General and Physical
CHEMISTRY

Dr. A. Ghoshal

CELEBRATING 55 YEARS 1960-2015

Since BOOKS AND ALLIED 1960

[For Undergraduate students of Chemistry (CBCS) and Engineering.]

General and Physical CHEMISTRY

Dr. Amalendu Ghoshal, M.Sc., Ph.D.

Reader, Dept. of Chemistry,
Midnapore College, Midnapore (W.B.).

BOOKS AND ALLIED (P) Ltd.

No.1-E(1) "Shubham Plaza" (1st Floor)
83/1, Beliaghata Main Road, Kolkata 700010
Tel : (033)23740062 • 8274085530
e-mail : booksandallied1960@gmail.com • booksmktng@gmail.com

General and Physical
CHEMISTRY

First Published : 2000
Reprinted (with corrections) : 2009
Second Edition : 2013
Reprinted : 2015, 2017
Reprinted : 2019

Publisher : Ruba Sen
BOOKS AND ALLIED (P) LTD.
8/1 Chintamoni Das Lane, Kolkata 700009

Typesetter : BOOKS AND ALLIED (P) LTD.
8/1 Chintamoni Das Lane, Kolkata 700009

Printer : EUREKA
91/A, Baithakkhana Road, Kolkata 700009

ISBN 978-81-87134-67-1

Price : **₹315.00**

"When science embraces Religion
with meaning and means keenly
and fulfil each other
with kisses and riches—
Heaven comes down
through the royal road of knowledge
to flourish all
with elixir and bliss."

— Sri Sri Thakur

"Dedicated to my supreme-beloved,
***'Thakur Sri Sri Anukul Chandra'**,*
whose divine-presence was felt in
the darkest hour of my life."

Preface to the Second Edition

This is a thoroughly revised as well as enlarged edition. In recent years, syllabus of different universities have been changed and pattern of questions have undergone a radical change. Considering these two things, model answers to some present day questions of different universities for the last few years have been added in the appendix. Let me hope that the edition will continue to enjoy support of both teachers and students.

15th August, 2013 **DR. AMALENDU GHOSHAL**

Preface to the First Edition

When plenty of books are already available for the students of the B.Sc. level, was there any necessity to add one more? As author of this new addition I owe an explanation. I have written this book on ***General and Physical Chemistry*** for two basic reasons. First—my publisher Mr. Arunabha Sen wanted me to write this book. Second—my students (both past and present) who are fond of my teaching style were requesting me again and again to write a book following the way of my lecturing in class-room. Being encouraged by these two demands I took up the task of writing this book. Though the book is mainly for B.Sc. students of different Indian Universities, in many chapters additional discussions have been done so that even B.Sc. (Hons.) students may find it helpful as a preliminary book. Students, appearing at IIT, JEE will also get help from this book.

Since this is my maiden attempt in writing a text-book, many errors may creep in. I shall be grateful if my readers take the pain of writing me about the mistakes.

Finally, I would like to record my appreciation to Mr. Arunabha Sen, who always encouraged me to enter in the area of writing books and also to my friends in the department of printing who are unsung heros of a successful book.

New Millennium, 2000 **DR. AMALENDU GHOSHAL**
Midnapore.

Contents

Part I : GENERAL CHEMISTRY

Part II : PHYSICAL CHEMISTRY

Contents

Part I
GENERAL CHEMISTRY

Chapter 1

ATOMIC STRUCTURE — CLASSICAL AND WAVE-MECHANICAL APPROACH

1.1 ATOMS ARE NOT 'ATOM'

The term 'atom' is a Greek word; it means 'not divided'. When Dalton proposed his atomic theory (1805) he believed that all matters were composed of tiny discrete particles. These smallest particles were visualised as indivisible and that is why he called them atoms. But from the end of nineteenth century, evidences began to accumulate to prove that these 'atoms' are not really 'atom' i.e. they can be broken down to sub-atomic particles viz. electron, proton, neutron and a lot of many others.

1.2 CATHODE RAYS — ELECTRONS

Thomson (1896) carried out experiment on electric discharge with gases at very low pressure. His instrument and experiment was as follows —

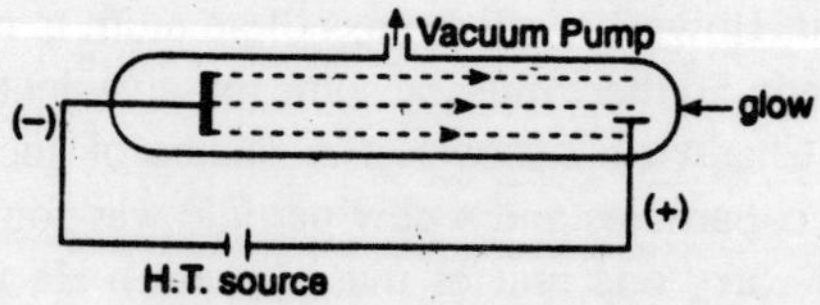

Fig. 1.1 Cathode ray tube.

He took a glass tube consisting of some gas at extremely low pressure. Two metal electrodes (cathode and anode) are fused at the two ends of the tube. Now they are connected to a high tension source so that electric discharge occurs. A series of negatively charged particles are found to be emitted from the cathode and they move towards anode. They produce a glow on the wall of the glass tube in their path. They can produce shadow when some object is placed in their path; they can also impart mechanical motion on a small wheel placed in their path. These particles were called cathode rays because of their origin. Lorentz named them as electrons. Its charge is to mass (*e/m*) ratio has been determined. Now we know that its charge is 1.60×10^{-19} coulomb and mass is 9.1×10^{-28} gm. Alternatively we can say that it bears one unit negative charge and its mass is (1/1836)th times of mass of hydrogen atom.

1.3 POSITIVE RAYS — DISCOVERY OF PROTONS

Discovery of electrons proved that atoms contain negative charge inside it. But as a whole an atom is electrically neutral. So an equivalent amount of positive charges are expected to be present inside it. The expectations were fullfilled by the discovery of positive rays. In the cathode ray tube if a perforated cathode is used then some glow on the glass wall just following the cathode (i.e. opposite to the anode) is found. These rays are called 'canal' rays (passing through a canal which is perforated cathode) or positive rays.

These rays have the following properties —

(a) They travel in straight line.

(b) They bear positive charge and their (*e/m*) ratio varies with the nature of gas.

(c) They cause fluorescence on ZnS.

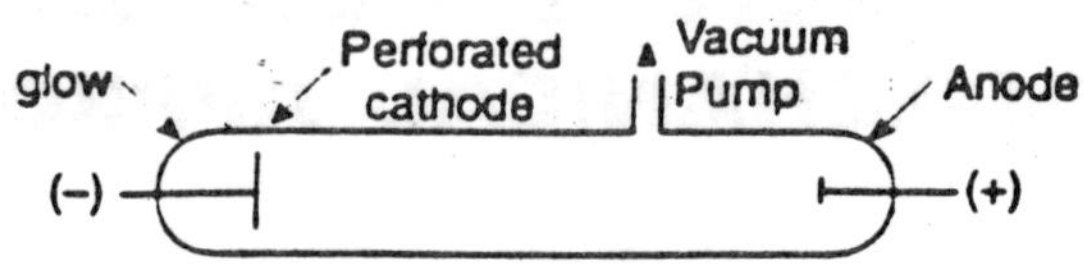

Fig. 1.2 Production of positive rays.

From the discovery of positive rays Aston (1913) concluded that atom consists of at least two parts which are (a) negatively charged electron and (b) a positive residue which is mainly responsible for atomic mass.

Thomson studied the nature of positive rays obtained with hydrogen gas. He found that in this case mass of positive particle is lowest and he named it proton. Its mass is almost equal to that of H-atom and is taken as one atomic mass unit (1 a.m.u.). Its actual mass is 1.672×10^{-24} gm and its charge is + 1.60×10^{-29} coulomb i.e. one unit positive charge.

1.4 NEUTRONS

Since electrons are of negligible mass, mass of an atom should be equal to total masses of all protons present inside it. However, often mass of an atom is much greater than total masses of all the protons present. So there must be some massive neutral particles to account for this difference. In 1932 Chadwick discovered existence of such particles. Beryllium metal was hit by a stream of α-particles and a new particle was found to be ejected. They bear no electrical charge but carry one unit of mass (1 amu). He named it neutron.

1.5 STRUCTURE OF ATOM — RUTHERFORD'S VIEW

After the discovery of electron and proton Rutherford performed experiment to know their relative positions. A stream of highly energetic α-particles were allowed to strike a thin gold foil which was surrounded by a screen coated with zinc sulphide. He found that most of the α-particles penetrate through the gold foil and produce flashes of the ZnS-screen behind it. To his astonishment he noticed that tiny flashes are observed on

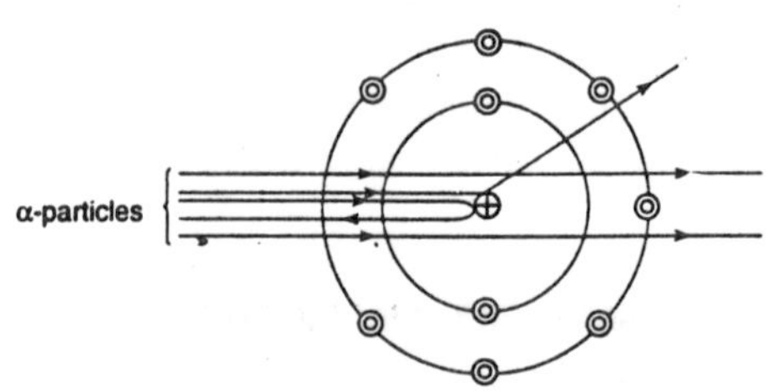

Fig. 1.3 Scattering of α-particles by gold foil.

other portion of the screen and very few flashes were noticed even in front of the gold foil. This led him to assume that atom has an inner core called nucleus where all positive charges are centrally located. Electrons are moving around the nucleus in circular orbits. So there are vacant spaces between electron and nucleus. This model is planetary model because movement of electrons around the nucleus is analogous to movement of planets around the sun.

1.6 ATOMIC NUMBER — MASS NUMBER-MOSELEY'S EXPERIMENT

During formation of periodic table for elements, the term atomic number was taken as a number showing the position of an element in the periodic table. So atomic number was position number. Moseley found that when cathode rays struck different elements used as anode in the discharge tube, X-rays of characteristic frequencies are emitted. He found that plot of atomic number of the anode element against the square root of frequency of X-ray emitted is a straight line. So atomic number is not merely a position number, it is related with the atomic structure. Moseley calculated the number of positive charges in the nucleus of the anode element and established that 'atomic number' of an element is equal to number of protons in the nucleus. Thus atomic number is actually proton number.

Mass number is regarded as the total number of proton and neutrons i.e. neucleons. Thus both atomic number and mass number are whole number. However, atomic weight is the weight of an atom in a. m. u. scale. It may be fraction. Its nearest whole number value is mass number. Thus atomic weight for Na is 22.9898 and mass number is 23.

1.7 WEAKNESS OF RUTHERFORD'S MODEL

The planetary model prescribed by Rutherford to explain position of nucleus and electrons is not in conformity with electrodynamics. We know that if a charged particle moves in a circular path it will begin to emit its energy. As electron is a charged particle, so during motion it should radiate and energy of it should be lowered. Thus its distance from nucleus should decrease and it should ultimately fall in the nucleus in a spiral path. But this thing does not happen.. So Rutherford's model needs modification. Modification was done by Niels Bohr with the help of Planck's quantum theory. Before discussing Bohr's model let us discuss the philosophy of quantum theory in brief.

1.8 QUANTUM THEORY — A NEW APPROACH

Prior to quantum theory all problems of mechanics were treated classically which believes in the idea of (i) continuity and (ii) certainty. Newtonian mechanics, which is based on laws of motion believes that past, present and future position of any object can be determined very accurately if we have informations regarding the forces which acted on it and are going to act on it. It also believes that all changes take place continuously. Thus by suitable manipulation of applied force we can give an object any position and any energy we like. These ideas worked well for macro particles. But Max Planck noticed that micro particles e.g. electron, proton, atom, molecules etc. do not obey classical i.e. Newtonian mechanics. In their cases we are to apply the idea of (i) uncertainty and (ii) discontinuity. Their positions energy etc. can never be determined accurately and changes that take place are always discontinuous. That is they can not have any position or any energy but they will have some selected position and so some selected energy values. Acquirance of such selected values is known as the idea of quantisation. Thus their position and energy become quantised. This quantisation concept is the philosophy of quantum mechanics.

1.9 PROBLEM BEFORE BOHR — FAILURE OF RUTHERFORD'S MODEL AND SPECTRA OF HYDROGEN ATOM

We have already discussed weakness of Rutherford's model. Niels Bohr, who tried to put forward an alternate model had the additional problem of explaining spectra of hydrogen atom. Hydrogen gas was taken in a discharge tube at low pressure and electric discharge is passed to excite H-atoms. The excited atoms begin to emit light of various frequencies which are examined by spectroscopes. Balmer found four lines in the visible region, now known as Balmer series. Besides these four other series, one in the UV region and the other three in the IR region are observed. They bear the name of their discoverers.

Table 1.1

Name	Region
Lyman Series	UV
Balmer Series	Visible
Paschen Series	IR
Brackett Series	IR
Pfund Series	IR

Classical theory is unable to account for such a discontinuous series of spectral lines.

1.10 BOHR'S MODEL — IDEA OF STATIONARY STATES

To explain the difficulty faced by Rutherford and also occurrence of discontinuous lines in the atomic spectra of hydrogen Bohr proposed the following model of atomic structure.

(a) In an atom electron travels around the nucleus in some selected orbits which will be called 'stationary-state' or 'stationary-orbit'.

(b) Such orbits should be such that, here angular momentum (*mvr*) of a moving electron, must be an integral multiple of $h/2\pi$ where h = Planck's constant i.e.

$mvr = \frac{nh}{2\pi}$ where 'n' is an integer having values 1,2,3,..... etc. and is called 'Principle Quantum Number'. We shall see later that 'n' denotes position and energy of an electron.

(c) While remaining in a stationary state electrons will neither radiate nor absorb energy. But if it changes orbit then it may absorb or emit energy depending on whether it is promoted or demoted. Thus if υ is the frequency of radiation emitted (or absorbed) during transition from orbit 1 to orbit 2 where electron possesses energy E_1 and E_2 respectively, then

$$E_1 \sim E_2 = h\upsilon.$$

1.11 WAVE CHARACTER OF ELECTRON — ESTABLISHMENT OF RELATION $mvr = \frac{nh}{2\pi}$

The term 'wave' means some sort of disturbance in a medium. When light passes through a medium it produces both electrical and magnetic disturbance in the medium. The pattern of this disturbance is wave. It is as follows.

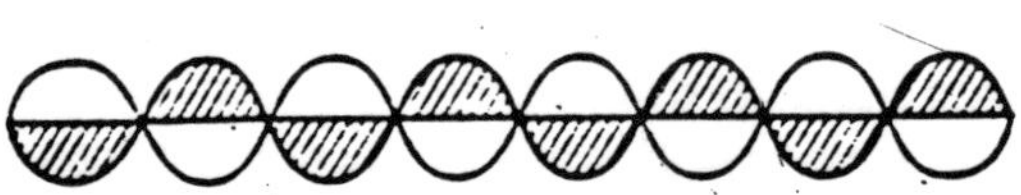

Fig. 1.4 Plane : Electrical disturbance pattern
Shaded : Magnetic disturbance pattern.

The electrical and magnetic disturbances occur in planes, which are perpendicular to each other. In space we can draw infinite number of planes through a point. Out of these, half of the planes are perpendicular to another half. If one half represents electrical disturbance, then another half is magnetic disturbance. Thus light propagation was treated as propagation of electromagnetic wave. Such wave treatment of light worked well till we discovered photoelectric effect and Compton effect. When light falls on some alkali metals e.g. Li, Na etc. the metals eject electrons, **the phenomena is photoelectric effect**. Compton found that when X-rays strike sample of graphite, some electrons are ejected and wavelength of the scattered X-ray becomes larger than incident ones i.e. its energy is lowered. **This effect is called Compton effect.** These two effects can not be explained by wave concept of light. On the contrary if we think that light rays consists of bundles of very small particles (now called **photons**) then photoelectric effect and compton effect can be explained with the help of idea of collision between photon and electron where the former looses energy and the latter gains sufficient energy to be ejected. Thus 'wave-particle' dual character of light was established. Louis-de-Broglie thought that if light (which till now was regarded as wave) behaves as a particle then electron which is recognised as particle may also show wave character. If a particle of mass '*m*' is converted to energy then according to Einstein energy $E = mc^2$ and according to Planck a light quanta (particle) of frequency υ possess energy $h\upsilon$ (h = Planck's constant).

So $$E = mc^2 = h\upsilon$$

Now $$\upsilon\lambda = c$$ [λ = Wave length, c = Velocity of light.]

So $$mc^2 = \frac{hc}{\lambda}$$

or $$\lambda = \frac{h}{mc} \qquad \text{......(1.1)}$$

Using above relation de-Boglie thought that if electron moves with velocity 'v' its wave-length will be

$$\lambda = \frac{h}{mv}. \qquad \text{......(1.2)}$$

Thus electron has also dual nature. The wave-particle duality may appear as ambiguous, because in our sense, concepts of wave and particle reside in opposite poles. Actually we do not know what electron is. In our dictionary we do not have appropriate word to describe

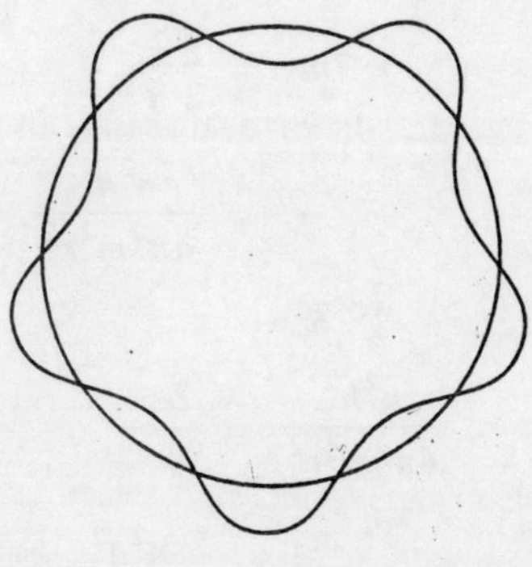

Fig. 1.5 de-Broglie wave in Bohr's orbit.

it. Quoting Heisenberg' it is like searching a black cat in a dark room when perhaps the cat is not there'. But the fact is that, if we want to know position of an electron then mathematics takes the help of particle concept but if we are eager in knowing its activity then wave concept serves better. Now if electron moves in a circular orbit then according to wave-concept it must be a closed wave and circumference of the circle must be some integral multiple of wavelength of electron.

So if 'r' is the radius of the circle then

$$2\pi r = n\lambda.$$

But $$\lambda = \frac{h}{mv}.$$

So $$2\pi r = \frac{nh}{mv}$$

or, $$mvr = \frac{nh}{2\pi}. \qquad \text{......(1.3)}$$

This is Bohr's postulate.

1.12 EXPRESSION FOR RADIUS OF A BOHR ORBIT AND ALSO ENERGY OF AN ELECTRON

Applying Bohr's idea we can now calculate radius of a Bohr orbit and also energy of an electron moving in such an orbit. Let us consider an atom having Ze units of positive charge (Z = at. no.) at the nucleus and also let r be the radius of an orbit where our electron is moving. The centrifugal force during movement is exactly balanced by the attractive force executed by the nucleus. From Coulomb's law attractive force = $\frac{Ze^2}{r^2}$ and centrifugal force for an electron of mass 'm' moving with a velocity v in an orbit of radius 'r' is $\frac{mv^2}{r}$.

So here $$\frac{mv^2}{r} = \frac{Ze^2}{r^2}$$

or $$v^2 = \frac{Ze^2}{mr}. \qquad \text{......(1.4)}$$

From Bohr's definition of stationary state

$$mvr = \frac{nh}{2\pi}$$

or $$v^2 = \frac{n^2h^2}{4\pi^2m^2r^2}. \qquad \text{......(1.5)}$$

Equating eqn. (1.4) and (1.5) we get,

$$\frac{n^2h^2}{4\pi^2m^2r^2} = \frac{Ze^2}{mr}$$

or $$r = \frac{n^2h^2}{4\pi^2Ze^2m}. \qquad \text{......(1.6)}$$

This is the expression for radius of a Bohr's orbit. For H-atom $z = 1$, so for radius of, 1st Bohr orbit $n = 1$ and so

$$r = \frac{h^2}{4\pi^2 e^2 m}.$$

Now $h = 6.626 \times 10^{-27}$ erg-sec; $\pi = 3.14$

$e = 4.8 \times 10^{-18}$ e.s.u. and $m = 9.109 \times 10^{-28}$ gm.

Putting these values we get $r = 0.529 \times 10^{-8}$ cm i.e. ~ 0.53 A°.

Total energy for the electron can now be calculated. Total energy = K.E. + P.E.

K.E. $= \frac{1}{2}mv^2$ and P.E. $= -\frac{Ze^2}{r}$ [work to be done to bring an electron from infinity to the given orbit]. But we have seen —

$$\frac{mv^2}{r} = \frac{Ze^2}{r^2}.$$

So $$mv^2 = \frac{Ze^2}{r}.$$

Hence $$\text{K.E.} = \frac{1}{2}mv^2 = \frac{Ze^2}{2r}.$$

So $$\text{Total energy } E = \text{K.E.} + \text{P.E.}$$

$$= \frac{Ze^2}{2r} - \frac{Ze^2}{r}$$

$$= -\frac{Ze^2}{2r}$$

Putting the value of r i.e. $r = \frac{n^2h^2}{4\pi^2 Ze^2 m}$

$$E = -\frac{2\pi^2 Z^2 e^4 m}{n^2 h^2}. \quad ......(1.7)$$

The expression for energy of an electron in the first Bohr's orbit for H-atom is —

$$E = -\frac{2\pi^2 e^4 m}{h^2}. \quad ......(1.8)$$

1.13 ENERGY CHANGE DURING ORBIT CHANGE — RYDBERG'S CONSTANT

If an electron changes orbit from n_1 to n_2 its energy will change from E_1 to E_2 where

$$E_1 = -\frac{2\pi^2 Z^2 e^4 m}{n_1^2 h^2} \text{ and } E_2 = -\frac{2\pi^2 Z^2 e^4 m}{n_2^2 h^2}.$$

So energy absorbed by the electron will be

$$E_2 - E_1 = \Delta E = \frac{2\pi^2 Z^2 e^4 m}{h^2}\left[\frac{1}{n_1^2} - \frac{1}{n_2^2}\right]$$

If υ is the frequency of the light absorbed, then $\Delta E = h\upsilon$

$$\upsilon h = \frac{2\pi^2 Z^2 e^4 m}{h^2}\left[\frac{1}{n_1^2}-\frac{1}{n_2^2}\right]$$

$$\text{or,}\quad \upsilon = \frac{2\pi^2 e^4 m}{h^3} Z^2\left[\frac{1}{n_1^2}-\frac{1}{n_2^2}\right]. \qquad \text{......(1.9)}$$

If $\bar{\upsilon}$ is wave number i.e. no. of waves/ unit length

then $\bar{\upsilon} = \frac{1}{\lambda} = \frac{\upsilon}{c}$ [λ = wave length; c = velocity of light].

So equation (1.9) becomes

$$\bar{\upsilon} = \frac{2\pi^2 e^4 m}{h^3 c} Z^2\left[\frac{1}{n_1^2}-\frac{1}{n_2^2}\right].$$

The term $\frac{2\pi^2 e^4 m}{h^3 c}$ is a constant and is known as Rydberg's constant R.

So
$$\bar{\upsilon} = RZ^2\left[\frac{1}{n_1^2}-\frac{1}{n_2^2}\right].$$

For H-atom Z = 1, so

$$\bar{\upsilon} = R\left[\frac{1}{n_1^2}-\frac{1}{n_2^2}\right] \qquad \text{......(1.10)}$$

value of 'R' is 109679 cm^{-1}.

With the help of this equation Bohr's was able to explain spectrum of H-atom. Following table may help to understand it.

Table 1.2

Series	n_1	n_2	Region	Wave-length (A°)
Lyman	1	2, 3, 4, 5, etc.	U.V.	920 – 1200
Balmar	2	3, 4, 5, 6, etc.	Visible	4000 – 6500
Paschen	3	4, 5, 6, 7, etc.	I. R.	9500 – 18750
Brackett	4	5, 6, 7, etc.	I. R.	19450 – 40500
Pfund	5	6, 7, etc.	I. R.	37800 – 75000

1.14 ZEEMAN EFFECT

In 1896 Zeeman observed that if the source, which is exhibiting emission spectra is placed under magnetic field then the spectral lines are split up into a number of fine lines. This effect is called Zeeman effect. It can be explained as follows. If we consider electron as charged particle then its movement around the nucleus is like the flow of electricity in a loop of conductor. And certainly it will create a magnetic field. If some external magnetic field is applied it will create a torque on the loop and so it will take some possible orientations each of which is associated with a definite amount of energy. Of course these energy values will be very close to each other. Number of such possible orientations is expressed by 'm' known as magnetic quantum number. Its possible values range from $-l$ to $+l$ including zero (so total number of orientations is $2l + 1$), where l = azimuthal quantum number.

1.15 THE QUANTUM NUMBERS

The complete state of an electron inside an atom is now described with the help of four numbers viz. 'n', 'l', 'm', and 's' known as quantum numbers. A brief discussion about their characteristics are as follows :

(a) **Principal quantum number 'n'** : This number describes energy of the electron and average size of the electron-cloud which measures radius of Bohr orbit. Its numerical values are 1, 2, 3, 4, 5, etc. Theoretically it may increase by integral values up to infinity, but in practice values from 1 to 7 have been established.

(b) **Azimuthal quantum number 'l'** : This quantum number, whose numerical value ranges from 0 to $n - 1$ (all integral values only) determines the spatial distribution of electron cloud i.e. shape of the orbital. Thus when $l = 0$ the orbital is spherically symmetrical and is called 's'-orbital. When $l = 1$, the orbital is 'p'-orbital and is dumbel shaped. The orbital with $l = 2$ is called 'd' and the one with $l = 3$ is called 'f' orbitals. Their shapes are complicated. The terms 's', 'p', 'd' and 'f' have been coined from spectral study; they represent sharp; principal; diffused and fundamental line.

(c) **Magnetic Quantum number 'm'** : From the study of Zeeman effect we have known that electronic orbitals occupy some selected positions under the influence of some magnetic field. Total number of possible orientations is indicated by magnetic quantum number, having values $-l$ to $+l$ including zero. Following figures will help us to understand.

Azimuthal Q. No.	Name	Magnetic q. nos. and shape
$l = 0$	s	$m = 0$
$l = 1$	p	$m = -1, 0, +1$
$l = 2$	d	$m = -2, -1, 0, +1, +2$

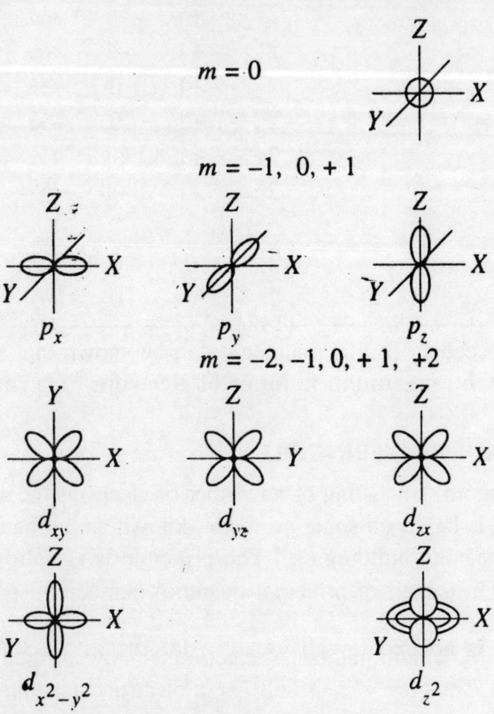

Fig. 1.6 Shapes of atomic orbitals.

In '*p*' and '*d*' orbitals different lobes are separated by planes where probability of finding electron is zero. These are known as nodal planes. When $l = 3$ orbital is '*f*' and there are seven magnetic quantum numbers −3, −2, −1, 0, +1, +2 and +3. Their shapes are very complicated.

(d) **Spin quantum number** (s) : This quantum number denotes the spin of the electron around its own axis. Since only two types of spin viz. clock-wise and anti clock-wise are possible there are two possible spin quantum numbers; $-\frac{1}{2}$ and $+\frac{1}{2}$. Their difference is '1' in accordance with other quantum numbers. The spins are indicated by arrows. The symbol '$\uparrow\uparrow$' indicates parallel spin whereas the symbol '$\uparrow\downarrow$' indicates opposite spin.

1.16 PAULI'S EXCLUSION PRINCIPLE

Pauli put forward a principle which puts some restriction on the assignment of four quantum numbers towards different electrons. It states that — no two electrons in an atom can have the same set of four quantum numbers. Thus if two electrons have same '*n*', '*l*' and '*m*' values then their '*s*' value must differ. We can now guess about possible orientations.

Table 1.3

Pr. Q. No.	Possible distributions					Total electrons present	
$n = 1$	n	l	m	s			
	1	0	0	+½	1*s*	2	($1s^2$)
	1	0	0	−½			
	n	l	m	s			
	2	+1	−1	+½			
	2	+1	−1	−½			
	2	+1	0	+½			
$n = 2$	2	+1	0	−½	2*p*	8	$2s^2$, $2p^6$
	2	+1	+1	+½			
	2	+1	+1	−½			
	2	0	0	+½	2*s*		
	2	0	0	−½			

Proceeding in this manner it can be shown that for an orbit with principal quantum number '*n*', **maximum number of electrons that can be accommodated is $2n^2$.**

1.17 AUFBAU PRINCIPLE

In atoms consisting of a number of electrons the arrangement of electrons in different orbitals is based on some principle, known as 'aufbau' principle. ['Aufbau' is a German word meaning building up.] The principle is as follows :—

(a) In a shell of principal quantum number '*n*' the maximum number of electrons is $2n^2$.

(b) Maximum number of electrons in any subshell is $2(2l + 1)$; l = azimuthal quantum number. Thus for '*s*' subshell $l = 0$; electrons accommodated = 2; for '*p*'; $l = 1$, electrons

(c) The electrons will always occupy lowest available energy.

(d) Electrons never pair until all available subshells with same energy [e.g. p_x ; p_y ; p or d_{xy}; d_{yz} ; d_{zx} ; $d_{x^2-y^2}$; d_{z^2}] are half filled.

(e) After half-filling all subshells electrons tend to pair with antiparallel spin.

The general pattern of filling up is —

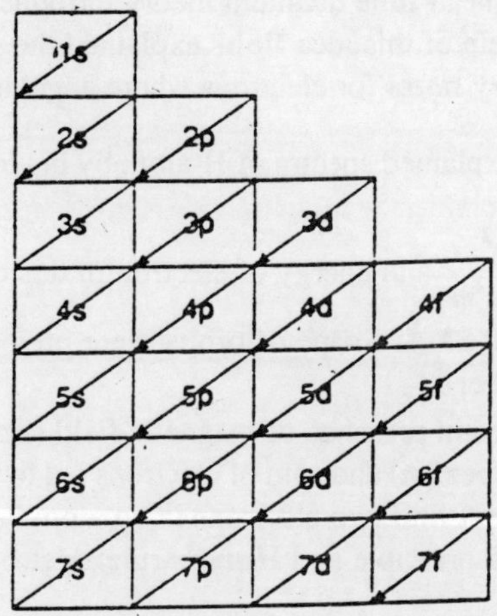

Fig. 1.7 Electronic configuration : arrow heads show direction of filling up.

1.18 SOME EXCEPTIONS

According to 'aufbau' principle 4s orbitals will be filled up by electrons before occupying 3d orbitals. Thus electronic configuration of Chromium (At. No. 24) should be —

$1s^2$
$2s^2\ 2p^6$
$3s^2\ 3p^6\ 3d^4$
$4s^2$

But from energetic point of view a half-filled or completely filled orbital gets extra stability. Since 'd' orbital can accommodate '10' electrons; $3d^5$ state is more stable than $3d^4$. And so its electronic configuration will be —

$1s^2$
$2s^2\ 2p^6$
$3s^2\ 3p^6\ 3d^5$
$4s^1$.

Similar is the case with Cu (At. No. - 29). In this case, expected configuration is $1s^2$, $2s^2$, $2p^6$, $3s^2$, $3p^6$, $3d^9$, $4s^2$, but actual configuration is $1s^2$, $2s^2$, $2p^6$, $3s^2$, $3p^6$, $3d^{10}$, $4s^1$. Such exceptions are also found in few more cases. These are Mo (At. No.- 42); Pd (At. No. - 46); Ag (At. No. - 47); Pt (At. No. - 78) and Au (At. No. - 79).

POINTS TO REMEMBER

Assuming them as '**indivisible**' smallest particle of matter (they were) termed as **atom.** However, now they have been broken and a number of subatomic particles have been discovered. Notable among them are **electron** (one unit negative charge and negligible mass), **proton** (one unit positive charge and one unit mass) and **neutron** (no charge but one unit mass). **Rutherford** first thought that electrons are rotating around a tiny nucleus (consisting of neutron and proton) in a **circular** path. This planetary model was rejected by electrodynamics. In the mean-time quantum theory introduced the idea of discontinuity and uncertainty. With the help of this idea **Bohr** explained the structure of hydrogen atom. He proposed some *stationary states* for electrons where angular momentum (***mvr***) must be $\frac{nh}{2\pi}$ (n = + *ve* integer). He explained spectra of H-atom by his formula $E_1 \sim E_2 = h\nu$ Radius of orbit was found $\frac{n^2h^2}{4\pi^2Ze^2m}$ and energy of electron in this orbit is $-\frac{2\pi^2Z^2e^4m}{n^2h^2}$. First Bohr orbit has a radius of **0.53 Å.** Shape of orbits were predicted by a quantum number l (azimuthal quantum number).

Splitting of spectral lines in presence of magnetic field (**Zeeman effect**) gave the idea of **magnetic quantum number** (m) and spin of electrons led to **spin quantum number** (s). According to Pauli we can not find two electrons in an atom having same 'n', 'l', 'm' and 's'. With the help of **aufbau principle** and **Hund's** rule electronic configurations of atoms are obtained.

QUESTIONS

SHORT TYPE

1. What does azimuthal quantum number indicate ?
2. Which quantum number was prescribed after the discovery of Zeeman effect and how ?
3. Draw the shapes of five '*d*' - orbitals ?
4. What is the reason behind failure of Rutherford's model ?
5. State whether in the ground state electron is nearest or furthest from the nucleus.
6. Name the series associated with the transition of an electron from higher orbit to 'L' orbit.
7. Write down the electronic configuration for an atom with atomic number 24 and 29.

ESSAY TYPE

1. Explain Rutherford's model to explain atomic structure. What is the limitation of this model ? How Bohr modified this.
2. State and explain Bohr's theory to elucidate structure of an atom. How atomic spectra of H can be explained on the basis of this theory ?
3. Starting from Bohr's postulate derive an expression for the energy of an electron and radius of Bohr's orbit.
4. What is Zeeman effect ? State and explain Pauli's exclusion principle.
5. Describe in details the characteristics of different quantum numbers.
6. Explain 'aufbau' principle. State at least two cases where the principle is not rigidly followed.
7. Prove that maximum number of electrons that can be accommodated in an orbit is $2n^2$ where n is principal quantum number.

PROBLEMS

1. A particle of wavelength 6.6×10^{-4} c.m. is moving with a velocity of 10^6 cm/sec. Find out its mass.

   [***Hints*** : $\lambda = \frac{h}{mv}$; $h = 6.62 \times 10^{-27}$ erg sec.]

2. What is the wavelength associated with a particle of mass 0.1 gm moving with a speed of 1×10^5 cm/sec.

   [***Hints*** : Same as before.]

3. If radius of first Bohr's orbit is 0.528 Å; find out the radius when $n = 10$.

   [***Hints*** : $r = \frac{n^2h^2}{4\pi^2 Ze^2 m}$ $\therefore r \propto n^2$; when $n = 1$

   $r = 0.528$; when $n = 10$; $r = 52.8$ Å.]

4. The energy of the first Bohr orbit for H-atom is -2.168×10^{-11} erg. Find the energy in 3rd Bohr orbit.

5. Calculate the wavelength of emitted radiation when an electron in H-atom jumps from 3rd to 2nd orbit. Given $E_2 = -5.42 \times 10^{-12}$ and $E_3 = -2.41 \times 10^{-12}$ ergs.

[*Hints* : $\Delta E = (E_3 - E_2) = h\upsilon$; $\upsilon = \frac{c}{\lambda}$; $c = 3 \times 10^{10}$ cm sec^{-1}.]

6. A spectral line in Lyman series of H-atom has a wave number of 82,200 cm^{-1}. What transition is responsible for this emission ? [R = 109679 cm^{-1}.]

[*Hints* : $\bar{\upsilon} = R\left[\frac{1}{n_1^2} - \frac{1}{n_2^2}\right]$. For Lyman series $n_1 = 1$, $n_2 \approx 2$.]

OBJECTIVE TYPE

1. Fill in the blanks.

(a) When an electron jumps from —— to —— orbit it emits energy.
(b) By passing α-particles through a thin gold particles —— discovered nucleus.
(c) The —— quantum number represented as —— indicates shape of an orbital.
(d) —— effect led to the concept of magnetic quantum number.
(e) —— principles means a building up process.
(f) Possible number of '*d*' orbitals is —— but it is —— for '*f*' orbitals.
(g) —— quantum numbers are required to describe an electron fully.
(h) Spectral lines in U.V. region (for H-atom) are called ——.
(i) Pfund series show spectral lines in —— region.
(j) Radius of first Bohr orbit is ——

2. Write T (for true) or F (for false) in the boxes besides following statements.

(a) Cathode rays are stream of electrons. ☐
(b) Neutron was discovered by Rutherford. ☐
(c) Bohr introduced the idea of stationary state. ☐
(d) Quantum theory was proposed by Pauli. ☐
(e) There are exceptions to aufbau principles. ☐

3. Match terms of column A with those of column B.

A	B
(a) Moseley	(a) Classical theory
(b) Planck	(b) Quantum theory
(c) Newton	(c) Expt. on atomic number.
(d) Heisenberg	(d) Exclusion principle
(e) Brackett	(e) Multiplicity Rule
(f) Pauli	(f) I.R. region
(g) Hund	(g) Uncertainty principle
(h) Planck	

Answers :

1. (a) higher; lower (b) Rutherford (c) azimuthal; *l* (d) Zeeman (e) Aufbau (f) 5; 7 (g) Four (h) Lyman (i) I.R. (j) 0.53 Å
2. (a) T (b) F (c) T (d) F (e) T
3. a – c; b – b; c – a; d – g; e – f; f – d; g – e; h – b.

Chapter 2

CHEMICAL BOND-LEWIS THEORY-VALENCE BOND THEORY M. O. CONCEPT AND V S E P R IDEA

2.1 BOND — DEFINITION AND TYPES

We know that hydrogen and oxygen molecules are written as H_2 and O_2 but helium and neon as He and Ne. Carbon and oxygen combines to give CO and CO_2 but C_2O; C_2O_7 etc. are not found. Na^+ and Cl^- remain rigidly attached in common salt. Millions of atoms of a metal exist as a united species together. All these suggest that some sort of force act to keep a selected number of atoms together. In chemistry we call this force as chemical bond. So bond is a force, which holds the atoms together. Depending on the nature of the force different types of bonds have been recognised. These are (i) Ionic or electrovalent bond (ii) Covalent bond (iii) Co-ordinate co-valent bond (iv) Metallic bond and (v) Hydrogen bond.

2.2 ROLE OF ELECTRONS IN FORMING A BOND

During formation of chemical bond, the most important role is played by electrons. Lewis and Kossel observed that noble gases having stable electronic configuration are inert and highly stable. Since in nature every substance tends to have stability (excepting human being !) by lowering its free energy, so they tend to acquire electronic configuration of noble gases. Now every noble gas, excepting He has eight electrons in the outermost shell, known as valence shell. The following table will illustrate.

Table 2.1

Electronic configuration of inert gases

Name	At. No.	Distribution of electrons in different shells.
He	2	$1s^2$
Ne	10	$1s^2$; $\underbrace{2s^2\ 2p^6}_{8}$
Ar	18	$1s^2$; $2s^2\ 2p^6$ $\underbrace{3s^2\ 3p^6}_{8}$
Kr	36	$1s^2$; $2s^2\ 2p^6$; $3s^2\ 3p^6\ 3d^{10}$; $\underbrace{4s^2\ 4p^6}_{8}$
Xe	54	$1s^2$; $2s^2\ 2p^6$; $3s^2\ 3p^6\ 3d^{10}$; $4s^2\ 4p^6\ 4d^{10}$; $\underbrace{5s^2\ 5p^6}_{8}$

In 1916 Lewis and Kossel suggested that during formation of chemical bond, atoms interact to acquire stability i.e. to have electronic configuration of inert gas. In trying to do so atoms may loose; gain or share electrons. Depending on how they are acting bonds will be classified in to various types.

2.3 IONIC BOND

Elements of groups IA, IIA and IIIA have 1, 2 and 3 valence electrons; whereas elements of groups VA; VIA and VIIA have 5, 6 and 7 valence electrons. So if elements of groups IA, IIA and IIIA combine with elements of groups VA; VIA and VIIA then the former can give up their electrons to attain inert gas structure whereas the latter may take up those electrons to attain inert gas structure. After giving up electrons former will have positive charge whereas the latter will bear negative charge. In the compound the ions will remain bound by electrostatic force. Such a bond is electrovalent or ionic bond. Some examples are given below.

Table 2.2

1.	Na	+	Cl	$\longrightarrow$	Na^+	+	Cl^-	;	$Na^+ — Cl^-$
	2, 8, 1		2, 8, 7		2,8		2, 8, 8		
2.	Ca	+	O	$\longrightarrow$	Ca^{2+}	+	O^{2-}	;	$Ca^{2+} — O^{2+}$
	2,8,8,2		2, 6		2, 8, 8		2, 8		
3.	Mg	+	Cl	+ Cl	$\longrightarrow$	Mg^{2+}	$+ Cl^- + Cl^-$	;	$Cl^- — Mg^{2+} — Cl^-$
	2,8,2		2,8,7	2,8,7		2,8	2,8,8 2,8,8		

In NaCl crystals Na^+ and Cl^- are present. Since any positive charge can attract any negative charge and vice-versa, so each Na^+ can attract as many Cl^- ions as possible. Actually in crystal lattice each ion is surrounded by six ions of opposite charge. So a single $Na^+ Cl^-$ molecule has no existence. That is why, for ionic compounds we call formula weight instead of molecular weight. Characteristic properties of ionic compounds are —

They are solid crystal having high melting points. The solids are hard, brittle and easily soluble in water because of its strong polar character. Aqueous solution is an electrolytic conductor.

2.4 COVALENT BOND

In case of homonuclear diatomic molecules e.g. H_2, O_2, Cl_2 and in many compounds (majority of which are organic) the idea of electrostatic nature of chemical bond does not work. In such cases Lewis introduced the idea of electron-pair sharing. Such a sharing helps both the combining atoms to attain inert gas configuration. And sharing a common pair act as a chain holding the atoms together. Such type of bond has been termed as co-valent bond. Some examples are given in the following table. Here a dash (—) indicates a co-valent bond consisting of the electron pair jointly shared by both atoms.

Table 2.3

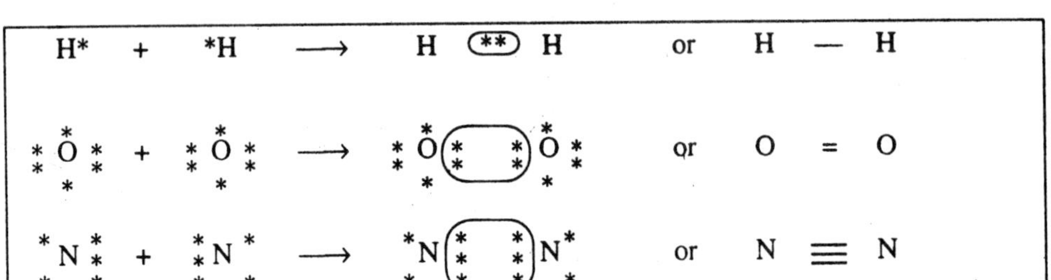

(only valence electrons are shown as *)

Co-valent compounds are mostly gases and liquids and low melting point soft solids. They are soluble in organic solvents and can not conduct electricity.

2.5 CO-ORDINATE BOND

In a normal covalent bond the electron pair, which is jointly shared by the combining atoms has been made by equal contribution of the two atoms. But there are some co-valent bonds where the shared pair has been given by one atom (donor), the other simply shares without contributing. Such a co-valent bond is called co-ordinate co-valent bond or dative bond. The compound thus produced is very often called as co-ordinate compound. Some examples are given in the following table.

Table 2.4

Ammonium ion formation

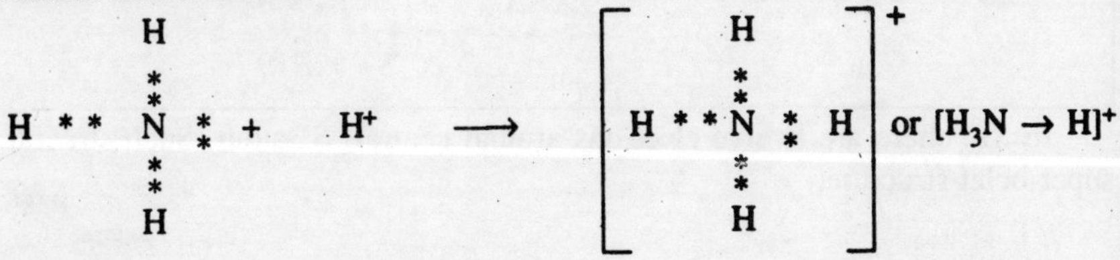

Addition compound between NH_3 and BCl_3

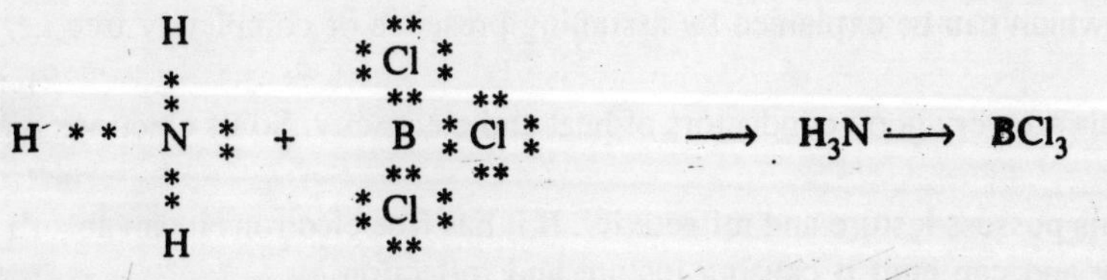

In NH_4Cl there are three N – H covalent bonds; one N → H dative bond and one ionic bond between NH_4^+ and Cl^-.

2.6 EXCEPTIONS TO OCTET RULES

To explain formation of ionic covalent and co-ordinate bond we have assumed that atoms tend to acquire inert gas share by donating; accepting or sharing electrons. Excepting He all other inert gas has eight electrons in the outermost orbit. That is why we call it 'octet' rule or rule of eight. However, in course of time we found many molecules, which do not obey this rule. Some such exceptions are given in the next page —

Table 2.5

Name of the compound	Lewis structure	No. of electrons around central atom
$BeCl_2$	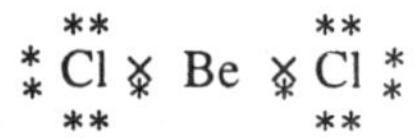	4
BF_3	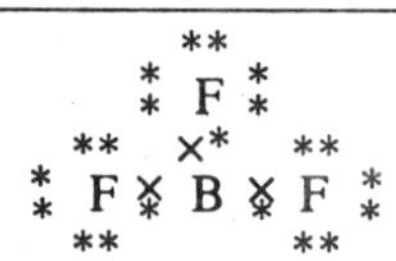	6
ClO_2	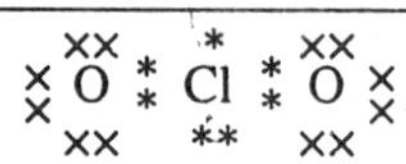	7
PCl_5	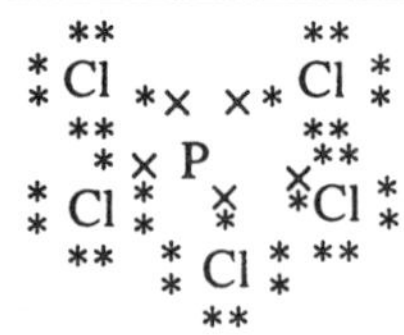	10

In SF_6 there are twelve electrons around central 'S'-atom. Such structure are called super-oclet-structure.

2.7 METALLIC BOND

If bond is a force holding the atoms together then quite naturally the question that will come in our mind is — "What type of force is holding the atoms of a metallic crystal together ?". The answer is given on the basis of metallic bond. The idea of metallic bond is based on '**the electron sea model**'. Before discussing this let us analyse some properties of metal which can be explained by assuming presence of completely free i.e. delocalised electrons.

Metals are very good conductors of heat and electricity. So its electrons are very much mobile.

Metals possess lusture and reflectivity. If it has free electrons it can absorb light of any frequency and can emit it causing lusture and reflection.

On heating or during absorption of light some metals emit electrons (thermionic and photoelectric emission). So electrons are loosely bound and are mobile.

Based on these observations the idea of metallic bonds was developed as follows. We can assume that electrons of metal atoms are removed and accumulated together to form an 'electron-sea'. In this 'electron-sea' the nuclei (produced after robbing off the electrons) are dipped one by one. Thus the electrons no longer remain bound to a particular nucleus but they become capable of moving any where inside the lattice. Thus they become free i.e. completely delocalised. Hence a metallic bond is the electrostatic attraction between metallic ions and the delocalised electrons.

2.8 HYDROGEN BOND

When hydrogen is bonded with a much more electronegative atom by covalent bond, then the electron pair is not shared equally but it is pulled towards more electronegative atom making it slightly negative charged. Automatically H-atom will acquire slight positive charge. This slightly charged atoms will attract oppositely charged atoms of their immediate neighbour and thus a weak electrostatic force will come in to play. This force is called hydrogen bond.

Two types of hydrogen bond formation are known. These are (i) intermolecular hydrogen bond i.e. hydrogen bond between neighbouring molecules of the same substance and (ii) intramolecular hydrogen bond i.e. hydrogen bond between atoms inside a single molecule.

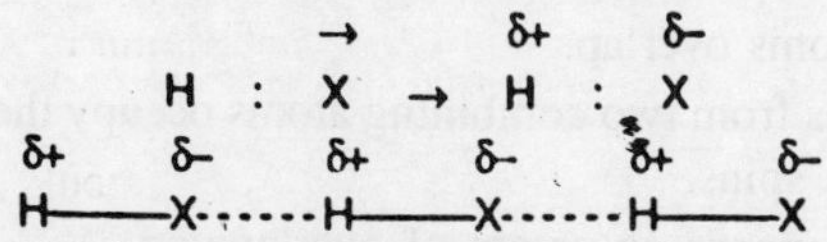

Fig. 2.1 Hydrogen bond formation.

Examples of intermolecular H-bonding are HF, H_2O etc. molecules.

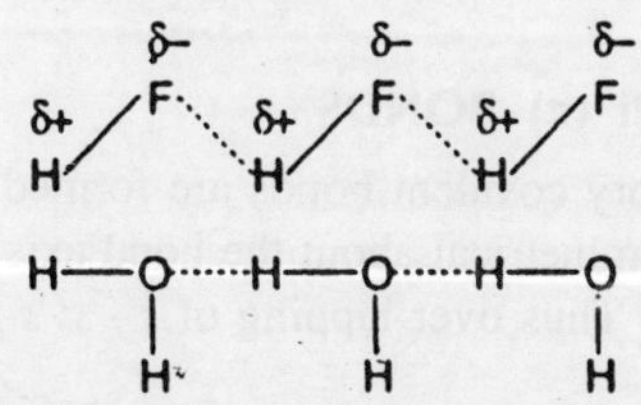

Fig. 2.2 Example of intermolecular H-bonding.

Examples of intramolecular H-bonding in 2-nitrophenol.

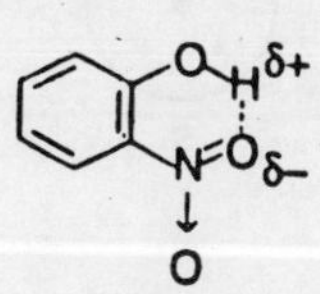

Fig. 2.3 Intramolecular H-bonding.

Compounds, possessing H-bond are generally characterised by high boiling and melting points. Since H-bond has directional property, it generally helps to form three-dimensional crystal lattice e.g. ice crystal.

2.9 LIMITATIONS OF LEWIS THEORY FOR COVALENT BOND

The idea of sharing of electron pairs during covalent bond formation (Lewis concept) may lead to many confusions in our mind. These are —

(a) What happens to the orbit (or orbital) of the electrons after sharing ?

(b) If the two electrons coming from two atoms form a pair, then is it a static pair remaining in between the two nuclei ?

(c) What would be the geometry of a molecule ?

Answers to these questions are not found in the theory of covalent bond formation.

2.10 MODERN THEORIES OF BOND FORMATION

To answer questions mentioned in section 2.9, two modern theories based on quantum mechanics have been proposed. These are

(i) **Valence-Bond theory** proposed by **Heitler** and **London** and (ii) **Molecular Orbital theory** proposed By **Mullikan**, **Huckel** and **Hund.**

2.11 VALENCE-BOND THEORY

This theory was proposed by Heitler and London in 1927 and later extended by Pauling and Slater in 1931. The fundamental assumptions are —

(a) A covalent bond is formed when orbitals of valence electrons (consisting of only one electrons) of two atoms overlap.

(b) The two electrons from two combining atoms occupy these overlapped orbitals and they must have opposite spins.

(c) Bond strength depends on extent of overlapping.

(d) Directionless orbitals viz. '*s*' orbital do not show any directional preference but '*p*', '*d*' etc. orbitals having directional property will tend to form bond along their axis where electron density is maximum.

(e) Bonds between orbitals with directional properties are stronger.

2.12 SIGMA (σ) AND Pi (π) BONDS

According to V.B. theory covalent bonds are formed by overlapping atomic orbitals. Bond orbitals, which are symmetrical about the bond axis (i.e. line joining the two nuclei) are called sigma (σ) bonds. Thus over-lapping of s - s; s-p; p_x-p_x form σ bond. These are head-on overlapping.

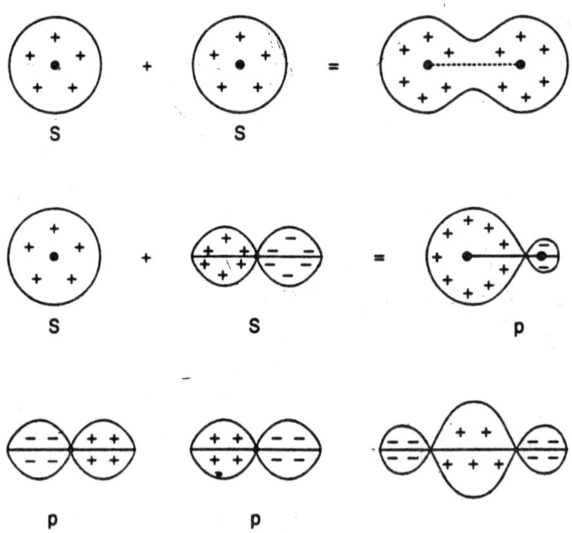

Fig. 2.4

[Overlapping of *s-s*; *s-p* and *p-p* (along axis) orbitals. Sign of wave function is same on the both sides of bond - axis. So these are symmetrical and hence bonds are σ-type.]

On the contrary bond-orbitals, which are anti-symmetrical around the bond axis are called pi (π) bonds. These are produced by lateral overlapping of *p*-orbitals.

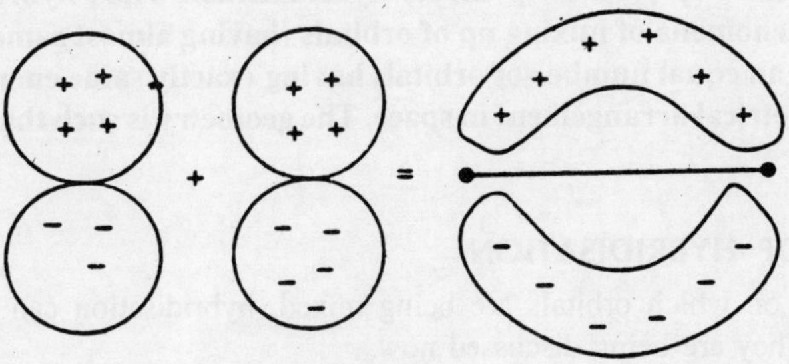

Fig. 2.5

[Lateral overlapping produces an orbital, where sign of wave functions are different above and below the bond axis. These are antisymmetrical and hence bonds are π-type.]

2.13 SOME EXAMPLES

We shall now consider some examples of such bond formation.

(a) **Hydrogen molecule :** Hydrogen atom has $1s^1$ configuration. During bond formation $1s$ orbitals of two atoms overlap to produce a σ-bond between them.

(b) **HF :** Electronic configuration of hydrogen and fluorine are $1s^1$ and $1s^2$, $2s^2$, $2p_x{}^2$, $2p_y{}^2$, $2p_z{}^1$. Here $1s$ orbital of hydrogen overlap with $2p_z{}^1$ orbital and *s-p* overlap forms one σ-bond.

(c) **NH_3 :** Electronic configuration of hydrogen and nitrogen are $1s^1$ and $1s^2$, $2s^2$, $2p_x{}^1$, $2p_y{}^1$, $2p_z{}^1$. So overlapping of three $1s$ orbitals of three H-atoms with p_x, p_y and p_z produces three σ-bonds (s-p overlap).

(d) **O_2-molecules :** Electronic configuration of oxygen atom is $1s^2$, $2s^2$, $2p_x{}^2$, $2p_y{}^1$, $2p_z{}^1$. During combination of two atoms two *p*-orbitals overlap along axis (σ-bond) and the remaining two overlap laterally. So in O_2 molecule we have one σ-bond and one π-bond.

(e) **N_2-molecule :** Electronic configuration of nitrogen atom is $1s^2$, $2s^2$, $2p_x{}^1$, $2p_y{}^1$, $2p_z{}^1$. During combination of two atoms two *p*-orbitals (let $2p_x{}^1$) overlap along axis and the remaining orbitals ($2p_y{}^1$, $2p_z{}^1$) overlap laterally. So it has one σ-bond and two π-bonds.

[Note : During bond formation we have considered overlapping between those orbitals only where we get unpaired electrons. This is characteristic of V.B. theory. We shall see later that M.O. theory differs in this regard.]

2.14 CONCEPT OF HYBRIDISATION

Electronic configuration of carbon is $1s^2$, $2s^2$, $2p_x$, $2p_y$. Since it has two unpaired electron it should be bivalent. But in most of its compounds it exists as tetravalent. One way to make it tetra-valent is to unpair $2s^2$ state and promote one electron to the empty $2p_z$ orbital. The electronic configuration will be then $1s^2$, $2s$, $2p_x$, $2p_y$ $2p_z$. However in this situation C will produce three bonds at right angles to each other (overlapping of three *p*-orbitals with those of combining atoms) and the remaining bond will have no preference for direction. In actual practice we never find this. On the contrary, we find four bonds to be exactly equivalent. That is why it has been assumed that one '*s*' ($2s$) and three '*p*' ($2p_x$, $2p_y$, $2p_z$) orbitals have been mixed (i.e. hybridised) to produce a set of four orbitals, which are exactly identical. Condition of lowest energy demands that these four orbitals will be directed to the four sides of a regular tetrahedron, where bond angle is 109° 28'. The phenomena is called sp^3. **(read *s-p* three) hybridisation. Thus hybridisation may be defined as a phenomena of mixing up of orbitals (having almost same energy) of same atom to obtain an equal number of orbitals having exactly same energy and following a definite geometrical arrangement in space. The geometry is such that energy becomes minimum.**

2.15 TYPES OF HYBRIDISATION

Depending on which orbitals are being mixed, hybridisation can be classified into several types. They are being discussed now.

(a) **sp^3 (*s-p* three) hybridisation :** If one *s* and three *p* orbitals having the same principal quantum number values are mixed, then the result is sp^3 hybridisation. As has already been mentioned C acquires tetravalency in this manner. The orbitals are directed along four sides of a regular tetrahedron. And they can now overlap with four orbitals of other atoms to produce four covalent bonds. Formation of methane occurs in this manner.

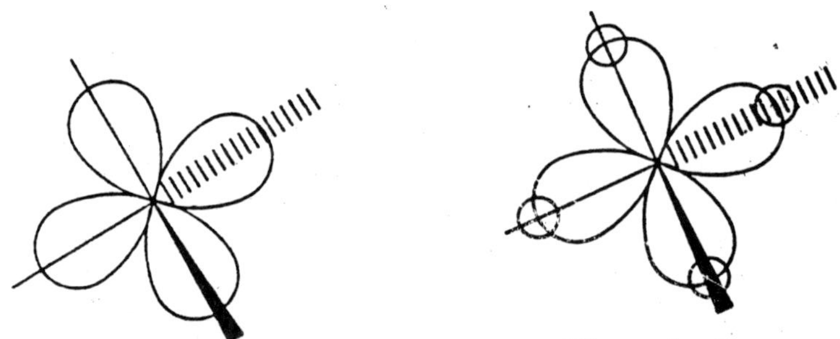

Fig. 2.6

[sp^3-orbitals. Solid lines are in the plane of the paper. Dotted line is above the plane and (thick) line is below the plane. Regular tetrahedron. Circles are *s*-orbitals of H-atoms. All are σ-bonds]

(b) **sp^2 — hybridisation :** Let us consider the molecule BCl_3. Electronic configuration of boron is $1s^2$, $2s^2$, $2p_x$. So it should be univalent. But it can be made trivalent by promotion one of two $2s$ electrons to empty $2p_y$ orbitals. Thus the configuration will be $1s^2$, $2s$, $2p_x$, $2p_y$. Now one $2s$ and two $2p$ orbitals hybridise and three hybrid orbitals will lie in the same plane at an angle 120° with each other to have lowest energy. This is a case of sp^2 (read *s-p* two) hybridisation. These three orbitals can now overlap with orbitals of chlorine to give BCl_3. Example of sp^2 hybridisation is also found in carbon compounds. Thus in ethylene molecule C shows sp^2 hybridisation. Two of these three orbitals overlap with $1s$ orbitals

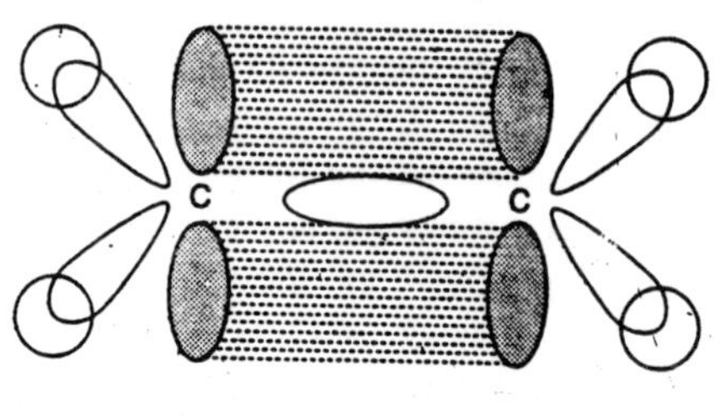

Fig. 2.7

of H-atoms the remaining orbital overlaps with identical orbital of another 'C' atom. These are σ-bonds. The unhybridised p_z orbitals of each 'C' remain at an angle 90° with these. Two such p_z orbitals of two C-atom overlap laterally. Thus one π-bond is also produced.

[Circles are 1*s* orbitals of H-atom. Shaded orbitals are unhybridised p_z orbitals of two C-atoms. They overlap laterally to form π-bond. There are five σ and one π-bond.]

(c) ***s-p* hybridisation :** When one '*s*' orbital mix up with only one '*p*' orbital, then it is a case of *sp*-hybridisation. In acetyelene molecule we find such case. $2p_y$, and $2p_z$ orbitals of carbon do not hybridise. Thus they overlap laterally to produce π-bond. In *s-p* hybridisation since only two hybride orbitals are produced, condition of lowest energy demands that they are at an angle of 180°.

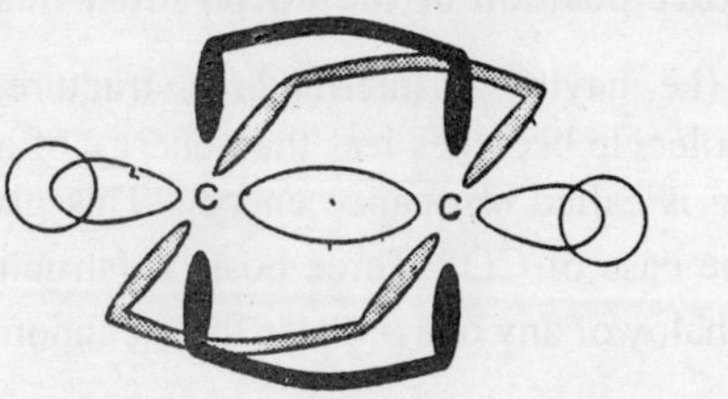

Fig. 2.8

[Circles are 1*s* orbitals of H-atom. Unhybridised p_y and p_z orbitals are shown as  and  shapes. They overlap to produce π-bond. So in C_2H_2 there are three σ-bonds and two π-bonds.]

(d) **Hybridisation involving *d*-orbitals :** Involvement of '*d*' orbitals in hybridisation process may lead to several possibilities. Most important among them are dsp^3; d^2sp^3 and dsp^2 type. The names indicate the number and types of orbitals used in hybridisation purpose. The geometries are trigonal bipyramid for dsp^3; octahedral for d^2sp^3 and square plannar for dsp^2.

2.16 RESONANCE OR MESOMERISM

It has been found that there are many compounds for which a single structural formula can not explain all its properties. A number of structural formulas can be proposed and all have the capability of explaining many of its properties. Actual structure is therefore some sort of a peculiar mixture of these structural formulas. As a crude comparison we may consider visible light as a mixture of seven different colours. **Ingold (1933)** called this phenomena **mesomerism** (meaning intermediate structure). **Heisenberg** gave some theoretical explanation of this phenomena from quantum mechanics. He called it **resonance** and that name is widely used. Let us consider the structure of carbonate ion (CO_3^{2-}). Since all three oxygen atoms are found to be in identical positions, the actual structure is a mixture of following three structures in equal proportions.

Above three structures are called **cannonical** forms or resonating structures of CO_3^{2-}. Since in the formation of CO_3^{2-} these three structures have equal contribution, we can consider every oxygen atom getting a negative charge of $\frac{2}{3}$ units. The three C — O bond lengths are same and it is $\frac{2 \times 1.43 + 1.20}{3} \approx 1.35$ A° [as C ==O bond length is 1.20 A° as C — O bond length is 1.43 A°].

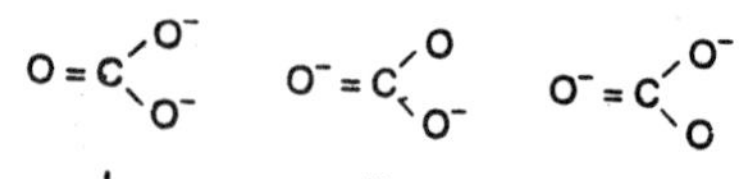

I II III

Fig. 2.9 Resonating structures of CO_3^{2-} ion.

Condition for observing resonance are —

(a) **Resonating structure must have approximately same energy.**

(b) **Number of unpaired electrons in each structure must be same and**

(c) **In each structure position of the nuclei must be same.**

Due to resonance (i.e. having an intermediate structure among the cannonical forms) actual energy of the molecule becomes less than energy of any of the proposed structures. The extent of lowering is called resonance energy. This gives some extra stability.

Let us consider the case of CO_2. Three possible structure are proposed —

The calculated enthalpy of any one of these three cannonical forms is ~ 1590 KJ mole^{-1}

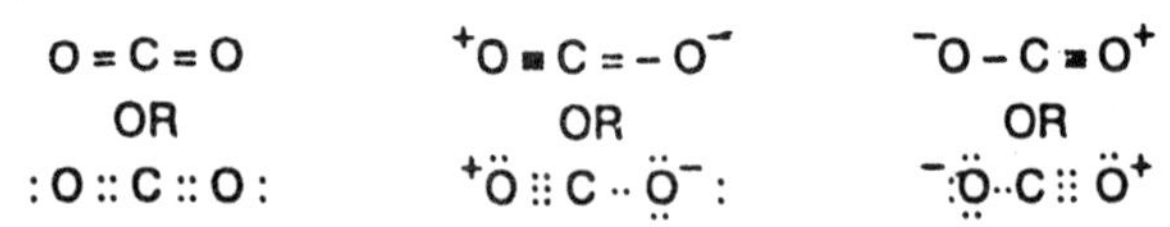

I II III

Fig. 2.10

but the experimental value calculated from heat of formation is ~ 1464 KJ mole^{-1}. The difference ~ 126 KJ mole^{-1} is resonance energy, which makes the real structure more stable than cannonical forms.

Resonance is widely found in organic compounds. Benzene may be represented by any one of the following structures —

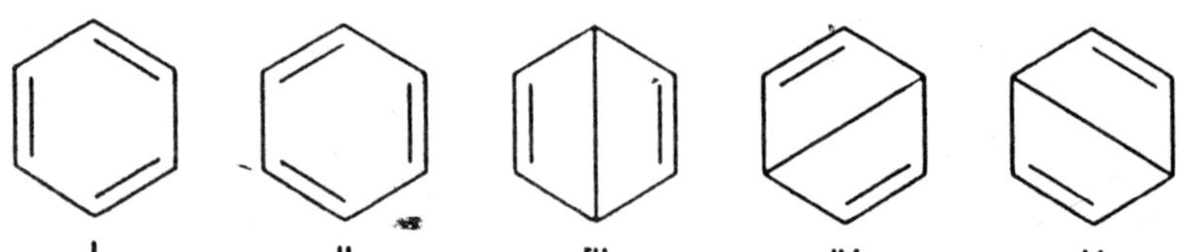

I II III IV V

Fig. 2.11

Resonance energy of C_6H_6 molecule is ~ 167.4 KJ mole^{-1}, which means that actual structure of C_6H_6 is a hybride of these and is more stable.

2.17 V S E P R THEORY

The idea of valence-bond theory and the concept of hybridisation inspired **Gillespie** to put forward valence shell electron pair repulsion theory (abbreviated as **V S E P R** but pronunciated as **vesper**) to explain geometry of molecules. It states that electron pairs of valence orbits (bonded pair or lone pair) surrounding the central atoms will be arranged in space in such a manner that electrostatic repulsion will be minimum. Thus if there are only two pairs they will form bond at an angle of 180°. If there are three pair they will remain at an angle 120° apart. When four such pairs are involved they will try to accommodate along four sides of a tetrahedron at an angle 109°28'. However, there may come distortion of shape and widening or shortening of bond angles due to various reasons. These will be discussed on the basis of following assumptions —

(a) **Multiple bond will be treated as a single electron pair bond and**

(b) **Order of repulsion follow the sequence —**

$$lp\text{-}lp >> lp\text{-}bp > bp\text{-}bp$$

[lp = lone pair; bp = bonded pair]

Thus presence of lone pair will push bonded pair closer and bond angle will be shortened.

Some examples may now be considered :

(i) CO_2 Carbon dioxide has two bonded pair — so bond angle is 180°; structure is linear.

Fig. 2.12

(ii) BF_3 In BF_3 molecule there are three bonded pairs. So bond angle is 120°. Structure is trigonal planar.

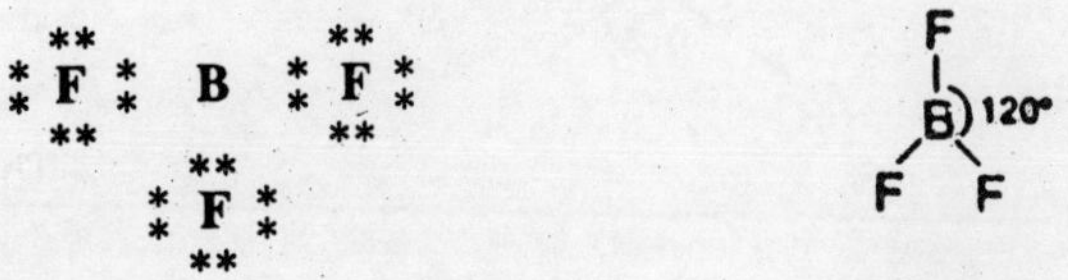

Fig. 2.13

(iii) CH_4 In CH_4 molecule there are four bonded pairs. So it will have the structure of a regular tetrahedron. Bond angle = 109°28'.

Fig. 2.14

(iv) NH_3 In NH_3 we have three bonded pairs and one non-bonded pair. Because of four pairs it should have a tetrahedral arrangement with bond angle 109° 28'. But because of greater repulsive force exerted by non-bonded pair bond angle is shortened and NH_3 has a pyramidal shape with bond angle 107.3°.

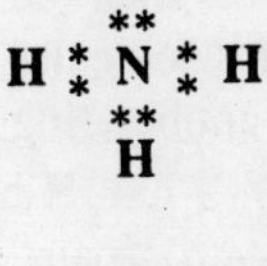

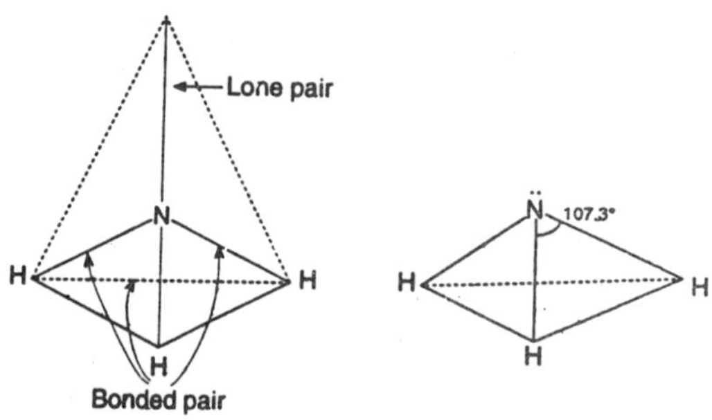

Fig. 2.15

(v) H_2O

$$H \overset{**}{\underset{**}{\overset{*}{\underset{*}{}}O\overset{*}{\underset{*}{}}}} H$$

In water there are two bonded and two non-bonded pairs. Because of four pairs tetrahedral arrangement with 109° 28' bond angle is anticipated. But with a much greater (more than NH_3) repulsion exerted by two non-bonded pairs, bond angle is considerably shortened to 105° Thus water has an angular structure.

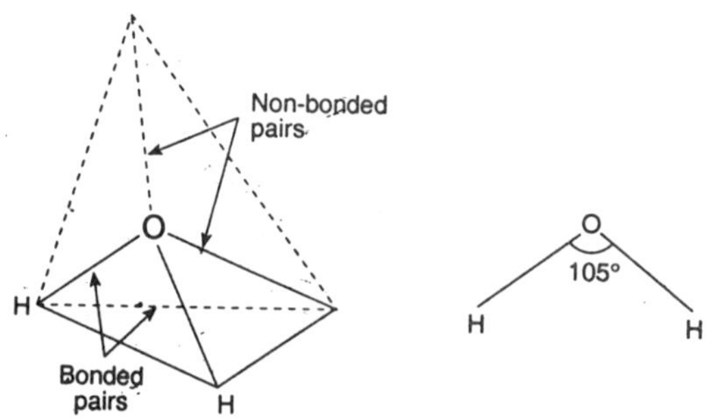

Fig. 2.16

2.18 MOLECULAR ORBITAL (M. O.) THEORY

Valence -Bond (V. B.) theory assumes that during bond formation between atoms, only valence electrons (i.e. electrons of outermost orbital with unpaired spin) participate. Their orbitals overlap and form covalent bond. Remaining electrons of the combining atoms (i.e. electrons of inner orbitals) remain intact and they do not participate in bond formation. Molecular orbital (M. O.) theory which was developed by Mullikan, Huckel and Hund explains bond formation in an entirely different manner. It assumes that just like an atom (which has atomic orbitals to accommodate extranuclear electrons) a molecule has molecular orbitals to accommodate all the electrons of a molecule. These molecular orbitals are formed by linear combination of atomic orbitals (L. C. A. O). Actually orbitals of combining atoms mix up to give rise an equivalent number of new orbitals, which now belong to the molecule. Quantum mechanical calculation shows that when two atomic orbitals combine they can produce two molecular orbitals, one having energy less than the total energies of two A. O's and the other having energy more than this total. Molecular orbital having less energy is called bonding and the other having more energy is antibonding (having a '*' mark).

Formation of some molecular orbitals are shown in the following table —

Table 2.6

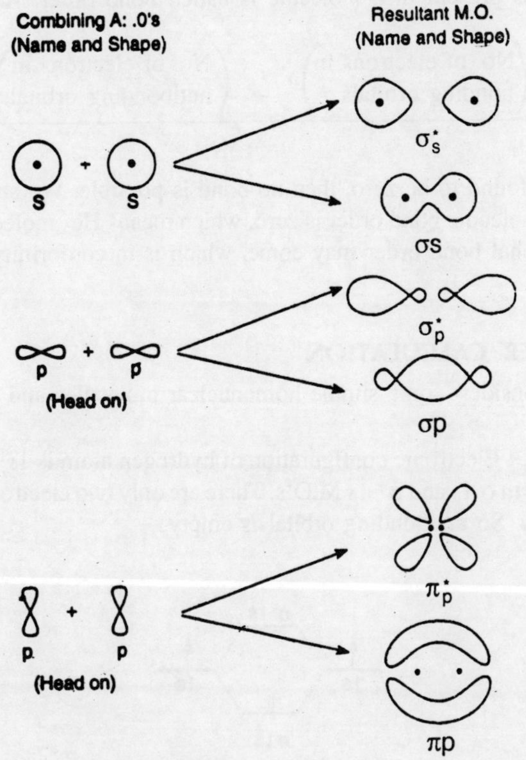

Energy-values of the M.O.s follow the order —

$$\sigma\,(1s) < \sigma\,*\,(1s) < \sigma\,(2s) < \sigma\,*\,(2s) < \sigma\,(2p) < \begin{matrix}\pi\,2p\\ \pi\,2p\end{matrix} < \begin{matrix}\pi^*\,2p\\ \pi^*\,2p\end{matrix} < \sigma\,*\,(2p)$$

Filling up these orbitals will follow 'aufbau' principle.

Following figures is the general representation of different M.O.s arranged in order of their energies.

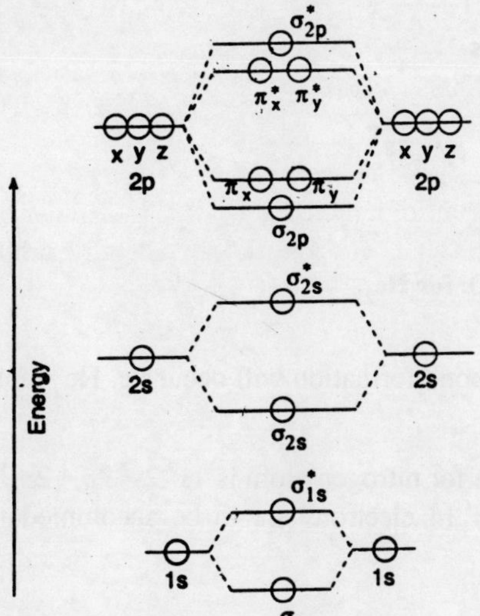

Fig. 2.17 (a)

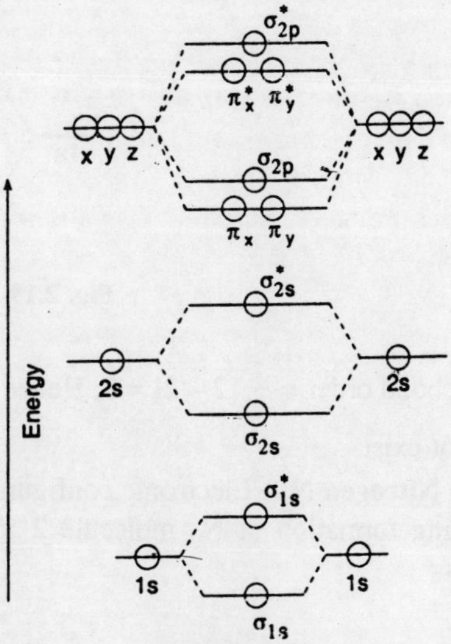

Fig. 2.17 (b)

2.19 BOND ORDER

Number of bonds present in a molecule is called bond order. According to M. O. theory —

$$\text{Bond order} = \frac{\begin{pmatrix}\text{No. of electrons in}\\ \text{bonding orbitals}\end{pmatrix} - \begin{pmatrix}\text{No. of electrons in}\\ \text{antibonding orbitals}\end{pmatrix}}{2}$$

If bond order is found to be zero, then no bond is possible. We shall see in the next section that for He_2 molecule bond order is zero, which means He_2 molecule does not exist. In many cases fractional bond order may come, which is in conformity with the concept of resonance.

2.20 BOND ORDER CALCULATION

We shall now consider some simple homonuclear molecules and calculate the bond order.

(a) **Hydrogen H_2 :** Electronic configuration of hydrogen atom is $1s^1$. So two $1s$ atomic orbitals combine to form σ $1s$ and σ^* $1s$ M.O's. There are only two electrons and they occupy bonding orbitals σ $1s$. So antibonding orbital is empty.

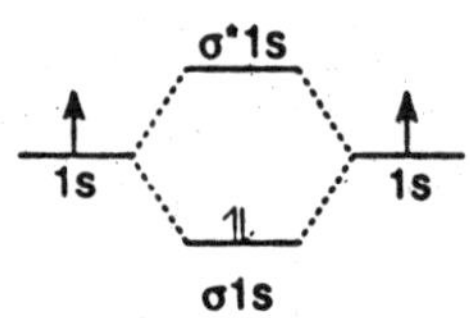

Fig. 2.18 M. O. for H_2.

Hence bond order $= \frac{1}{2}$ [2 – 0] = 1. i.e. one covalent bond exists.

(b) **Helium He_2 :** Electronic configuration of He atom is $1s^2$. So if there is any He_2 molecule it will have four electrons distributed as below —

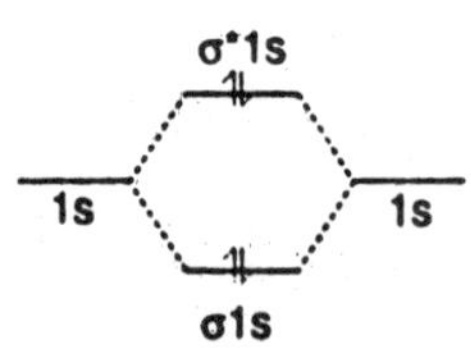

Fig. 2.19 M. O. for He_2.

So bond order $= \frac{1}{2}$ [2 – 2] = 0. Hence no bond formation will occur i.e. He_2 molecule does not exist.

(c) **Nitrogen N_2 :** Electronic configuration for nitrogen atom is $1s^2\,2s^2\,2p_x^1\,2p_y^1\,2p_z^1$. So during formation of N_2 molecule 2 × 7 = 14 electrons are to be accommodated as follows :

Figure shows that number of electrons present in bonding orbitals (σ_{1s} ; σ_{2s} ; σ_{2p} and $2\pi_{2p}$) is 10 and number of electrons present in anti-bonding orbitals (σ^*_{1s} ; σ^*_{2s}) is 4. So bond order = $\frac{1}{2}$ [10 – 4] = 3. Hence two nitrogen atoms are bonded by three covalent i.e. a tripple bond.

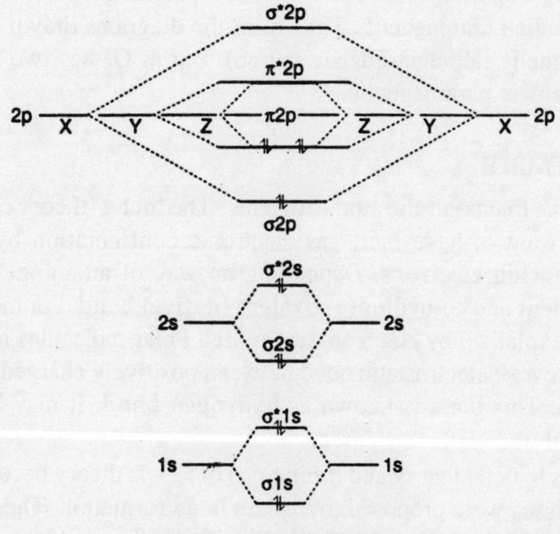

Fig. 2.20 M. O. for N_2.

(d) **Oxygen O_2** : Oxygen has the electronic configuration $1s^2$, $2s^2$, $2p_x^2$, $2p_y^1$, $2p_z^1$. So in O_2 molecule 2 × 8 = 16 electrons are to be accommodated in various bonding and antibonding orbitals. This can be done as follows —

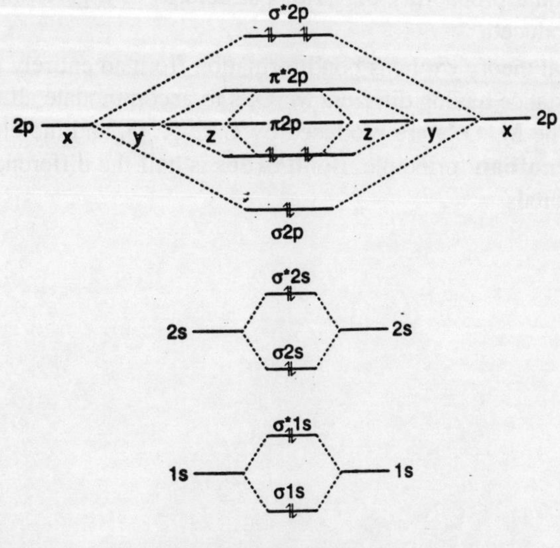

Fig. 2.21 M. O. for O_2.

Here number of electrons present in bonding orbitals (σ_{1s} ; σ_{2s} ; σ_{2p} and $2\pi_{2p}$) is 10 and number of electrons in antibonding orbitals (σ^*_{1s} ; σ^*_{2s} ; π^*_{2p} ; π^*_{2p}) is 6. So bond order $= \frac{1}{2}$ [10 − 6] = 2. So there are two covalent bonds i.e. one double bond in oxygen molecule.

2.21 PARAMAGNETISM AND DIAMAGNETISM

Molecules having unpaired electrons are called paramagnetic whereas those having paired electrons are called diamagnetic. Thus from the diagrams drawn in section 2.20 we see that N_2 is diamagnetic (all electrons are paired). But as O_2 has two unpaired electrons in two π^*_{2p} orbitals it is paramagnetic.

POINTS TO REMEMBER

A '**bond**' is a force that holds the atoms together. The '**octet**' theory conceived by **Lewis** assumes that atoms want to have inert gas electronic configuration by **(i) donating (ii) accepting or (iii) sharing electrons**. Depending on way of attaining 'octet' bonds were named as **ionic**; **covalent** and **co-ordinate covalent (dative) bond**. For metals bond between constituent atoms is explained by **electron-sea model**. Polar molecules involving hydrogen atom possess another weak electrostatic bond between positively charged H -atom and more electronegative atom. This bond is known as **hydrogen bond**. It may be **intermolecular** or even **intramolecular.**

When exceptions to octet theory and limitations of Lewis theory become more and more prominent, other theories were proposed to explain bond formation. These are (a) **Valence-bond theory** and (b) **Molecular-orbital theory**. V. B. theory believes that orbitals of unpaired valence electrons overlap to form bond. For **head-on overlap (*s-s*; *s-p* and *p-p*) bond orbitals are symmetrical and these are σ-bonds. For lateral overlap (*p-p*) bond orbitals are antisymmetrical and they are π-bonds**. Concept of **hybrid orbitals** was introduced to explain why four C—H bonds in CH_4 are exactly equivalent. Different types of **hybridisation viz.** sp^3; sp^2; sp; dsp^3; d^2sp^3 etc. explain geometry of the molecules very well. Considering equal probability of various structures of a given molecule concept of **resonance** was introduced.

Molecular orbital theory explains bond formation from an entirely new angle. It treats a molecule as a substance having different M. O.'s to accommodate all the electrons of the combining atoms. The **M. O.'s** are produced by **L. C. A. O.** and the electrons are accommodated following '**aufbau**' principle. **Bond order** is half the difference betweenbonding and antibonding orbitals.

QUESTIONS

SHORT TYPE

1. Name the type of bond (or bonds) present in the following molecules : (a) $CaCl_2$ (b) NH_4 Cl (c) HCl (d) 2-nitrophenol (e) Ag-metal.
2. Discuss the nature and shapes of sp^3 ; sp^2 and sp hybrid orbitals.
3. 'Covalent bond is directional but ionic bond is not' — why ?
4. The term 'Molecular weight of NaCl' is meaningless — why ?
5. 'H_2S is a gas but H_2O is a liquid' — why ?
6. 'Phosphorous can form pentachloride but nitrogen can not' — why ?
7. Explain the following order of bond lengths $C - C > C = C > C \equiv C$.
8. Why BF_3 has a plannar structure but NH_3 pyramidal ?
9. Name some molecules where 'octet rule' has been violated.
10. What are σ and π bonds ?
11. How many σ and π bonds are present in C_2H_2 molecule ?
12. What are cannonical structure ?
13. C – O bond length in CO_3^{2-} is some what in between C – O and C = O bond lengths' — why ?
14. Draw different mesomeric varieties of C_6H_6.
15. Bond angle in NH_3 molecule is less than 109° 28' — why ?
16. Establish the angular structure of O_2 molecule.
17. Show that Be_2 (At. No. 4) can not exist.
18. N_2 is diamagnetic; but O_2 is paramagnetic — why ?
19. What is bond order ?
20. Arrange different molecular orbitals in order of increasing energy.

ESSAY TYPE

1. What is a bond. Explain Lewis theory for different types of bond formation.
2. Describe the 'electron-sea model' for metallic bond and discuss how it explains electrical and thermal conductivity and also lusture of metals.
3. What is a hydrogen bond ? How many types of H-bonds are possible ? What are the characteristics of compounds having H-bond ?
4. Discuss the fundamental assumption of valence-bond and Molecular-orbital theories. Explain the structure of N_2-molecules from both these theories.
5. What is hybridisation ? What will be the shapes of (1) sp^3 hybridised (ii) dsp^2 hybridised and (iii) d^2sp^3 hybridised molecules ?
6. Briefly discuss the concept of resonance to explain geometry of covalent compounds. Mention the essential conditions for resonance. What is resonance energy ?
7. Discuss the formation of sigma and pi bonds with necessary diagrams. Which bond is stronger ?
8. Explain the important features of V S E P R theory. With its help explain the geometries of BF_3; NH_3 and H_2O molecules.

9. What are σ and π molecular orbitals and what are bonding and antibonding orbitals ?
10. What type of magnetism will be shown by O_2 molecule according to (i) Valence-bond theory (ii) Molecular orbital theory. What is the experimental observation and what is your comment on this regard ?

OBJECTIVE TYPE

1. Select the correct answer from the list in the following below.

(a) Weaker bond among the following types is
  (i) ionic bond (ii) H-bond (iii) Covalent bond.

(b) The correct term for NaCl is
  (i) Atomic weight (ii) Molecular weight (iii) Formula weight.

(c) K^+ is iso-electronic with
  (i) NO_3^- (ii) Cl^- (iii) Na^+.

(d) sp^3 hybride orbitals are
  (i) square-plannar (ii) trigonal (iii) tetrahedral

(e) Carbon-Carbon bond in acetylene contains
  (i) 3σ bonds (ii) 2σ and 1π bonds (iii) 1σ and 2π bonds.

(f) Bond order of Ne_2 molecule is — (i) zero (ii) 2 (iii) Negative.

(g) Resonance is also known as
  (i) Isomerism (ii) Mesomerism (iii) Hyperconjugation.

2. Match terms of column A with these of column B.

A	B
a) Lewis	a) Resonance
b) Heitler and London	b) M. O. theory
c) Ingold	c) VSEPR theory
d) Gillespie	d) Octet theory
e) Hund and Mullikan	e) Valence Bond theory.

3. Write T for true and F for false statements in the boxes.

(a) Number of covalent bonds is the difference between number of bonding and antibonding electrons. ☐

(b) Covalent bonds have directional properties. ☐

(c) Lone-pair — loan pair repulsion > lone pair — bonded pair repulsion > bonded pair — bonded pair repulsion. ☐

(d) A pi bond is formed by lateral overlapping of two pi-orbitals. ☐

(e) Overlapping of a '*s*' and a '*p*' orbital form pi bond. ☐

(f) Resonance energy is a measurement of stability. ☐

(g) VSEPR theory predicts a linear structure for H_2O but angular for CO_2. ☐

(h) O_2-molecule exhibits diamagnetism. ☐

(i) M. O. theory is more successful than V. B. theoy to explain magnetic character. ☐

(j) Energies of antibonding orbitals are more than those of bonding orbitals. ☐

Answers :

1. (a) (ii) ; (b) (iii) ; (c) (ii) ; (d) (iii) ; (e) (iii) ; (f) (i) ; (g) (ii) ;

2. a – d ; b – e ; c – a ; d – c ; e – b

3. (a) F (b) T (c) T (d) T (e) F (f) T (g) F (h) F (i) T (j) T.

Chapter 3

RADIOACTIVITY AND NUCLEAR CHEMISTRY

3.1 THE DISCOVERY

Henry Becquerel was studying phosphorescence phenomena of certain minerals. Among his substance was a piece of mineral containing uranium. Accidently he kept this substance over a photographic plate which was thoroughly wrapped by thick black paper and was kept in his desk which was dark. On developing the photographic plate he found a spot in the position over which the mineral was kept. This led him to conclude that some **radiation** is being emitted by the mineral and this radiation is capable of penetrating. The phenomena was termed as **radioactivity.** It was concluded that heavy metal uranium is responsible for this. Madam Curie and Piere Curie noticed that radioactivity of pitch-blende (a mineral of uranium) is much more than that of equal amount of pure uranium. Hectic research led to the discovery of another radioactive metal, who they named as **Polonium** in honour of their motherland **Poland**. However, uranium and polonium can not quantitatively explain radioactivity of pitch-blende. Pain-staking work showed that almost a **tonne** of pitch-blende contains **few milligrams** of a pure substance, whose radioactivity is almost a **million times more** than that of uranium. This new element was named as **radium.**

3.2 NATURE OF RADIATIONS

Rutherford carried out following experiment to know the nature of radiations emitted by radioactive substances. Incidently we define **radioactivity as the spontaneous disintegration of unstable atoms accompanied by emission of various radiations. The disintegration process is never influenced by change of any external factor viz. temperature, pressure, cosmic ray etc.**

Inside a lead box (thick lead can absorb radioactive radiations) a piece of radioactive material was taken. There is an opening in the box through which radiations come out. Two charged plates are kept in their path and there is a ZnS-screen where the radiations cause luminescence. From the positions of the luminescence it was concluded that radioactive radiations consist of three types, one of which is not deflected by electrical field (γ-rays); another showing little deflection towards negative plate (α-rays) and the remaining showing a very large deflection towards positive plate (β-rays).

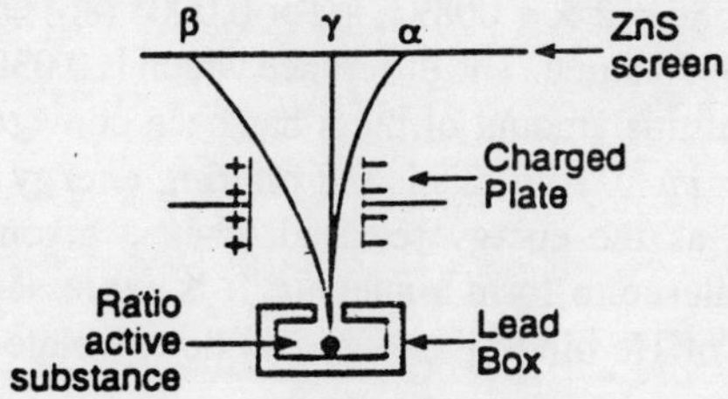

Fig. 3.1

3.3 PROPERTIES OF RADIATIONS

Properties of radioactive radiations (viz. α, β and γ rays) can be best studied under the headings — (i) Nature (ii) velocity (iii) penetrating capacity and (iv) ionising power. We are discussing it in a tabular form.

Table3.1

	α–rays	β–rays	γ–rays
1. Nature	These are charged particles having two units of positive charge and four units of mass i.e. double positive He- atom $^4He_2$	Same as electrons i.e. having one unit of negative charge and $\frac{1}{1827}$ a.m.u. mass.	These are nothing but electromagnetic radiation.
2. Velocity	About $\frac{1}{10}$th of the velocity of light.	Nearly same as the velocity of light.	Exactly equal to the velocity of light.
3. Penetrating capacity	Because of large size andmass penetrating capacity is small. Can be stopped by 0.01 mm thick aluminium.	Penetrating power is 100 times more than α-particle; can be stopped by 1 c.m. thick aluminium.	Penetrating power very very high; can pass through 5 cm thick lead or several meteres of concrete layer.
4. Ionising Power	Because of strong attraction for electrons it can cause intense ionisation.	Ionising capacity is $\frac{1}{100}$th of that of α-particles.	Very weak ionisers.

3.4 REASON FOR RADIOACTIVITY — NUCLEAR STABILITY

Since in an atom all the protons reside at the nucleus a question may come in our mind about the stability of nucleus in spite of strong repulsive forces among positively charged protons. To account for the stability of the nucleus various ideas have been proposed. These are (a) Idea of binding energy (b) Meson theory (c) Liquid drop model of the nucleus and (d) Nuclear-shell model.

(a) **Binding Energy and Packing Fraction :** It has been found that mass of an atom is slightly less than total masses of all subatomic particles present in it. Let us make some calculation for He atom. It consists of 2 protons (each of mass 1.00758 amu); 2 neutrons (mass = 1.00893 e.m.u.) and 2 electrons (mass = 0.0005486 a.m.u.). So its exact mass should be = $(2 \times 1.00758 + 2 \times 1.00893 + 2 \times 0.0005486) = 4.0341172$ a.m.u. But the actual mass of He is 4.00390 a.m.u. The difference which is 0.0302172 a.m.u. is called **mass defect**. It is believed that this amount of mass has been converted to energy, according to Einstein's equation ($E = mc^2$). And this is the **binding energy** of the nucleus. So binding energy may be defined as the energy released when a given number of nucleons (i.e. neutrons & protons) coalesce to form a nucleus. It is expressed as energy per neucleons. Thus in the above case of He binding energy can be calculated as follows —

We first calculate energy equivalent to 1 a.m.u.

1 a.m.u. = 1.66×10^{-27} kg. and $c = 2.998 \times 10^8$ metre sec^{-1}.

So $E = mc^2 = 1.66 \times 10^{-27} \times (2.998 \times 10^8)^2 = 1.4922 \times 10^{-10}$ Joule.

Now one million electron volt i.e. 1 MeV $\equiv 1.602 \times 10^{-13}$ Joule.

$$\text{So energy equivalent to one e.m.u.} = \frac{1.4922 \times 10^{-10}}{1.602 \times 10^{-13}} = 931.5 \text{ MeV.}$$

In case of He mass defect = 0.03021 a.m.u.

$$\text{So binding energy per neucleon} = \frac{0.03021 \times 931.5}{4} \text{ MeV.}$$

$\because$ there are two neutrons and two protons). It is 7.03 MeV. Binding energy per neucleon is small for lighter elements. It gradually increases and rises to a maximum of 9 MeV and then falls slowly to about 7.5 MeV for heavy radioactive elements such as U. These are unstable neucleus.

We know that mass number of an element is the total number of neutrons and protons. So it is a whole number. Now, since mass of either proton or neutron is little more than 1 a.m.u. so actual mass should always be greater than mass number. But just now we have learnt about mass defect i.e. lowering of total mass due to conversion of some mass in to energy. Thus actual mass is again little lower than expected. **Aston** defined a term **packing fraction**, which is —

$$\textbf{Packing Fraction} = \frac{\text{Actual mass} - \text{Mass number}}{\text{Mass Number}} \times 10^4.$$

$$\text{For He it will be} = \frac{4.0039 - 4}{4} \times 10^4 = +\,9.75.$$

Mass number for Nickel is 58; but actual mass = 57.942 due to mass defect.

$$\text{So its packing fraction} = \frac{57.942 - 58}{58} \times 10^4 = -\,10.$$ Packing fraction may be positive or negative. Negative value means greater mass defect and more stability.

(b) **Meson field theory :** In 1935 H. YuKawa predicted existence of a particle, which he called 'meson' in the neucleus. Their mass is about 275 times that of electron and they may bear one unit positive charge (π-meson) or one unit negative charge (μ-meson) or no charge (k-meson). They act as a bridge between proton and neutron to hold them together. Thus if a proton ejects a π-meson it becomes neutron and if neutron absorbs the π-meson it becomes proton. Similarly if neutron ejects a μ-meson it becomes proton and if proton absorbs that meson it becomes neutron i.e.

$$p \rightleftharpoons n + \pi\text{-meson}$$

$$\text{and} \quad n \rightleftharpoons p + \mu\text{-meson.}$$

Rapid exchange of particles is cause for binding force of the nucleus.

(c) **Liquid- drop model :** According to this idea nucleus is supposed to be a liquid drop having unbelivable high density (~ 10^{14} gms/c.c.) and very very high surface tension (~ 9.3 $\times 10^{18}$ Newton metre^{-1}). It is this very high surface tension which keeps the particle united inside a very very small sphere. With increase in mass number deformity in shape begins to come. And when mass number is quite high e.g. uranium atom, the deformed nucleus breaks down. The theory is particularly suitable to explain nuclear fission.

(d) **Nuclear-shell model :** It is believed that just like electrons, neueleons have also different energy levels. The idea gets support from the fact that neuclei having 2, 8, 20,

28, 50, 82 or 126 neutrons or protons are very much stable (cf. electronic configuration of stable inert gases). These numbers are now regarded as **magic numbers**. These particles are arranged in shells as paired nucleons (with paired spin). **Thus nucleus with even number of protons and neutrons are quite stable; those with odd number are unstable.** The stability of the neucleus seems to depend on neutron, proton ratio (***n* : *p* value). When it is 1 the neucleus is quite stable.** For uranium $n : p = 1.56$. It accounts for its unstability. The following $n : p$ ratio determining curve speaks about stability.

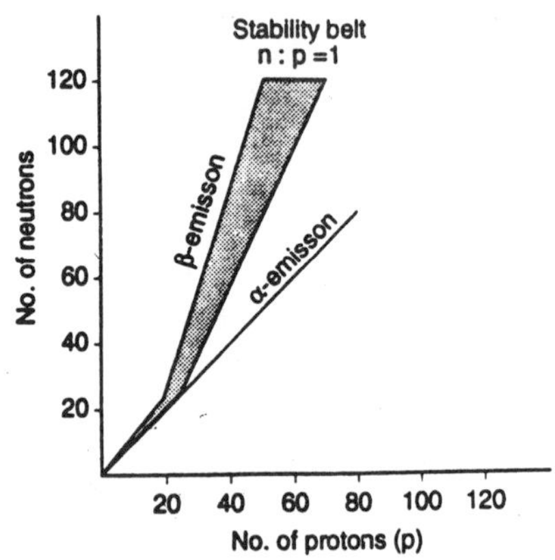

Fig. 3.2

Points to be noted from this curve are —

(i) Up to atomic number 20 (i.e., no. of proton = 20) elements are stable having $n : p = 1$.

(ii) Shadded region is stability belt, i.e. elements whose number of neutrons and protons fall in this belt are quite stable. Elements having '*n*' and '*p*' values outside this belt are radioactive and unstable.

(iii) Nuclei whose *n* and *p* values lie above the shaded belt usually emits β-particles and thus increases number of protons (see displacement law section : 3.5) and nuclei whose *n* and *p* values lie below this belt emits α-particles to increase $n : p$ ratio.

3.5 GROUP DISPLACEMENT LAW

We have seen that radioactive elements emit α and β particles and also γ-ray. Now emission comes from neucleus, so such emission will cause change in number of proton and neutron in the neucleus. The result is a change in nature of element i.e. after emission the radioactive elements will be converted to a new element. Hence their position in the periodic table will change. Soddy and Fajan stated the following law regarding this. '**An element formed by the emission of α-particle by a radioactive element occupies a position two groups left to the position of parent element in the periodic table; while an element produced by β-particle emission occupies a position, one group right to the parent element in the periodic table.**'

Explanation : α-particles are $^4He_2$ nucleus, so emission of α-particles decreases mass number by '4' and atomic number by 2. Thus the daughter element will have atomic number less by 2, so its position will be two groups left in the periodic table. Thus emission of α-particle from $^{238}_{92}U$ gives $^{234}_{90}Th$.

β-particle i.e. electrons are ejected from nucleus from neutron to (proton plus electron) conversion. Thus number of proton and hence atomic number increases by one. So the new element occupies a position, one group right to the parent element. Thus $^{234}_{90}Th$ becomes $^{234}_{91}Pa$ after emitting β-particle.

3.6 RADIOACTIVE DISINTEGRATION SERIES

Number of naturally occurring radio isotopes is about forty. They belong to one of three successive disintegration chains. These are known as radioactive disintegration series. The series are $4n$ (or Thorium) series; $4n + 2$ (or Uranium) series and $4n + 3$ (or Actinium) series. The term $4n$ means mass number is divisible by 4; whereas $4n + 2$ and $4n + 3$ means mass number when divided by 4 gives a remainder of 2 and 3 respectively.

In the $4n$ series **starting element is $^{232}_{90}Th$** and through a chain of successive disintegration it gives **ultimately $^{208}_{82}Pb$**.

Starting element of ($4n + 2$) series is $^{238}_{92}U$ and end product is $^{206}_{82}Pb$.

For ($4n + 3$) series starting element is $^{235}_{92}U$ and end product is $^{207}_{82}Pb$.

An artificially made disintegration series is Neptunium series ($4n + 1$). Its starts with $^{237}_{93}Np$ and ends at $^{200}_{83}Bi$.

Details of disintegration chart of ($4n + 2$) series is given below —

$$^{238}_{92}U \xrightarrow{\alpha} {}^{234}_{90}Th \xrightarrow{\beta} {}^{234}_{91}Pa \xrightarrow{\beta} {}^{234}_{92}U \xrightarrow{\alpha} {}^{230}_{90}Th \xrightarrow{\alpha}$$

$$^{226}_{88}Ra \xrightarrow{\alpha} {}^{222}_{86}Rn \xrightarrow{\alpha} {}^{218}_{84}Po \xrightarrow{\alpha} {}^{214}_{82}Pb \xrightarrow{\beta} {}^{214}_{83}Bi \xrightarrow{\beta}$$

$$^{214}_{84}Po \xrightarrow{\alpha} {}^{210}_{82}Pb \xrightarrow{\beta} {}^{210}_{83}Bi \xrightarrow{\beta} {}^{210}_{84}Po \xrightarrow{\alpha} {}^{206}_{82}Pb$$

3.7 LAW OF RADIOACTIVE DISINTEGRATION

Radioactivity is a spontaneous disintegration of the nucleus. It is not affected by any external factor and even chemical change. Number of nuclei disintegrating per unit time is called rate of disintegration. It is expressed in the differential form as $-\frac{dN}{dt}$. According to the law of radioactive disintegration **'rate of disintegration at any moment is directly proportional to number of radioactive atoms present at that moment.'** Thus

$$\frac{dN}{dt} \propto N$$ [N = No. of atoms present at the time of measuring rate.]

or, $$-\frac{dN}{dt} = \lambda N$$ [λ = Proportionality constant, known as decay constant]

or, $$-\frac{dN}{N} = \lambda\, dt$$

If N_0 is the number of atoms present at the beginning ($t = 0$) and N is the number present after time 't'

then $$-\int_{N_0}^{N} \frac{dN}{N} = \lambda \int_0^t dt$$

or, $$\ln \frac{N_0}{N} = \lambda t. \qquad \text{......(3.1)}$$

Also $N = N_0 . e^{-\lambda t}$. ......(3.2)

Thus number atoms (radioactive) will decrease exponentially with time. A plot of N vs t will be any asymptote. [See : 1st order reaction in chemical kinetics].

3.8 HALF-LIFE

Time required for decay of half the radioactive atoms present initially is called half-life. Thus in the equation (3.1), if 't' is half life ($t_{1/2}$), then $N = (N_0/2)$. So equation 3.1 becomes

$$\ln\left(\frac{N_0}{N_0/2}\right) = \lambda t_{1/2}$$

or, $$t_{1/2} = \frac{\ln 2}{\lambda} = \frac{2.303 \log 2}{\lambda} = \frac{2.303 \times 0.3010}{\lambda}$$

or, $$t_{1/2} = \frac{0.693}{\lambda}. \qquad \text{......(3.3)}$$

So half-life is a constant.

3.9 AVERAGE LIFE

In a sample of radioactive elements all the atoms present are potentially unstable. But yet all do not disintegrate at a time. Some decay earlier, others survive longer. Statistical average of life of all atoms present at any time is average life.

So $$\textit{Average life} = \frac{\textit{Summation of life-periods of all radioactive atoms}}{\textit{Number of such atoms}}$$

Let in a specimen N_0 is number of radioactive atoms present initially. After time 't' number of atoms become N and rate becomes $-\frac{dN}{dt}$. So now in between times t and $t + dt$, dN atoms are decaying. Since 'dt' is infinitesimally small, life period of these 'dN' number of atoms can be taken as t. So summation of their life periods become $dN.t$. Since life period of a particular radioactive element may have any value between zero to infinity, so

summation of life periods of all atoms $= \int_0^\infty dN.t$.

Hence average life $$\Gamma = \frac{\int_0^\infty dN.t}{N_0}. \qquad \text{......(3.4)}$$

But from law of disintegration —

$$-\frac{dN}{dt} = \lambda N$$

or, $-dN = \lambda N\, dt.$

From equation 3.1

$$N = N_0 . e^{-\lambda t}$$

So $$-dN = \lambda N_0 \; e^{-\lambda t} . dt .$$

Hence from equation (3.4) average life becomes

$$\text{Average life } \Gamma = \int_0^{\infty} \lambda . e^{-\lambda t} . t . dt$$ [$-ve$ sign is ignored. As $-ve$ time is meaning less]

$$\Gamma = \lambda \int_0^{\infty} e^{-\lambda t} . t . dt$$

$$= \lambda \left[\frac{t . e^{-\lambda t}}{-\lambda} - \int \frac{e^{-\lambda t}}{-\lambda} . dt \right]_0^{\infty}$$

$$= \lambda \left[-\frac{t . e^{-\lambda t}}{\lambda} - \frac{e^{-\lambda t}}{\lambda^2} \right]_0^{\infty}$$

$$= \lambda \frac{1}{\lambda^2} \left[(\lambda t + 1) . e^{-\lambda t} \right]_0^{\infty} = \frac{1}{\lambda} \qquad \text{......(3.5)}$$

Since half-life $t = \frac{0.693}{\lambda}$.

So $t_{1/2} = 0.693\ \Gamma$

or, $\Gamma = 1.44\ t_{1/2}$. ......(3.6)

3.10 UNITS OF DISINTEGRATION RATE

Disintegration rate is generally d p s (disintegration per second) . It means how many nuclei are disintegrating per second. Since the number is very high we use 'Curie' as practical unit.

1 Curie = 3.7×10^{10} dps.

Another unit is 'Rutherford' (rd).

1 rd = 10^6 d.p.s.

So 1 c = 3.7×10^4 rd.

3.11 RADIOACTIVE EQUILIBRIUM

Let us consider a radioactive element A disintegrating to give another radioactive substance B which again decays to C.

$$A \longrightarrow B \longrightarrow C$$

Let λ_A and λ_B be respective decay constant. Now rate of formation of B = rate of disintegration of A and rate of formation of C = rate of disintegration of B. If rate of disintegration of A becomes equal to rate of disintegration of B, then number of radioactive

B atoms will remain unchanged with time. This state is called radioactive equilibrium. So at this state

disintegration rate of A = disintegration rate of B

or $\lambda_A N_A = \lambda_B N_B$ (from law of disintregration)

or $\dfrac{\lambda_A}{\lambda_B} = \dfrac{N_A}{N_B}$. [$N_A$ and N_B are number of atoms of *A* and *B*]

But $t_{1/2} \propto \dfrac{1}{\lambda}$.

So $$\frac{N_A}{N_B} = \frac{t_{1/2}(B)}{t_{1/2}(A)}. \qquad ......(3.7)$$

This is the condition for existence of radioactive equilibrium.

3.12 ARTIFICIAL RADIOACTIVITY

Many stable neucleus, when bombarded by high speed particles (like protons, deuterons, neutrons or alpha particles) are converted to unstable radioactive neucleus. This phenomena known as artificial radioactivity was discovered by Irene Curie (daughter of Madam Curie) and her husband F. Joliot in 1934. They found that certain light elements e.g. B; Al: Mg etc. When bombarded with high speed α - particle cause emission of proton; neutron and positrons. Emission of proton and neutron ceases when bombardment is stopped but emission of positron continues for sometimes after bombardment has been stopped. This indicates formation of some radioactive element. Nuclear changes may be written as —

$${}^{27}_{13}\text{Al} + {}^{4}_{2}\text{He} \rightarrow + {}^{30}_{14}\text{Si} + {}^{1}_{1}\text{H}$$

and $${}^{27}_{13}\text{Al} + {}^{4}_{2}\text{He} \rightarrow + {}^{30}_{15}\text{P} + {}^{1}_{0}\text{n}.$$

Thus emission of proton (${}^{1}_{1}\text{H}$) and neutron (${}^{1}_{0}\text{n}$) occurs during bombardment. But ${}^{30}_{15}\text{P}$ is radioactive. So it disintegrates —

$${}^{30}_{15}\text{P} \rightarrow + {}^{30}_{14}\text{Si} + e^{+} \text{ (half-life 3 min. 15 sec).}$$

Thus emission of positron (e^+) continues even for sometimes after bombardment has been stopped.

Certain other examples are —

$${}^{27}_{13}\text{Al} + {}^{1}_{0}\text{n} \rightarrow {}^{24}_{11}\text{Na} + {}^{4}_{2}\text{He} \quad \text{and} \quad {}^{24}_{11}\text{Na} \rightarrow {}^{24}_{12}\text{Mg} + e^{-} \; (\beta\text{-particles})$$

$${}^{19}_{9}\text{F} + {}^{1}_{1}\text{H} \rightarrow {}^{19}_{10}\text{Ne} + {}^{1}_{0}\text{n} \quad \text{and} \quad {}^{19}_{10}\text{Ne} \rightarrow {}^{19}_{9}\text{F} + e^{+}$$

3.13 NUCLEAR REACTION

In section 3.11 we have seen that when high speed particles (now knows as **projectiles**) are bombarded on neucleus (called as **target**) certain new elements are produced. Such changes are called nuclear reaction. *So nuclear reaction is a process that causes change of the nucleus and hence a new element is produced.* It is called **transmutation**. Following **five** types of nuclear reactions are known.

(a) *Projectile is captured and no emission occurs.*

Examples : ${}^{12}_{6}\text{C} + {}^{1}_{1}\text{H} \rightarrow {}^{13}_{7}\text{N} + \gamma$

(b) *Projectile is captured and emission of particle occurs.*

Examples : $^{11}_{5}B + ^{1}_{1}H \rightarrow ^{12}_{6}C + ^{1}_{0}n$

(c) *High speed projectile breaks neucleus of medium weight in to smaller fragments.* The process is **spallation.**

Examples : $^{63}_{29}Cu + ^{4}_{2}He \rightarrow ^{37}_{17}Cl + 14\,^{1}_{1}H + 16\,^{1}_{0}n$

(d) *Projectiles break down very heavy nuclei in to two or more fragments.* The process is **nuclear fission.**

Examples : $^{235}_{92}U + ^{1}_{0}n \rightarrow ^{141}_{56}Ba + ^{92}_{36}Kr + 2 \text{ to } 3\,^{1}_{0}n$

(e) *Light nuclei fuse together to give big nuclei.* This is **nuclear fusion.**

Examples : $^{2}_{1}H + ^{3}_{1}H \rightarrow ^{4}_{2}He + ^{1}_{0}n$

3.14 COMPARISON BETWEEN CHEMICAL REACTION AND NUCLEAR REACTION

A comparison between nuclear and chemical reactions can now be made.

(a) In chemical reactions, extranuclear electrons play the key role, nuclei do not participate.

In nuclear reactions nuclei are the target and they are changed.

(b) In chemical reaction, no new element is produced.

But nuclear reactions produce new elements.

(c) Chemical reactions are greatly influenced by external factors viz. pressure, temperature, catalyst etc.

Nuclear reactions are not at all influenced by such factors.

(d) Chemical reactions are associated with small amount of energy change.

Nuclear reactions are associated with huge amount of energy change.

(e) All chemical reactions are reversible.

Nuclear changes are irreversible.

3.15 FISSION-FUSION

Out of the different types of nuclear reactions discussed in section 3.13, special mention should be made about nuclear fission and nuclear fusion because of their role in construction of nuclear weapons or nuclear reactors; the first leads to destruction whereas second brings development. Let us discuss nuclear fission first. Otto Hann and S. Strassmann observed that when a target of $^{235}_{92}U$ is bombarded by slow neutron then it breaks down in to $^{141}_{56}Ba$ and $^{92}_{36}Kr$. Along with these 3 neutrons are produced. It was found that total masses of all the products is somewhat less than the mass of $^{235}_{92}U$ and the difference (about 0.215 amu) is converted to energy (about 200 MeV) according to Einstein's equation. Now the three neutrons produced will break three more $^{235}_{92}U$ nuclei and produce 9 neutrons; which in their turn will break 9 uranium nuclei and produced 27 neutrons. Thus a **chain reaction** will begin and on unimaginable amount of high energy will be produced. Temperature will rise up to millions of degree celsius and deadly γ-rays will be released. This is the principle behind **atomic bomb**. However, if this reaction is brought under control, as is done in a nuclear reactor then large amount of electricity can be produced.

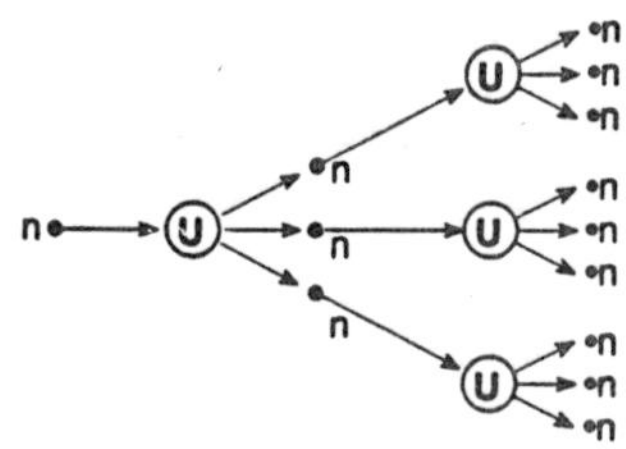

Fig. 3.3

A chain reaction continues till most of original nuclei are fissioned. However, all neutrons released may not cause fission if amount of fissionable substance is not above a certain mass, known as **critical mass.** The exact value of this critical mass and the process of trigerring is a top military secret. Very few countries have this knowledge. India entered the nuclear club in 1974.

Fusion reaction is opposite to fission. Nuclei of light elements can combine together to form heavy nucleus along with some mass loss, which again is converted to energy. However, during fusing light nuclei we are to overcome very strong repulsive force between the positive charges. For this purpose kinetic energy of the nuclei to be fused should be very high and that is done by supply of heat. That is why fusion reactions are also known as **thermonuclear reaction**. Such reactions occur in the stars where high temperature can initiate such a process. In fact energy of the sun and other stars comes from nuclear fusion.

The following reactions are believed to be occurring inside the sun.

$$^{12}_{6}C + ^{1}_{1}H \rightarrow ^{13}_{7}N + \gamma$$

$$^{13}_{7}N \rightarrow ^{13}_{6}C + e^{+} + \gamma$$

$$^{13}_{6}C + ^{1}_{1}H \rightarrow ^{14}_{7}N + \gamma$$

$$^{14}_{7}N + ^{1}_{1}H \rightarrow ^{15}_{8}O + \gamma$$

$$^{15}_{8}O \rightarrow ^{15}_{7}N + e^{+} + \gamma$$

$$^{15}_{7}N + ^{1}_{1}H \rightarrow ^{12}_{6}C + ^{4}_{2}He + \gamma$$

Summing up $4\,^{1}_{1}H \rightarrow ^{4}_{2}He + \text{Energy}$

The principle of thermonuclear or **hydrogen bomb** is based on this fusion reaction. While exact reaction is a strictly military secret, it is believed that H-bomb consists of a small plutonium fission bomb and isotopes of hydrogen. The fission bomb first explodes to supply necessary energy for fusion reaction.

3.16 NUCLEAR REACTORS

Power plants where nuclear fission or fusion reactions can be carried out under controlled conditions and electricity can be generated are called nuclear reactors. At present nuclear reactors all over the world carry out fission reaction. However, there are many risk factors. A number of serious accidents and leakage of radioactive material had occurred both in the U.S.A. and erstwhile USSR. Problem of disposal of radioactive waste material is causing headache. It is therefore most likely that in the coming twenty-first century nuclear reactors will be fusion reactors because of abundance of fusion fuel deuterium and non-radioactive nature of fusion products.

Present day fission reactors consist of fissionable $^{235}_{92}U$ or $^{239}_{94}Pu$ as fuel. They are stacked with heavy water or graphite rod, known as moderator. Their function is to slow down the projectiles, i.e. neutrons. Neutrons produced by fission are captured by the moderators and they loose some of their energies. Thus the rate of the chain reaction can be controlled. The energy released during reaction appear as heat and can be removed by circulating water around the reactor. Steam produced thus is used in steam turbines to generate electricity.

3.17 USES OF RADIO ISOTOPE

Since radioactive nuclei always indicate their presence by emitting radiation, it is quite easy to detect the site where they exist. This phenomena has made radio isotopes important tools in analytical; medicinal and also varieties of other works. Their presence is detected by electrometer; electroscope or Geiger-Muller Counter (GMC). Some of their applications are discussed below.

(a) **As tracer :** Radio isotopes are used to know mechanism of chemical reaction e.g. in the hydrolysis of ester RCOOR if we add water with radioactive oxygen (labelled O_2) then this oxygen is found in the resultant acid, which means – OR group has been replaced by – O*H group.

$$R-\overset{\displaystyle O}{\overset{\|}{C}}-OR + HO^*H \longrightarrow R-\overset{\displaystyle O}{\overset{\|}{C}}-OH + ROH$$

Radio isotopes are also used in tracing the location of brain tumour. A small quantity of radioactive iodine is injected, it is strongly adsorbed by the tumour from where it begins to emit radiation and show its presence.

Radioactive phosphorous is introduced in phosphate fertilisers which are adsorbed in plants and which show how phosphorous influences growth of plants.

(b) **Radiotherapy :** Radioactive Cobalt-60 is used for treating cancer growths. Radioactive phosphorous is used for treatment of leukemia.

(c) **Radio-Carbon dating :** Age of earth; fossils and of many evidences of lost civilisation can be determined by measuring the ratio of $^{12}C_6$ and $^{14}C_6$ and comparing it with normal value.

Few worked out problems

1. *1 gm of Ra - 226 emits 3.7 × 10^{10} α - particles per second. What is its half-life?*

Solution : Rate of disintegration = Rate of α-emission = 3.7×10^{10} = ($-dN/dt$)

$$\text{No. of atoms of Ra in 1 gm} = \frac{6.023 \times 10^{23}}{226} = N$$

$$-\frac{dN}{dt} = \lambda N.$$

$$\text{So} \qquad \lambda = \frac{3.7 \times 10^{10}}{6.023 \times 10^{23}} \times 226$$

$$t_{1/2} = \frac{0.693}{\lambda} = \frac{0.693 \times 6.023 \times 10^{23}}{3.7 \times 10^{10} \times 226} \text{ seconds}$$

$$\equiv 1583 \text{ years.}$$

2. *A piece of wood taken from an old coffin shows* ^{14}C : ^{12}C *ratio as* 0.477 *times a living plant. What is the approximate age of the coffin ?*

[$t_{1/2}$ for ^{14}C *is 5730 yrs]*

Solution : We know if N_0 is the number of radioactive atoms at the beginning and N_t is that after disintegration occurred for a time 't', then

$$\ln \frac{N_t}{N_0} = -\lambda t$$

Also
$$\lambda = \frac{0.693}{t_{1/2}} = \frac{0.693}{5730}.$$

So
$$\ln (0.477) = -\frac{0.693}{5730} \times t$$

or, $t = 6100$ yrs.

POINTS TO REMEMBER

Nucleus of heavy metals e.g. U; Po; Ra; Th etc. are unstable. They spontaneously disintegrate and emit various rays. The emitted rays are **α-rays (double positive helium); β-ray (electron) and γ-ray (electro megnetic radiation of very high frequency)**. α-ray has least penetrating capacity but highest ionising capacity whereas γ-ray has exactly opposite property.

Stability of the nucleus is explained by the concept of (i) mass defect and binding energy (ii) meson-field theory (iii) liquid-drop model and (iv) nuclear-shell model. Nuclei with neutron (n) : proton (p) ratio 1 are very much stable. **A n : p stability curve can be drawn and a stability belt has been identified. Nuclei having n : p value outside the belt are unstable. Those nuclei whose position is above the belt emits β-particles and those below the belt emit α-particle.**

Group displacement law predicts that **emission of α-particle** decreases **atomic number by 2**, whereas **β-emission** raises the **atomic number by 1.**

Radioactive disintegration is a first order process and obeys the first order rate constant.

Half-life is $\frac{\mathbf{0.693}}{\lambda}$ and **average life is** $\frac{1}{\lambda}$ where λ is decay constant. Curie is the unit of measuring disintegration rate; **1 curie = 3.7 × 10^{10} dps.**

Though natural radioactive substances are all heavy metals; light radioactive nuclei can be produced artificially by heating suitable targets with particles (projectiles) like neutron; α - particle; protons etc. Converting one nuclei into another is called nuclear reaction. They are of five types, notable among them are **spallation, fision and fusion.** The last two are used in nuclear reactors and construction of atom bomb and hydrogen bomb. At present radioactive isotopes have found plenty of applications.

QUESTIONS

SHORT TYPE

1. What is radioactivity ? What type of nuclei are radioactive ?
2. What is mass defect ? How is it related with binding energy of the nucleus ?
3. What do you understand from the word 'Packing-fraction' ? Packing fraction for one element is positive and for another it is negative; which one is more stable?
4. $^{14}_{7}N$ is stable but $^{17}_{7}N$ is unstable — what type of emission will occur from here and why ? [*Hints* : $n : p$ ratio is above stability belt.]
5. Complete the following reactions :

   (a) $^{9}_{4}Be + ^{1}_{1}H \rightarrow \ldots\ldots\ldots + ^{4}_{2}He$

   (b) $^{14}_{7}N + ^{14}_{2}He \rightarrow ^{17}_{8}O + \ldots\ldots\ldots$

   (c) $^{238}_{92}U \rightarrow ^{234}_{90}Th + \ldots\ldots\ldots$

   (d) $^{27}_{13}Al + \ldots\ldots\ldots \rightarrow ^{27}_{12}Mg + ^{1}_{1}H$

   (e) $^{37}_{17}Cl + \ldots\ldots\ldots \rightarrow ^{38}_{18}Ar + ^{1}_{0}n$

   (f) $^{214}_{82}Pb \rightarrow ^{214}_{83}Bi + \ldots\ldots\ldots$
6. What are half-life and average life of radioactive elements ? How are they related ?
7. What are 'Curie' and 'Rutherford' units ? State their relation.
8. Write down the fusion reaction occurring inside the stars.

ESSAY TYPE

1. Briefly explain different ideas to account for the stability of the nucleus.
2. State and explain (with specific examples) group-displacement law.
3. State the law of radioactive disintegration and derive expressions for half-life and also average life.
4. Write notes on (a) Artificial radioactivity, (b) Different types of nuclear reaction, (c) Distinction between nuclear and chemical reaction, (d) Radioactive equilibrium.
5. What are 'fission', 'fusion' and 'spallation' ? Explain each with example.
6. Describe some of the uses of radio-isotopes.

PROBLEMS

1. Calculate the packing fraction for $^{16}O_8$ atom given masses of proton and neutron are 1.007277 and 1.008665 a.m.u.

   [*Hints* : Ans = $\dfrac{Actual\ mass - 16}{16}$.]

2. Calculate the mass of ^{14}C having activity 1 Ci and half-life 5770 yrs.

   [*Hints* : $\lambda = \dfrac{0.693}{t_{1/2}}$ and $-\dfrac{dN}{dt} = \lambda N$;

   Here $-\dfrac{dN}{dt} = 3.7 \times 10^{10}$ and $N = \dfrac{x}{14} \times 6.023 \times 10^{23}$; x = gm. of ^{14}C.]

3. Activity of ^{14}C in a living tree is 15.3 disintegration per minute per gram. A piece of ancient wood is found to have ^{14}C activity of 2.42 disintegration per minute per gram. Calculate its approximate age. [$t_{1/2}$ for ^{14}C = 5720 yrs.]

[*Hints* : $t = \frac{2.303}{\lambda} \log \frac{N_0}{N_t}$; Here $\lambda = \frac{0.693}{5720}$; $\frac{N_0}{N_t} = \frac{15.3}{2.42}$.]

4. On analysis of an uranium ore ratio of Pb : U was found to be 15 : 100. Calculate approximate age of earth. Given decay constant for U is 1.52×10^{-10} per year.

[*Hints* : ^{238}U ultimately give ^{206}Pb. So 206 gm Pb comes from 238 gm U i.e. 15 gm came from $\frac{238}{206} \times 15 \approx 17$. So original amount of uranium = 100 + 17 = 117 gm. Now $t = \frac{2.303}{\lambda} \log \frac{N_0}{N_t}$; $\lambda = 1.52 \times 10^{-10}$; $\frac{N_0}{N_t} = \frac{117}{100}$]

5. Half-life of ^{221}Fr is 6.93 minutes. If we start with 1 m.g. of it how much will be left after 23 minutes ?
[**Ans.** = 0.1 mg]

OBJECTIVE TYPE

1. Write T (for true) and F (for false) against the following statements.

(a) An α-particle shows greater deflection than β-particles in electrical field. ☐
(b) A β-particle has the highest penetrating capacity. ☐
(c) γ-ray has the highest velocity. ☐
(d) Negative packing fraction means more stable then positive packing fraction. ☐
(e) Radioactive disintegration follows first order kinetics. ☐
(f) One rutherford unit is greater than one curie unit. ☐
(g) Artificial radioactivity was discovered by Becquerel. ☐
(h) Fusion reaction occur inside stars. ☐
(i) In atom-bomb fission reaction occurs. ☐
(j) Fusion reactors are free from hazards. ☐

2. Fill up the gaps.

(a) Radioactivity is a ——— phenomena.
(b) Nuclear reactions are different from ——— reactions.
(c) ——— particles account for stability of nucleus.
(d) Nuclei having odd number of neutrons and protons are ———.
(e) 2, 8, 20, 28, 50 etc. number of protons and (or) neutrons are called ———.
(f) Group displacement law was given by ———.
(g) Minimum mass required to initiate chain reaction in fission process is ———.
(h) Breaking of medium weight nuclei in to number of smaller particles is called ———.
(i) Radio carbon dating can measure ——— of the earth.
(j) Cobalt - 60 is used for treatment of ———.

3. Match terms of column A with those of column B.

A	B
(a) Soddy	(a) Natural radioactivity
(b) Curie	(b) Nuclear fission

(c) Becqurell (c) Artificial radioactivity

(d) Otto Hann (d) Group displacement law

(e) Irene Curie. (e) Radioactivity unit

Answers :

1. (a) F (b) F (c) T (d) T (e) T (f) F (g) F (h) T (i) T (j) T

2. (a) nuclear (b) chemical (c) meson (d) unstable (e) magic number (f) Soddy (g) critical mass (h) spallation (i) age (j) Cancer

3. a – d; b – e; c – a; d – b; e – c.

Part II
PHYSICAL CHEMISTRY

Chapter 1

IDEAL GAS — KINETIC THEORY

1.1 GAS — A TYPE OF MATTER

If we look at our universe we find and feel existence of only two things. These are — matter and energy. Matter is something, which occupies space and energy is something, which is responsible for any change. Now our world consists of three types of matter viz. solid, liquid and gas. Solid is something having definite mass, volume and geometry. On heating it looses the geometry and becomes liquid. This on heating looses the volume and becomes gas. Thus gas is a state of matter. Besides these three states, there is yet another state, known as plasma, which is a highly ionised state. This state, known as fourth state of matter is found in stars, where temperature is very very high.

1.2 GAS LAWS — A RECOLLECTION

One interesting aspect of properties of gases is that, irrespective of their nature all gases obey certain laws. These are (i) Boyle's law, (ii) Charle's law and (iii) Gay-Lussac's law.

(a) **Boyle's law :** Temperature remaining constant, volume of a definite mass of any gas varies inversely with pressure. So

$$V \propto \frac{1}{P} \text{ [when } T \text{ and mass are constant].}$$

$$\therefore \; PV = \text{constant.}$$

At constant temperature plot of P vs V will be a rectangular hyperbola. Such a plot is called *isothermal.*

(b) **Charle's law :** At constant pressure volume of a definite amount of gas increases (or decrease) by $\frac{1}{273}$th fraction of its original volume for every degree centigrade rise (or fall) of temperature.

Thus if V_0 is volume at 0°C; then volume at + t°C and – t°C will be

$$\left[V_0 + V_0 \cdot \frac{t}{273}\right] \text{ and } \left[V_0 - V_0 \cdot \frac{t}{273}\right]$$

Hence at – 273°C volume will be $\left[V_0 - V_0 \cdot \frac{273}{273}\right] = 0$. The temperature is called 0°A.

Now volume at t_1°C $\quad V_1 = V_0\left[1 + \frac{t_1}{273}\right]$

$$\text{or, } V_1 = V_0\left[\frac{273 + t_1}{273}\right]$$

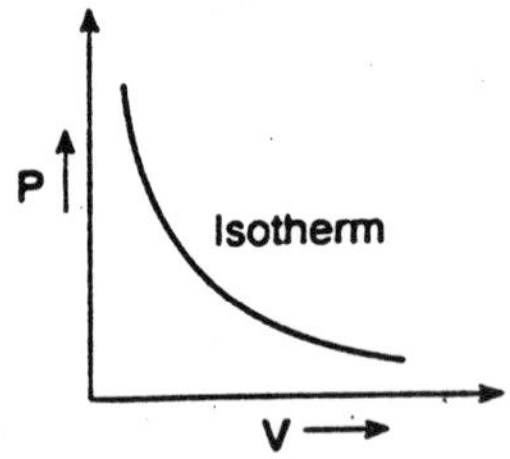

Fig. 1.1

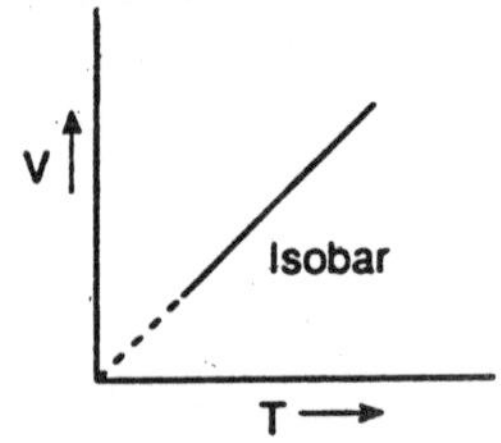

Fig. 1.2

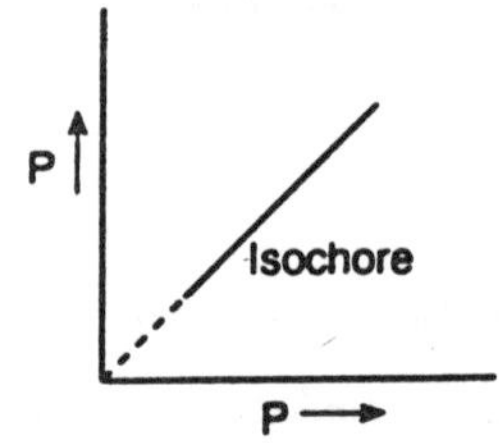

Fig. 1.3

and volume at t_2°C $\quad V_2 = V_0\left[\dfrac{273 + t_2}{273}\right]$.

Now $\quad \dfrac{V_2}{V_1} = \dfrac{273 + t_2}{273 + t_1} = \dfrac{T_2}{T_1}$ [where T = absolute temp.].

So $\quad V \propto T$.

Hence Charle's law can also be stated as — *At constant pressure volume of a definite amount of gas is directly proportional to absolute temperature.* Plot of V against T at constant P will be a straight line through origin. Such a plot will be called **isobar**.

(c) **Gay-Lussac's law :** At constant volume pressure of a definite amount of gas varies inversely with absolute temperature.

Thus $P \propto T$ [at constant volume and mass]. Thus plot of P vs T at constant volume will be a straight line through origin. Such a plot will be called **isochore**.

1.3 COMBINATION OF GAS LAWS

From Boyle's law $V \propto \dfrac{1}{P}$ [T = constant] and from Charle's law $V \propto T$ [P = constant].

So when all varies $\quad V \propto \dfrac{T}{P}$

i.e., $\quad PV$ = constant × T.

1.4 UNIVERSAL GAS CONSTANT–R

From above equation

$$\frac{PV}{T} = \text{constant.}$$

Now from Avogadro's hypothesis at a given pressure and temperature, one mole of *every* gas has the same volume. So we see that the value of the constant is same for each and every gas. Since it is same for all gases, we call it universal gas constant and express it by the symbol 'R'. So universal gas constant $R = \dfrac{PV}{T}$. The equation $PV = RT$ is thus for one mole gas. Gases, which obey this equation are called **ideal**. We shall later see that every gas obeys this equation *under certain conditions* and show departure from this equation under different conditions. When they obey this equation, we call them ideal but when they show deviation, we say that they have become real. Anyway, as volume is proportional to number of moles (n); so for 'n' moles gas the ideal gas equation is

$$PV = nRT. \qquad .....(1.1)$$

For 'g' gm gas with molecular weight M, we have $n = \left(\frac{g}{M}\right)$ and so

$$PV = \frac{g}{M} . RT. \qquad .....(1.2)$$

1.5 DIMENSION AND VALUES OF R

For one mole ideal gas

$$R = \frac{PV}{T}.$$

Now, $P = \text{pressure} = \frac{\text{force}}{\text{area}} = \frac{\text{force}}{(\text{distance})^2}$

$V = \text{volume} = (\text{distance})^3.$

So $R = \frac{PV}{T} = \left(\frac{\text{force} \times \text{distance}}{\text{degree}}\right) \Big/ \text{mole} = \frac{\text{work}}{\text{deg.}} \times \frac{1}{\text{mole}}.$

In S.I. unit work is expressed as joule, in traditional unit it is expressed as ergs and sometimes as calories also. Sometimes work is expressed as litre-atmosphere also. So different units are (i) joule deg^{-1} $mole^{-1}$ (ii) erg deg^{-1} $mole^{-1}$ (iii) calorie deg^{-1} $mole^{-1}$ and (iv) litre-atm deg^{-1} $mole^{-1}$. Its values in different units are —

(a) S.I. unit :

$$R = \frac{PV}{T}.$$

At N. T. P. $P = 1.013 \times 10^5$ pascal [i.e., newton/metre^2]

$V = 0.0224$ metre^3; $T = 273°$A

$$R = \frac{1.013 \times 10^5 \times 0.0224}{273} = 8.311 \text{ joule/deg/mole.}$$

(b) Traditional unit :

$$R = \frac{PV}{T}.$$

At N. T. P. $P = 76$ cm of Hg $\equiv 1.013 \times 10^6$ dynes/cm^2

$V = 22{,}400$ c.c.; $T = 273°$A

$$R = \frac{1.013 \times 10^6 \times 22{,}400}{273} = 8.31 \times 10^7 \text{ ergs/deg/mole.}$$

Now 4.2 joules $\equiv$ 1 calorie,

As $R = 8.311$ joule/deg/mole.

So $R = \frac{8.311}{4.2} = 1.987 \ (\simeq 2)$ cal/deg/mole.

(c) **Litre-atm unit :**

At N.T.P. $P = 1$ atm; $V = 22.4$ litre.

So $$R = \frac{1 \times 22.4}{273} = 0.082 \text{ litre-atm deg}^{-1} \text{ mole}^{-1}.$$

1.6 KINETIC THEORY — POSTULATES

One of the difficulties in understanding properties of gases is its invisibility. Barring few gases, almost all gases are colourless and invisible. Since we can not use our important sense organs to study their behaviour; we are to take the help of our imagination. Of course such imagination should be based on sound logic and reason. Maxwell, Boltzmann, Clausius and many others used logic and reason to put forward some forecast regarding properties of gases. These are popularly known as postulates of kinetic theory. Their assumptions are as follows —

(a) Every gas consists of tiny discrete particles called **molecules**.

(b) Molecules are always in random motion.

(c) During movement they collide with each other and also with the wall of the containing vessel.

(d) Collision is perfectly elastic.

(e) Pressure of a gas is due to the force on the wall, which is felt due to bombardment of the molecules on it.

(f) Molecules are point masses; their actual volume is negligible in comparison to that of the vessel.

(g) There is no force of attraction and repulsion among gas molecules.

(h) Kinetic energy of the gas is directly proportional to absolute temperature.

1.7 EXPLANATION

We have already said that there are some logic behind such assumptions. Let us see what led the scientists to put forward these postulates. The idea of tiny atoms/molecules as smallest part of any substance was already there. In case of gases, the property of diffusion in all possible direction gave the idea of random motion. And if the motion is random, there is every possibility of collision among molecules. If such a collision is inelastic there will be a fluctuation of total momentum of the gas molecules; which means some force will constantly operate on it. But that never occurs. So collision must be elastic in nature. Since almost entire vessel is available for movement of gas molecules, there actual volume is negligible. Because of very small amount of mass and comparatively large intermolecular distance, there is the absence of any attraction or repulsion among the molecules. Finally, as temperature increases molecular motion and hence kinetic energy; so K. E. must be proportional to temperature.

1.8 EXPRESSION FOR PRESSURE

One of the ways to test the validity of these postulates is to derive an expression for easily measurable parameter (say pressure) assuming the postulates as true. We are now going to do this.

Let us assume a cubical vessel, each side of which is of length 'l'.

[**Note.** Though we are assuming our vessel as cubical, it can be shown that same expression is reached for any type of vessel; for cubical vessel calculation is simple.]

Let the vessel consists of 'n' molecules of a gas, each of which is of mass 'm'. Since different molecules may move with different velocities, we take the root mean square (r.m.s.) velocity as 'c'.

[**Note.** Why we consider root mean square and not ordinary mean will be discussed in the next section.]

This velocity can be resolved in to three components u, v and w along three co-ordinate axes X, Y and Z respectively.

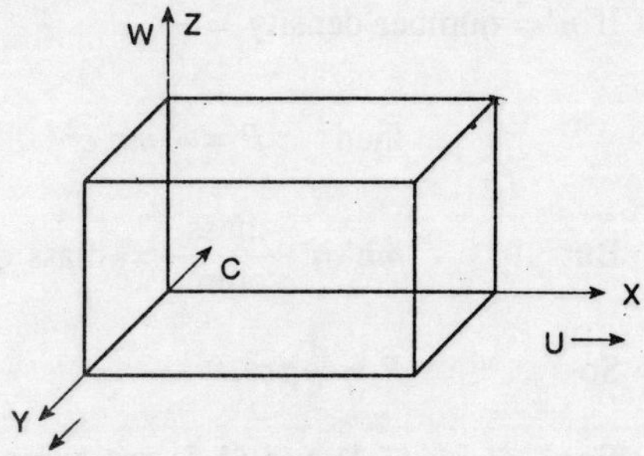

Fig. 1.4

From dynamics $c^2 = u^2 + v^2 + w^2$.

Now let us consider a molecule moving along X-direction with uniform velocity u. Since there is no force of attraction it will strike the wall facing it with momentum 'mu'. As wall does not move total momentum of colliding partners = $mu + 0 = mu$. Now collision is perfectly elastic. So total momentum after collision must be $-mu$. And this is the momentum of the molecule, as the wall does not move. So momentum change per collision $= mu - (-mu) = 2mu$. As length of the cube is 'l', the molecule must move 'l' distance for a single collision. But it moves 'u' distance in one second. So number of collisions suffered by it in one second $= \frac{u}{l}$. Again momentum change per collision $= 2mu$. So momentum change per second for one molecule moving along X-direction $= 2mu \times \frac{u}{l}$ $= \frac{2mu^2}{l}$. Similarly if it moves in Y-direction momentum change/sec will be $\frac{2mv^2}{l}$ and in Z-direction it will be $\frac{2mw^2}{l}$. So, during movement in any arbitrary direction, resultant momentum change will be

$$\left[\frac{2mu^2}{l} + \frac{2mv^2}{l} + \frac{2mw^2}{l}\right] \text{ per second}$$

$$= \frac{2m}{l}\left(u^2 + v^2 + w^2\right) = \frac{2mc^2}{l}.$$

This is for a single molecule. But our vessel has 'n' molecules. So total momentum change per second due to collision with wall is $\frac{2mnc^2}{l}$. From Newton's law, this is force on the wall (F). But

$$F = \text{pressure} \times \text{total area of wall} = P \times 6l^2$$

So $$P \times 6l^2 = \frac{2mnc^2}{l}$$

or $$P = \frac{1}{3}\frac{mnc^2}{l^3} = \frac{1}{3}\frac{mnc^2}{V} \quad [\text{as } l^3 = \text{volume } V]$$

So $$PV = \frac{1}{3}mnc^2. \quad \text{.....(1.3)}$$

If n' = number density $= \frac{n}{V}$

then $P = \frac{1}{3}mn'c^2$. .....(1.4)

But $mn' = \frac{\text{mass}}{\text{volume}}$ = mass density (ρ).

So $P = \frac{1}{3}\rho c^2$. .....(1.5)

Eqns. (1.3), (1.4) and (1.5) are expression for pressure.

1.9 USE OF THE POSTULATES IN DERIVATION OF PRESSURE

During derivation of the expression for pressure, we have used almost all the postulates. Firstly, we assumed existence of molecules, which are moving freely. We also assumed their random collision. During calculation of momentum change, we assumed the collision to be elastic. Also as we assumed that the molecule is moving with uniform velocity, so it is indirectly believed that there was no attraction or repulsion among gas molecules. Because if any such force exists, it will decrease or increase molecular velocity, just at the time of its impact with the wall, as only then such force becomes unbalanced. We also treated pressure as an outcome of the force on the wall. Finally we assume that the molecule is free to move anywhere in the vessel. This means we think that their existence inside the vessel is not putting any obstacle for free movement i.e., they have no volume. Only the postulate regarding K. E. has not been used. All other postulates have been used.

1.10 ROOT MEAN SQUARE VELOCITY : ITS IMPORTANCE

During derivation of the equation $PV = \frac{1}{3}mnc^2$, we have taken C as root mean square velocity and not as ordinary mean or average velocity. Thus if $C_1, C_2, C_3, ..., C_n$ are velocities of different molecules then

C_{av} i.e., $$\bar{C} = \frac{C_1 + C_2 + C_3 + ... + C_n}{n}$$

and $C_{r.m.s.}$ i.e., $$\sqrt{\overline{C^2}} = \sqrt{\frac{C_1^2 + C_2^2 + C_3^2 + ... + C_n^2}{n}}$$

The advantages of $C_{r.m.s.}$ over C_{av} is that only the former can give us exact value of kinetic energy. To illustrate this point, let us calculate a gaseous system having only two molecules moving with velocities 4 and 6 units respectively. Its average velocity $\bar{C} = \frac{4+6}{2} = 5$ unit. If mass of a molecule m, then average K. E. $= \frac{1}{2} m (5)^2 = 12.5m$. For two molecules total K. E. $= 25m$. If we calculate individual K. E.'s then for first molecule, it is $\frac{1}{2} m (4)^2 = 8m$ and for second molecule, it is $\frac{1}{2} m (6)^2 = 18m$, so that the total $= 8m + 18m = 26m$. Hence the two results do not agree. But if we find $C_{r.m.s.}$ then $C_{r.m.s.}$

$= \sqrt{\dfrac{4^2 + 6^2}{2}} = \sqrt{26}$. Average K. E. $= \dfrac{1}{2}\ m\ (C_{r.m.s.})^2 = 13m$ and total for two molecules $= 13m \times 2 = 26m$. So results agree. Hence we see that to get correct value of K. E. from average calculation we are to use $C_{r.m.s}$ and not C_{av}.

1.11 EXPRESSION FOR K. E. AND ALSO $C_{r.m.s.}$

We have already seen that $PV = \frac{1}{3}mnc^2$. Now for 1 mole gas n = Avogadro number N; also for 1 mole gas $PV = RT$.

So $$\frac{1}{3}mNc^2 = RT$$

$$\text{or,}\quad \frac{2}{3}N \cdot \frac{1}{2}mc^2 = RT$$

$$\text{or,}\quad \frac{2}{3}\ N \times \text{K. E. of one molecule} = RT$$

$$\text{or,}\quad \frac{2}{3} \times \text{K. E. per mole} = RT.$$

So K. E. per mole $= \dfrac{3}{2}\ RT,$ for ideal gas. .....(1.6)

For one molecule K. E. $= \dfrac{3}{2} \cdot \dfrac{R}{N} T = \dfrac{3}{2} kT$ .....(1.7)

where k = Boltzmann constant.

As mN = mass/mole i.e., molecular weight M.

So $$\frac{1}{3}\ MC^2 = RT$$

$$\text{or,}\quad C_{\text{r.m.s.}} = \sqrt{\frac{3RT}{M}}. \qquad .....(1.8)$$

Problem

1. Calculate the $C_{\text{r.m.s}}$ for H_2 at 27°C.

Solution :

$$C_{\text{r.m.s.}} = \sqrt{\frac{3RT}{M}} = \sqrt{\frac{3 \times 8.31 \times 10^7 \times 300}{2}}\ \text{cm./sec}$$

$$= 1.93377 \times 10^5\ \text{cm./sec.}$$

2. What is the K. E. of 2 moles of a perfect gas at 25°C ?

Solution :

$$\text{K. E.} = 2 \times \frac{3}{2} \times 8.313 \times 10^7 \times 298\ \text{ergs}$$

$$= 7.4318 \times 10^{10}\ \text{ergs.}$$

1.12 DERIVATION OF GAS LAWS FROM $PV = \frac{1}{3}mnc^2$

(a) **Boyle's law and Charle's law** : We have proved $PV = \frac{1}{3}mnc^2$. Now $PV = \frac{2}{3}n \times \frac{1}{2}mc^2$. But $\frac{1}{2}mc^2$ = K. E./molecule. So $PV = \frac{2}{3}$ × total K. E. Now one of the postulates states that K. E. ∝ T i.e., K. E. = constant × T. So PV = constant × T. Hence we see that if T is constant, then PV = constant; this is Boyle's law. Also if P is constant, $V \propto T$; this is Charle's law. Thus the postulates lead to well-established laws.

(b) **Avogadro's law** : To derive Avogadro's law from the above expression let us assume that V volume of two different gases at same pressure (P) and temperature (T) has number of molecules n_1 and n_2 respectively. To arrive at Avogadro's law, we are to prove $n_1 = n_2$. If mass of each molecule of two gases are m_1 and m_2 and r.m.s. velocities are C_1 and C_2 at this temperature, then —

$$PV = \frac{1}{3}m_1n_1C_1^2 \text{ for first gas and}$$

$$PV = \frac{1}{3}m_2n_2C_2^2 \text{ for second.}$$

So $$m_1n_1C_1^2 = m_2n_2C_2^2. \qquad .....(1.9)$$

Again as their temperatures are same, so from postulate K. E./molecule must be same, i.e.,

$$\frac{1}{2}m_1C_1^2 = \frac{1}{2}m_2C_2^2$$

or, $$m_1C_1^2 = m_2C_2^2. \qquad .....(1.10)$$

On comparison of eqns. (1.9) and (1.10) we get $n_1 = n_2$. This is Avogadro's law.

(c) **Graham's law of diffusion** : Graham's law of diffusion can also be derived from the eqn. $PV = \frac{1}{3}mnc^2$. Graham's law states that rate of diffusion is inversely proportional to square root of density (or molecular weight) of the gas. Now diffusion rate is dependent on velocity of the gas. We see that

$$C^2 = \frac{3PV}{mn}; \text{ for 1 mole gas } mn = M \text{ (mol. wt.)}$$

also $$\frac{M}{V} = \text{density } (\rho).$$

So for 1 mole gas $$C^2 = \frac{3PV}{M} \text{ or, } C = \sqrt{\frac{3PV}{M}} = \sqrt{\frac{3P}{\rho}}.$$

Hence $$C \propto \frac{1}{\sqrt{M}} \text{ or, } C \propto \frac{1}{\sqrt{\rho}}.$$

This is Graham's law.

(d) **Dalton's law of partial pressure** : According to Dalton, total pressure of a gas mixture present in a vessel is the summation of their partial pressures. The term partial pressure of a gas means the pressure that would have been felt, when that gas is present alone in the entire volume. Thus, let n_A moles A occupies a volume V and shows pressure P_A; we have then $P_A V = n_A RT$

If the same volume is occupied by n_B moles of B and it shows a pressure P_B, then $P_B V = n_B RT$. These P_A, P_B are partial pressure. On adding $(P_A + P_B)V = (n_A + n_B)RT$. Again when n_A moles of A and n_B moles of B are present together, then it total pressure is P, we write $PV = (n_A + n_B)RT$. On comparison we get $P = P_A + P_B$. Thus total pressure is the summation of partial pressure. This is Dalton's law.

[**Note.** $P_A V = n_A RT$; $PV = \Sigma nRT$. So $\frac{P_A}{P} = \frac{n_A}{\Sigma n}$ = mole fraction; so partial pressure = mole fraction × total pressure.]

Dalton's law of partial pressure can be derived from eqn. $PV = \frac{1}{3}mnc^2$. From this equation we have seen that K.E. for n mole of a gas $= \frac{3}{2}nRT = \frac{3}{2}PV$. Let n_A moles of A, n_B moles of B are present together and their K. E.'s are E_A and E_B. Then total K. E.

$$E = E_A + E_B.$$

If P is total pressure and P_A and P_B are partial pressures, then

$$\frac{3}{2}PV = \frac{3}{2}P_A V + \frac{3}{2}P_B V$$

$$\text{or,} \quad P = P_A + P_B.$$

This is Dalton's law.

1.13 DEGREE OF FREEDOM; EQUIPARTION OF ENERGY

We know that position of a particle in space can be indicated with the help of three co-ordinates. During translatory motion these co-ordinates change. Now a particle may translate in such a manner that it can change any one co-ordinate, keeping the other two unchanged. Thus we see that the three co-ordinates are independent of each other. We say, that there are three translational degrees of freedom. Thus degree of freedom can be described as the number of co-ordinates, whose values can be changed independently during motion. It is numerically equal to total number of quadratic terms required to express the energy due to that motion. Thus we see that translational K. E. $= \frac{1}{2}mc^2$ can be expressed as summation of three quadratic terms $\left(\frac{1}{2}mu^2 + \frac{1}{2}mv^2 + \frac{1}{2}mw^2\right)$ ['u', 'v', 'w' are three components of 'c']. So translational degree of freedom = 3.

One of the most important properties of gas is that it is perfectly isotropic in nature. *Isotropicity* means probability of movement of gas molecules in any direction is exactly same. Thus total K. E. may be thought to be made by equal contribution of three co-ordinate axes. Since total K. E. /mole $= \frac{3}{2}RT$. Therefore contribution of each axes $= \frac{1}{2}RT$. Hence we can write that K. E./mole/degree of freedom $= \frac{1}{2}RT$ and K. E./molecule/degree of freedom $= \frac{1}{2}kT$. This is known as principle of equipartition of energy.

1.14 MOLAR HEAT CAPACITY OF GASES

We know that specific heat of any substance is equal to amount of heat (in cals) required to raise the temperature of 1 gm of the substance through 1°C. Now if a substance is heated at constant volume, all the heat is used to raise its temperature. But if same substance is heated at constant pressure, then some expansion work is done and so a part of heat raises its temperature and another part does work. Hence, if want to increase temperature by 1°C, then at constant pressure it requires more heat than at constant volume. So we see that there are two specific heats (c_p and c_v); one at constant pressure and another at constant volume. The former is defined as the amount of heat required to raise the temperature of 1gm substance through 1°C at constant pressure and the latter as amount of heat required to do the same at constant volume. Molar heat capacities (C_p and C_v) are for 1 mole substance. Thus if molecular weight = M then $C_p = M.c_p$ and $C_v = M.c_v$. According to above discussion $C_p = C_v +$ work done during heating through 1°C at const. pressure.

For solids and liquids volume change due to 1°C rise in temperature is negligible, so $C_p \simeq C_v$; but for gases volume change is much and expansion work is large; so $C_p > C_v$.

[**Note.** For water during heating in the range 0°C — 4°C there is contraction in stead of expansion. So work is negative. And hence $C_p < C_v$ for water in this range. This is a strange case.]

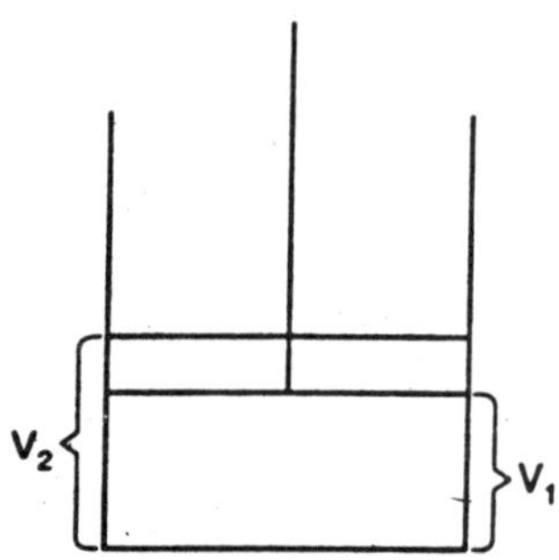

Fig. 1.5

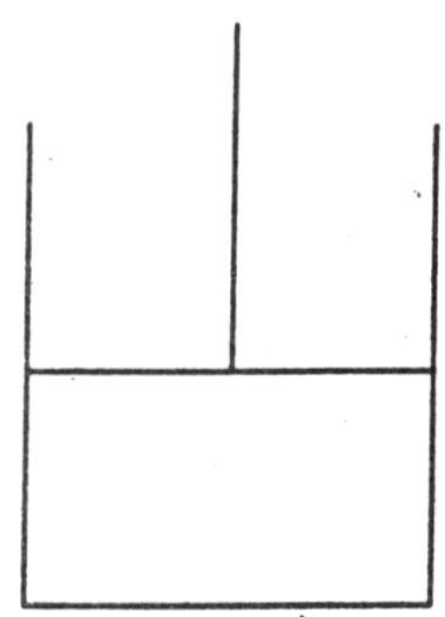

Fig. 1.6

1.15 C_p - C_v FOR IDEAL GAS

To derive a relation between C_p and C_v for ideal gas, let us consider one mole of an ideal gas enclosed in a cylinder fitted with a weightless and frictionless piston. Let its pressure be P_1; volume V_1 and temperature T_1. Now certain amount heat is supplied to it so that its temperature becomes $(T_1 + 1)$. When this is done at constant volume (fig. 1.6), no expansion work is done and by definition all the heat is used to increase temperature.

Rise in temperature increases K.E. of all types of gas molecules; it also increases rotational and vibrational energies of poly-atomic molecules. We call these two energies as intramolecular energy. Let its increase per degree rise in temperature be 'a'. So

C_v = (Increase in K.E. + Increase in intramolecular energy) per degree rise in temperature.

Now, K. E/mole before heating = $\frac{3}{2}RT_1$ and K.E./mole after heating = $\frac{3}{2}R\left(T_1 + 1\right)$.

So
$$C_v = \frac{3}{2}R(T_1 + 1) - \frac{3}{2}RT_1 + a$$
$$= \frac{3}{2}R + a.$$

If the same gas is heated from T_1 to $(T_1 + 1)$ at constant pressure, (fig. 1.5) then there will be expansion from volume V_1 to volume V_2. Now,

$$\begin{aligned}\text{Expansion work} &= \text{force} \times \text{distance changed}\\ &= \text{pressure} \times (\text{area} \times \text{distance changed})\\ &= \text{pressure} \times \text{volume change}\\ &= P(V_2 - V_1)\\ &= PV_2 - PV_1 = R(T_1 + 1) - RT_1\\ &= R.\end{aligned}$$

So in this case

Heat supplied = C_p = Increase in K. E. + Increase in intramolecular energy + work done.

$$\text{or, } C_p = \frac{3}{2}R\left(T_1 + 1\right) - \frac{3}{2}RT_1 + a + R$$

$$\text{or, } C_p = \frac{5}{2}R + a.$$

So
$$C_p - C_V = \left(\frac{5}{2}R + a\right) - \left(\frac{3}{2}R + a\right)$$

$$= \text{R for one mole ideal gas.}$$

1.16 HEAT-CAPACITY RATIO

The ratio $C_p : C_v$ is denoted by γ. So

$$\gamma = \frac{C_p}{C_v} = \frac{\frac{5}{2}R + a}{\frac{3}{2}R + a}.$$

For monoatomic gas there is no rotation, vibration; so $a = 0$ and $\gamma = \frac{5}{3} = 1.66$. For diatomic gas molecule γ is found to be 1.41 and for triatomic gas molecule γ is 1.29.

1.17 DISTRIBUTION OF MOLECULAR VELOCITY : MAXWELL'S LAW

In an assembly of gas molecules, different molecules may move with different velocities. In order to know the fraction of molecules moving with an arbitrary velocity lying between c and $c + dc$ Maxwell derived the following equation, known as Maxwell's law. The equation is

$$\frac{dn_c}{n} = \sqrt{\frac{2}{\pi}}\left(\frac{m}{kT}\right)^{\frac{3}{2}}.e^{-\frac{mc^2}{2kT}}.c^2dc \qquad .....(1.11)$$

where dn_c is the number of molecules, out of a total number 'n' moving with a velocity lying between c and $c + dc$ at T °A. 'm' is the mass of a molecule and T is absolute temperature $\frac{dn_c}{n}$ is also the probability of finding molecules having velocity in between c and $c + dc$.

Basic assumptions involved during derivation of above equation were (i) gas is free from any external and internal force field, so that number density is same everywere i.e., gas is ideal (ii) gas is isotropic and (iii) there is steady-state i.e., number of molecules moving with a given velocity is always constant, though an individual molecule may change its velocity.

Plot of $\frac{dn_c}{n}$ vs c is as follows —

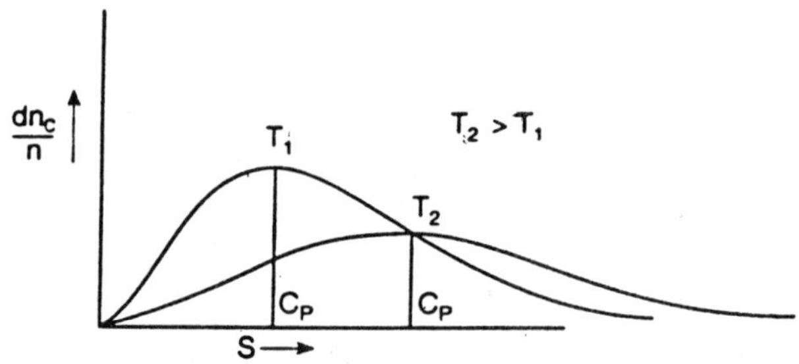

Fig. 1.7

Characteristic features of the curve (Fig.1.7) are as follows.

(a) No molecule is at rest and also no molecule move with infinite velocity. $[\frac{dn_c}{n} = 0$ at either $c = 0$ or $c = \infty]$

(b) Curve has a maxima at $c = C_p$. This is the velocity with which most of the molecules move. Since probability of finding molecules, moving with this velocity is maximum, it is called most probable velocity.

(c) At high temperature curve flattens, i.e., most probable velocity increases, though comparatively less molecules move with this velocity now.

1.18 THREE TYPES OF VELOCITIES : INTER RELATIONS

We have already learnt about two types of molecular velocities, which are r.m.s. and average velocity. Now we have learnt about most probable velocity. Their values, as calculated from Maxwell's law are — $C_{r.m.s.} = \sqrt{\frac{3RT}{M}}$; $C_{av} = \sqrt{\frac{8RT}{\pi M}}$ and $C_p = \sqrt{\frac{2RT}{M}}$.

So $$C_{r.m.s.} : C_{av} : C_p = \sqrt{3} : \sqrt{8/\pi} : \sqrt{2}$$
$$= 1 : 0.92 : 0.82.$$

So at any temperature $C_{r.m.s}$ is highest. R. M. S. velocity is exclusively used in K. E. calculation.

In all other calculations average velocity is used.

Most probable velocity is of academic interest i.e., to know velocity of maximum molecules. We should not forget that average K. E. /mole $= \frac{1}{2}M(C_{r.m.s.})^2 = \frac{1}{2}M.\frac{3\pi}{8}.(C_{av})^2$.

1.19 COLLISION FREQUENCY : MEAN FREE PATH

In an assembly of gas molecules, continuous collision is occurring among the gas molecules due to their random motion. Number of collisions suffered by a single molecule in one second has been calculated and this value comes as $\sqrt{2}\pi\sigma^2 C_a.n$, where σ = collision diameter i.e., twice the radius of a molecule ($2r$); C_a is average velocity and n is number density i.e., number of gas molecules present per c.c. of the container. Number of collisions occurring per c.c. per second is called collision frequency (Z). Is is equal to

$$Z = \frac{1}{\sqrt{2}}.\pi\sigma^2 C_a.n^2.$$

Between two successive collisions, a molecule moves certain path. This path is called **free path**. Average of all such free paths, executed in one second is called **mean free path.**

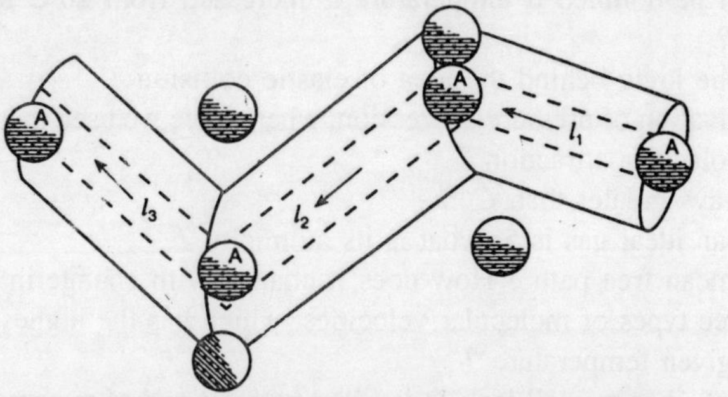

Fig. 1.8 Path covered by molecule a during collisions.

Average of l_1, l_2, l_3, ...etc. is mean free path.

$$\text{So,} \quad \text{mean free path} = \frac{\text{summation of all free paths in 1 sec.}}{\text{number of such free paths}}$$

$$= \frac{\text{distance moved in one second}}{\text{number of collisions per second}}$$

$$= \frac{C_a}{\sqrt{2}\pi\sigma^2 C_a n} = \frac{1}{\sqrt{2}\pi\sigma^2 n}.$$

On increasing temperature or decreasing pressure, mean free path will increase.

POINTS TO REMEMBER

Gas is a state of 'matter.' It obeys certain laws. If it obeys empirical (i.e. based on observation) laws like Boyle's law : Charle's law; Avogadro's law; Graham's law etc.— then it will be called ideal. Regarding their behaviour some predictions (based on logic) have been made. These are called postulates. Based on these postulates an equation $PV = \frac{1}{3}mnc^2$ can be derived. 'c' is root mean square velocity. From this equation established laws can be obtained.

Behaviour of gas is greatly influenced by its **isotropic** character. Thus total K. E. is equally divided along three **degrees of freedom** (½ *RT per mole per degree of freedom*). Because of sensitivity of its volume towards temperature it has two types of molar heat capacities C_p and C_v (at constant pressure and constant volume). For ideal gas $C_p - C_v = R$.

Velocity of ideal gas molecules follows Maxwell's distribution law — according to which no molecule is at rest or is moving with infinite velocity. Majority of the molecules move with a particular velocity known as most probable velocity. During movement the average distance travelled by a molecule between two collisions is **'mean free path'**

QUESTIONS

SHORT TYPE

1. When a ballon is blown, both its volume and pressure inside increase. Does it violate Boyle's law ?
2. Charle's law states that $V \propto T$ at constant pressure. Does it mean that volume of a gas will be doubled if temperature is increased from 20°C to 40°C at constant pressure ?
3. What is the logic behind the idea of elastic collision ?
4. In the derivation of pressure expression, where have we used the idea about absence of intermolecular attraction ?
5. Is C_p always greater than C_v ?
6. If C_v for an ideal gas is 5, what is its atomicity ?
7. What is mean free path ? How does it change with change in pressure ?
8. Name three types of molecular velocities; which has the highest value for a given gas at a given temperature ?
9. A cycle tube has a small leak. It is filled with air and after sometime the air inside it is analysed. Will the composition i.e., relative amount of N_2 and O_2 alter ?
10. Why H_2 and He are not abundant in the atmosphere ?

ESSAY TYPE

1. What is universal gas constant ? Why it is called universal ? State its values in different units.
2. Explain the postulates of kinetic theory and hence arrive at an expression for pressure of an ideal gas.
3. Which postulates of kinetic theory have been utilised in deriving the expression for pressure of an ideal gas ?
4. Starting from the expression $PV = \frac{1}{3}mnc^2$, derive (a) Boyle's law (b) Charle's law (c) Avogadros's law (d) Graham's law and (e) Dalton's law for partial pressure.
5. What are molar heat capacities ? Show that for an ideal gas $C_p - C_v = R$.
6. Explain the terms *degree of freedom* and *principle of equipartition of energy.*
7. Write down Maxwell's law for the distribution of molecular velocities and explain the salient features of the curve.

PROBLEMS

1. It one litre of oxygen diffuses through a small crack in an apparatus in 1 hour, how it take for a litre of He (atomic wt. = 4) to diffuse through under the same conditions?

[*Hints.* rate $= \dfrac{\text{Volume diffused}}{\text{Time}}$ $\dfrac{r_1}{r_2} = \dfrac{t_2}{t_1}$. But $r \propto \dfrac{1}{\sqrt{M}}$

So $\dfrac{r_{H_2}}{r_{He}} = \sqrt{\dfrac{M_{He}}{M_{H_2}}} = \dfrac{t_{He}}{t_{H_2}}$.]

2. The specific heat for a gas at constant volume is 0.075 and that at constant pressure is 0.125. Calculate the mol. wt. of the gas.

[*Hints.* $C_p - C_v = R$ or, $M(c_p - c_v) = R$ or, $M(0.125 - 0.075) = 1.987$.]

3. The specific heat at constant volume of argon is 0.075, its molecular weight is 40. How many atoms are there in its molecule ?

[*Hints.* Find C_p following problem 2. Then find $\gamma = \frac{C_P}{C_V}$. It will be 1.66. So gas is monoatomic.]

4. Calculate the r.m.s. velocity of oxygen molecules at 15°C and 770 m.m. pressure.

[*Hints.* $C_{r.m.s.} = \sqrt{\frac{3RT}{M}}$. Put '*R*' in ergs; $M = 32$.]

5. What is root mean square velocity of methane molecules at 27°C ? At what temperature would ethane molecules have the same velocity as those of methane molecule ?

[*Hints.* First part same as above. Then

6. Calculate the r.m.s. velocity of oxygen molecules at 25°C.

[*Hints.* Same as Problem 4.]

OBJECTIVE TYPE

1. Choose the correct answer.

(a) Boyle's law is valid at — (i) const. T (ii) const. '*n*' [*n* = mole] (iii) at const '*T*' and '*n*'.

(b) According to Charle's law volume of a gas will be doubled it tempareture is raised from 20°C to 40°C — (i) True (ii) False.

(c) At constant '*p*' slope of *V* vs *T* Curve for 1 mole gas is — (i) *P* (ii) *R* (iii) $\frac{R}{P}$.

(d) Expression $PV = \frac{1}{3}mnc^2$ is valid for — (i) Cubilcal vessel (ii) Vessel of any shape.

(e) In the expression $C = \sqrt{\frac{3RT}{M}}$ if '*C*' is in cm sec^{-1} then '*R*' is in — (i) ergs deg^{-1} $mole^{-1}$ (ii) Joule deg^{-1} $mole^{-1}$ (iii) liter-atm deg^{-1} $mole^{-1}$.

(f) In question '*e*' if '*c*' is in metre sec^{-1} then *R* will be — (i) (ii) (iii).

(g) What is $C_p : C_v$ ratio for H_2 ? — (i) 1.66 (ii) 1.41 (iii) 1.29.

(h) 30 moles of CO_2 and 60 moles of N_2 are mixed; total pressure is 180 m.m. So partial pressure for CO_2 is (i) 120 m.m. (ii) 90 m.m. (iii) 60 m.m.

2. Match terms of column A with that of column B.

A	B
(a) Boyles	(a) Isobar
(b) Charles	(b) Diffusion
(c) Graham	(c) Isotherm
(d) Helium	(d) $\sqrt{3RT / M}$
(e) Maxwell	(e) $\sqrt{2RT / M}$
(f) R.M.S. velocity	(f) $r = 1.66$
(g) Most probable velocity	(g) Velocity distribution

Answers :

1. (a) (iii); (b) (ii); (c) (iii); (d) (ii); (e) (i); (f) (ii); (g) (ii); (h) (iii)

2. a – c; b – a; [illegible] d – f; e – g; f – d; g – e;

Chapter 2

REAL GAS

2.1 REAL GAS — AMAGAT'S CURVES

We have seen that postulates of kinetic theory lead to the equation $PV = RT$ for 1 mole ideal gas. Now if we determine PV values for 1 mole of a gas at different pressures, keeping temperature constant and plot these PV-values against P we would expect a straight line parallel to P-axis (dotted line in fig. 2.1). However, the curves that we really obtain are either of type III or type I. Amagat first carried out these experiments and that is why these curves are known as **Amagat's curves**. If experiment is done at room temperature then type I curve is obtained for H_2, He, N_2, and type III curve is for CO_2 and other easily liquefiable gases. Later experiments revealed that every gas can show type I, type II or type III curve. What type of curve will really be obtained that depends on experimental temperature. Every gas has a characteristic temperature at which it obeys Boyle's law. This temperature is called Boyle temperature. If PV is plotted against P at this temperature we shall get type II curve. If plot is done above Boyle temperature we get type type I curve and a plot below this temperature gives type III curve. For H_2 and He Boyle temperature is very low and so plotting at ordinary temperature gives type I curve, whereas for CO_2 ordinary temperature is less than Boyle temperature.

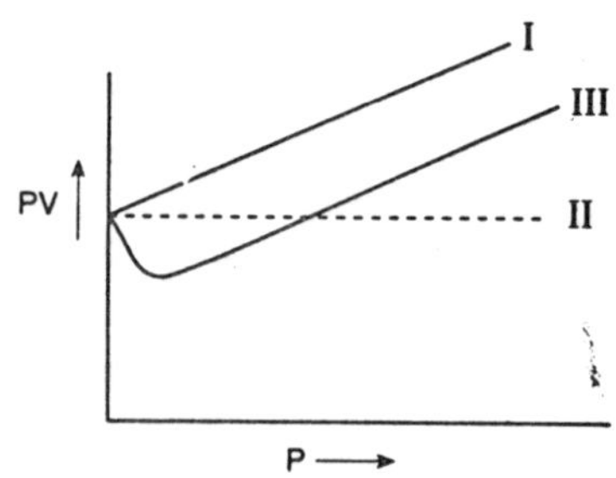

Fig. 2.1

2.2 COMPRESSIBILITY FACTOR — DEVIATION FROM IDEAL BEHAVIOUR

Amagat's curves indicates that for most gases at ordinary temperature $PV \neq RT$ for one mole. We write $PV = ZRT$ where Z, which is $\frac{PV}{RT}$ will be called compressibility factor. If $Z = 1$, the gas will be called ideal. From Amagat's curves we see that there are regions where $Z > 1$ (i.e., experimental PV -values are more than theoretically expected) and also $Z < 1$ [i.e., PV (exptl.) < PV (theo.)]. In region $Z > 1$, the gas is less compressible and in regions $Z < 1$ the gas is more compressible. Now less compressibility means, there are repulsive forces, which oppose during compression and more compressibility means there are attractive forces which favour compression process. Anyway the conclusion is that there are attraction and repulsion among gas molecules. Postulates of kinetic theory disregarded their presence. So ideal gas equation derived on the basis of these postulates needs modification.

2.3 EQUATION OF STATE FOR REAL GAS : VAN-DER-WAALS EQUATION

The term *states* of a gas mean those variables, on which its energy depends. For a given mole of gas 'P', 'V' and 'T' are its states. Any equation connecting the states of a gas will be called an equation of state. Thus $PV = nRT$ is an equation of state for ideal gas.

The first equation of state for real gas was proposed by Dutch chemist Van der Waals. According to him at least two postulates of kinetic theory require modification when the gas is at moderate pressure and temperature. These postulates are (i) consideration of gas molecules as point masses having negligible volume and (ii) assumption regarding absence of any intermolecular attraction and repulsion. Above postulates are approximately valid at very low pressure and high temperature. At these conditions number of molecules within the vessel is very small and so their volume and also mutual interaction can be disregarded. However, as pressure is raised and hence number density of the gas molecules increases, above factors become important. Van-der Waals considered these and brought necessary correction in both pressure and volume terms of ideal gas equation.

(a) Volume Correction : The term '*V*' in the ideal gas equation $PV = RT$ is really volume available for the movement of gas molecules. At very low pressure, when volume occupied by small number of tiny gas molecules is really negligible, entire volume of the vessel becomes available for movement and '*V*' becomes volume of the vessel. But at moderate to high pressure volume occupied by gas molecules is no longer negligible. A moving molecule will face obstacles due to repulsion by its neighbours during free movement. Thus a portion of the vessel (say '*b*') becomes unavailable for movement. This unavailable portion '*b*' is called **co-volume**. Available volume for movement thus becomes $(V - b)$. Thus the correction term '*b*' signifies repulsive interaction. Value of '*b*' can be calculated, It is found to be $4N.\frac{4}{3}\pi r^3$. [N = Avogadro Number; r = radius of a molecule].

(b) Pressure Correction : If we consider absence of any intermolecular attraction, then a moving gas molecule can move anywhere in the vessel with uniform velocity. But if intermolecular attraction is considered then we see that net attractive force on a molecule in the bulk is zero. But for a molecule, which is going to strike the wall there will be a unbalanced backward force, as there are no gas molecules on the wall side to neutralise backward attraction.This will retard the molecular velocity and it will strike the wall with less momentum. So force on the wall and hence pressure will be somewhat less than ideal gas pressure, where no such attraction is considered. If pressure of a real gas, having such attraction is P, then pressure of an ideal gas having no such attraction is (P + something). This 'something' can be calculated as follows. Attractive interaction is proportional to number of molecules striking the wall and number of molecules attracting them. Both these terms are proportional to density (d) of the gas.

So attractive interaction $\propto d^2$. But as $d \propto \frac{1}{V}$.

So interaction $\propto \frac{1}{V^2}$ i.e., $= \frac{a}{V^2}$ [where a = constant].

So ideal gas pressure $= \left[P + \frac{a}{V^2}\right]$.

Hence the modified equation will be $\left[P + \frac{a}{V^2}\right][V - b] = RT$.

This is Van der Waals equation for one mole real gas.

2.4 VAN-DER-WAALS EQUATION FOR '*n*' MOLE GAS

For one mole real gas Van-der-Waals equation is $\left[P + \frac{a}{V^2}\right][V - b] = RT$.

Let we have 'n' mole gas having volume v. Then molar volume $V = \frac{v}{n}$. Putting this in above equation, we get

$$\left[P + \frac{n^2 a}{v^2}\right]\left[\frac{v}{n} - b\right] = RT$$

$$\text{or,} \quad \left[P + \frac{n^2 a}{v^2}\right][v - nb] = nRT.$$

This is Van-der-Waals equation for 'n' mole gas.

2.5 UNITS OF 'a' AND 'b'

In Van-der-Waals equation $\frac{a}{v^2}$ is added with P. So it will have the same unit as P i.e., atmosphere. But v has the dimension of litre mole^{-1}.

So 'a' will have the dimension of litre^2 atm mole^{-2}.

Similarly 'b' is subtracted from V. So it will have same unit as V i.e., litre mole^{-1}.

2.6 MERITS AND DEMERITS OF VAN-DER-WAALS EQUATION

(a) **Merits :** Van-der-Waals equation successfully explains Amagat's curves. These curves show that during plot of PV vs P, the value of PV falls below the expected value (dotted line; $Z < 1$) in the low pressure range but becomes more than the expected value ($Z > 1$) at high pressure range for every gas, below Boyle temperature. Now in Van-der-Waals equation $\left[P + \frac{a}{V^2}\right][V - b] = RT$; '$a$' stands for attraction correction and 'b' for repulsion correction. At low pressure as number density is less repulsive interaction becomes less important and the equation becomes

$$\left(P + \frac{a}{V^2}\right)V = RT$$

$$\text{or,} \quad PV = RT - \frac{a}{V}.$$

Expected value of PV is RT. So here value of 'PV' is less than expected value.

Similarly when pressure is made high number density becomes also high and repulsion becomes more important than attraction we can then write

$$P(V - b) = RT$$

$$\text{or,} \quad PV = RT + bP.$$

Thus 'PV' value becomes more than expected value.

(b) **Demerits :** Van-der-Waals equation suffer from some defects also. These are —

The constant 'a' and 'b' are not truly constants. Their values decrease slightly at high temperature.

Critical volume (to be discussed later) V_c calculated from Van-der-Waals equation is $3b$, but its experimental value is '$2b$'.

The value of $\frac{RT_c}{P_c V_c}$ comes as 2.66 from Van-der-Waals equation; but its exact value is 3.69.

Also P-V isothermals drawn from Van-der-Waals equation below critical temperature, shows a S-shaped portion, which is absurd.

2.7 CONDITION FOR IDEAL BEHAVIOUR — BOYLE TEMPERATURE

For 1 mole real gas Van-der-Waals equation is $\left[P + \frac{a}{V^2}\right][V - b] = RT$

$$\text{or,}\quad P = \frac{RT}{V - b} - \frac{a}{V^2}.$$

Multiplying by 'V' $\quad PV = \frac{RTV}{V - b} - \frac{a}{V}$

$$\text{or,}\quad PV = \frac{RT}{\left(1 - \frac{b}{V}\right)} - \frac{a}{V}$$

$$\text{or,}\quad PV = RT\left(1 - \frac{b}{V}\right)^{-1} - \frac{a}{V}$$

$$\text{or,}\quad PV = RT\left[1 + \frac{b}{V} + \ldots\ldots\right] - \frac{a}{V}$$ (neglecting higher terms, which are small)

$$\text{or,}\quad PV = RT\left[1 + \frac{b}{V} - \frac{a}{RTV} + \ldots\ldots\right]$$

$$\text{or,}\quad Z = \frac{PV}{RT} = 1 + \left(b - \frac{a}{RT}\right)\frac{1}{V} + \ldots\ldots .$$

If pressure is very low and temperature very high, then 'V' is a large quantity and second term of R. H. S. becomes negligible. So $Z = \frac{PV}{RT} \simeq 1$ i.e., gas becomes ideal at very low pressure and high temperature.

We have seen that every gas behaves ideally at Boyle temperature (T_b). So when $T = T_b$, $Z = 1$.

$$\text{Hence}\quad b - \frac{a}{RT_b} = 0$$

$$\text{or,}\quad T_b = \frac{a}{Rb}$$

This is the value of Boyle-temperature.

2.8 OTHER EQUATIONS OF STATE — VIRIAL FORM

After Van-der-Waals many other scientists put forward equations of state for real gas. Dieterici, Berthelot are among them. They also did pressure and volume correction. Volume correction was done by the same method; but pressure correction techniques of them were different. For one mole real gas three different equations are —

(i) Van-der-Waals : $$\left(P + \frac{a}{V^2}\right)(V - b) = RT.$$

(ii) Dieterici : $$P = \frac{RT}{V - b}.e^{-\frac{a}{RTV}}.$$

(iii) Berthelot : $$\left(P + \frac{a}{TV^2}\right)(V - b) = RT.$$

All such equations can be expressed by the general form

$$Z = \frac{PV}{RT} = 1 + \frac{A}{V} + \frac{B}{V^2} + \frac{C}{V^3} + \ldots\ldots\ldots\ldots .$$

This equation is called virial equation. A, B, C are first, second, third virials. For Van-der-Waals and Dieterici equation first virial is same.

2.9 LIQUEFACTION OF GASES — ANDREW'S EXPERIMENT

The basic difference between a gas and liquid is that in the latter the molecules are in a condensed form. So one way to liquefy gas may be application of pressure on it to bring the molecule closer. But this compression has one side effect. Compression generally increases temperature of a gas. And increase in temperature is not conducive for liquefaction. So heat, equivalent to the work done during compression must be rejected by the gas. Such a compression is called isothermal compression. Andrew tried to liquefy gases by this technique. He first carried out experiment with CO_2 gas. The P-V isothermals of CO_2 drawn at various temperatures are as follows —

Below 31°C the curve has three distinctly separate portions, AB, BC and CD. The AB portion resembles rectangular hyperbola in accordance with Boyle's law. The BC portion is near horizontal; pressure at C is infinitesimally greater than that at B, but the difference in volume is very large. Actually at B liquefaction of the gas starts and at C it is complete. In between B to C the system is neither gas nor liquid or both liquid and gas. Since volume of a liquid does not change significantly with pressure the CD portion is almost vertical having slight inclination towards pressure axis.

When temperature is raised the length of horizontal portion of the isothermal gradually decreases and at 31°C it converts to a small hump. Above 31°C the isothermals are only rectangular hyperbola, which means liquefaction by applying pressure is not possible above 31°C.

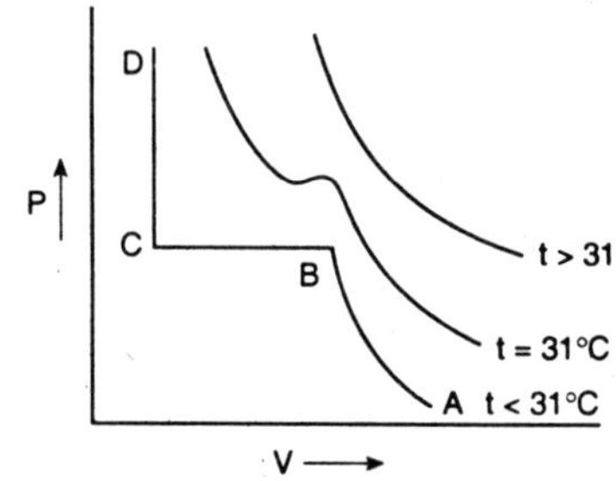

Fig. 2.2

2.10 CRITICAL PARAMETERS — T_C, P_C AND V_C

From Andrew's curve for CO_2 (Fig. 2.2) we see that for liquefaction of CO_2 by applying pressure, its temperature must be equal to or less than 31°C. For every gas there is such a temperature, at which or below which it can be liquefied by applying pressure; but above which it can never be liquefied by applying pressure. This temperature is called *critical temperature* (T_c). Minimum pressure required to liquefy the gas at this temperature is called *critical pressure* (P_c) and volume occupied by one mole gas at this state is called *critical volume* (V_c).

2.11 EVALUATION OF CRITICAL VALUES

Critical values P_c, V_c and T_c can be evaluated from equation of states for real gases. We have seen that at critical point the P-V isothermal has a hump. So calculus tells us that at critical point

$$\left(\frac{dP}{dV}\right)_T = 0 \text{ and also } \left(\frac{d^2P}{dV}\right)_T = 0.$$

Now if we want to find out critical values from Van-der-Waals equation, we write it as

$$P = \frac{RT}{V-b} - \frac{a}{V^2}.$$

So $$\left(\frac{dP}{dV}\right)_T = -\frac{RT}{(V-b)^2} + \frac{2a}{V^3}. \quad .....(2.1)$$

At critical state $$0 = -\frac{RT_c}{(V_c-b)^2} + \frac{2a}{V_c^3}$$

or, $$\frac{2a}{V_c^3} = \frac{RT_c}{(V_c-b)^2}. \quad .....(2.2)$$

Again from equation (2.1)

$$\left(\frac{d^2P}{dV^2}\right)_T = \frac{2RT}{(V-b)^3} - \frac{6a}{V^4}.$$

At critical state

$$0 = \frac{2RT_c}{(V_c-b)^3} - \frac{6a}{V_c^4}$$

or, $$\frac{6a}{V_c^4} = \frac{2RT_c}{(V_c-b)^3}. \quad .....(2.3)$$

Dividing equation (2.3) by equation (2.2)

$$\frac{3}{V_c} = \frac{2}{V_c - b}$$

or, $3V_c - 3b = 2V_c$

or, $V_c = 3b$. .....(2.4)

Putting this value in equation (2.2) we get

$$\frac{2a}{(3b)^3} = \frac{RT_c}{(3b-b)^2}$$

Now from Van-der-Waals equation

$$P_c = \frac{RT_c}{V_c - b} - \frac{a}{V_c^2}$$

or, $$P_c = \frac{R.\frac{8a}{27Rb}}{(3b - b)} - \frac{a}{(3b)^2}$$

or, $$P_c = \frac{4a}{27b^2} - \frac{a}{9b^2} = \frac{4a - 3a}{27b^2} = \frac{a}{27b^2}. \qquad .....(2.6)$$

Now every gas is ideal at very low pressure and $\frac{RT}{PV} = 1$ for one mole ideal gas. As pressure is gradually raised its deviation from ideal behaviour becomes more and more. At critical point, the deviation is highest, because any further increase in pressure will liquefy it and gas will no longer be present. So we may be curious to know the value of $\frac{RT_c}{PV_c}$ at this highest state of deviation. Putting the values of these parameters from Van-der-Waals equation we get

$$\frac{RT_c}{P_cV_c} = R \times \frac{8a}{27Rb} \times \frac{27b^2}{a} \times \frac{1}{3b} = \frac{8}{3} = 2.66$$

So compressibility factor $Z = \frac{P_cV_c}{RT_c} = \frac{3}{8}$.

Table 2.1

Gas	P_c (atm.)	V_c (c.c.)	T_c (°K)	$\frac{RT_c}{P_cV_c}$
H_2	12.8	69.7	33.2	3.05
He	2.25	61.5	5.2	3.08
CO_2	72.8	94.2	304.2	3.64
N_2	38.5	90	126	3.42

2.12 CONTINUITY OF STATE

If someone condenses a gas or heats a liquid then generally he finds that at a certain temperature and pressure transition from gas to liquid or liquid to gas is taking place. The change appears to be abrupt since the appearance of two phases are different. So one may think that gradual heating or cooling brings a discontinuous change at the point of transition. But this is not true. From Andrew's curve it can be shown that liquid state is the state of highly compressed gas and gaseous state is nothing but highly rarefied liquid; hence one is a continuation of other. In fact transition from one state to another can be done without any liquid-vapour co-existence state. These are understood from the following figure.

We have seen that in Andrew's curve there is a semi-horizontal portion below critical temperature. This portion indicates gas-liquid transition, which appears as discontinuous. Now if the two ends of different horizontal lines obtained at different temperatures (all below T_c) are joined by dotted line, we obtain a parabola with its apex at critical isotherm (at T_c).

Any point inside this parabola mean a gas-liquid co-existence, but outside it, we have either gas or liquid. If one can pass from gaseous line to liquid line avoiding this parabolic portion, he can change the state without any discontinuity. Let initially we are at point P (only gas). Now it is heated at constant volume till we reach point Q. If now it is cooled at constant pressure we can reach point R (only liquid) avoiding the parabolic zone (gas-liquid-transition part). Thus the transition is so smooth that one fails to distinguish between two phases. This is called continuity of state.

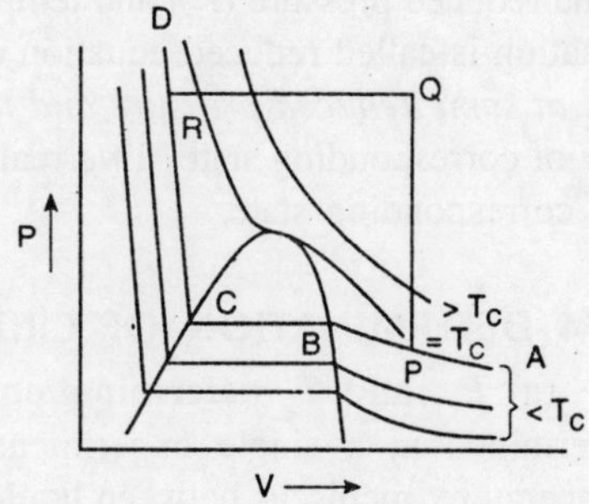

Fig. 2.3

2.13 REDUCED EQUATION OF STATE — LAW OF CORRESPONDING STATE

The ideal gas equation $PV = RT$ (1mole) is universal gas equation. Any two ideal gas (one mole) at same pressure and temperature will have same volume. However, any two real gas at same pressure and temperature will not have same volume; because real gas equation consists of gas dependent terms 'a' and 'b'. If eliminating these two terms we can derive real gas equation, which will be universally applicable for every real gas, then that equation of state will be called reduced equation of state. For deriving this equation, we first define reduced pressure; reduced volume and reduced temperature taking critical pressure (P_c), critical volume (V_c) and critical temperature (T_c) as unit pressure, unit volume and unit temperature. Thus if pressure = P in the ordinary scale, then in the reduced scale, reduced pressure $P_r = P/P_c$; similarly reduced volume $V_r = V/V_c$ and reduced temperature $T_r = T/T_c$.

Now Van-der-Waals equation for one mole real gas is

$$\left[P + \frac{a}{V^2}\right][V - b] = RT.$$

Putting $P = P_r.P_c$; $V = V_r.V_c$ and $T = T_r.T_c$ we get

$$\left(P_r.P_c + \frac{a}{V_r^2.V_c^2}\right)(V_r.V_c - b) = R.T_r.T_c.$$

But $P_c = \dfrac{a}{27b^2}$; $V_c = 3b$ and $T_c = \dfrac{8a}{27Rb}$.

So
$$\left(\frac{a}{27b^2}.P_r + \frac{a}{9b^2V_r^2}\right)(3b.V_r - b) = R.\frac{8a}{27Rb}.T_r$$

or,
$$\frac{a}{27b^2}\left(P_r + \frac{3}{V_r^2}\right).b(3V_r - 1) = \frac{8a}{27b}.T_r$$

or,
$$\left(P_r + \frac{3}{V_r^2}\right)(3V_r - 1) = 8T_r.$$

This equation does not contain any gas-dependent term and hence any two real gas at same reduced pressure (P_r) and temperature (T_r) will have same reduced volume (V_r). The equation is called reduced equation of state. The statement that *one mole of any two real gas at same reduced pressure and temperature will have same reduced volume* is called law of corresponding state. Two real gases at same P_r, T_r and V_r are said to be present in the corresponding state.

2.14 DETERMINATION OF CRITICAL VALUES — P_C, T_C AND V_C

(a) **P_c and T_c determination :** Critical pressure and critical temperature can be determined in a single experiment. The principle of the method is that, at critical temperature; meniscus between liquid and vapour vanishes and there is continuity of state. The instrument used is as shown in figure 2.4.

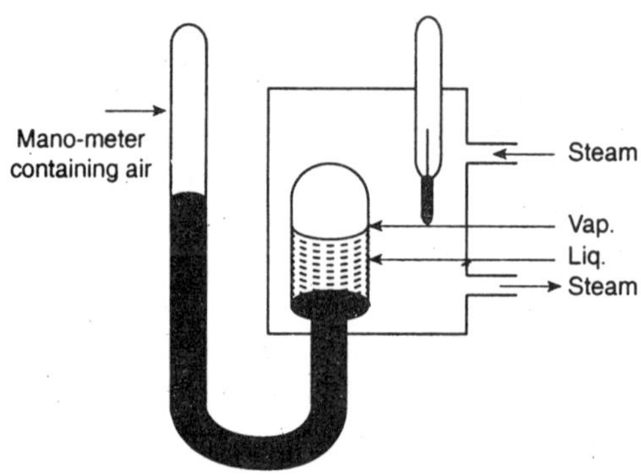

Fig. 2.4

The liquid under consideration is taken in a bulb over mercury kept in a mercury manometer; the other side of the manometer consists of air. The bulb is surrounded by a iacket, through which steam can be passed to heat the bulb. A thermometer is kept to note the temperature. Initially the meniscus between liquid and vapour is distinct. As temperature is raised the line of separation just disappears at a particular temperature. During cooling the meniscus appears almost at that temperature. The average of temperatures of appearance and disappearance of meniscus is critical temperature. Volume of the air at this temperature is noted from mercury meniscus. From this pressure of the vapour i.e., critical pressure (P_c) can be known.

(b) **Determination of V_c :** Mathias and Cailittet observed that if mean of densities of liquid and vapour (in equilibrium above it) is determined at various temperature and then it is plotted against temperature, we get a straight line. The nature of curves are as follows.

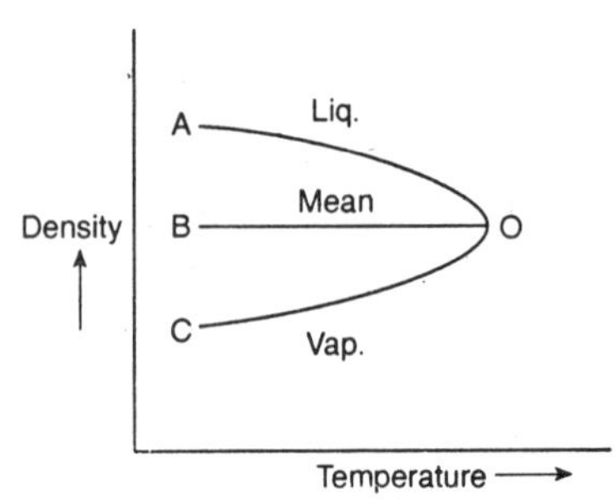

Fig. 2.5

In the above figure AO and CO are density vs temperature curves for liquid and vapour and BO is mean density vs temperature curve. The three met at 'O'. So at 'O' densities of liquid and vapour are same i.e., the system is both liquid and vapour. This happens at critical state. So density corresponding to point 'O' is critical density (d_c). If 'M' is the molecular weight of the substance, then critical volume (V_c) = M/d_c.

2.15 DETERMINATION OF VAN-DER-WAALS CONSTANTS

In section 2.14 we have discussed how critical parameters can be determined. Now $V_c = 3b$; so from critical volume, Van-der-Waals constant 'b' can be known. Also critical pressure $P_c = a/27b^2$. Thus knowing 'b' and determining 'P_c' Van-der-Waals constant 'a' can be determined.

POINTS TO REMEMBER

From Amagat's PV vs P isothermals it was found that no gas obeys ideal gas equation at ordinary pressure and temperature. Thus gases are not *'ideal'* but *'real'*. Ideal behaviour is found either at very low pressure and high temperature or at a given temperature, know as Boyle temperature $\left(T_b = \frac{a}{Rb}\right)$. Departure from ideal behaviour is due to existence of intermolecular attraction and repulsion, which become negligible at very low P and very high T. Considering this Van-der-Waals gave an equation of state for real gas, which is $\left(P + \frac{a}{V^2}\right)(V - b) = RT$ for one mole. 'a' and 'b' are constant and sta for attraction and repulsion correction. There are other equations of state for real gas. Generalised equation is virial equation and it is $Z = \frac{PV}{RT} = 1 + \frac{A}{V} + \frac{B}{V^2} + \frac{C}{V^3} + \ldots\ldots\ldots\ldots$

Reduced equation of state is an equation of state for real gases where gas-dependent parameters 'a' and 'b' have been eliminated.

Attempt to liquefy gas by isothermal compresson was made by Andrew. His observation is that - for this purpose temperature of the gas must be equal to or less than a particular value, known as critical temperature (T_c). Pressure required for liquefaction at this temperature is P_c and volume of the gas is V_c. For a Van-der-Waals gas $\frac{RT_c}{P_cV_c} = 2.66$.

Experimental methods are available for measuring P_c, T_c and V_c. No $P_c = \frac{a}{27b^2}$;

$T_c = \frac{8a}{27Rb}$ and $V_c = 3b$; thus 'a' and 'b' can be evaluated.

QUESTIONS

SHORT TYPE

1. What are compressibility factor (Z) and Boyle temperature (T_b) ? What is the value of Z at T_b ?
2. What are the units of '*a*' and '*b*' ?
3. What is co-volume ? Can you determine molecular radius from it ?
4. What are the defects of Van-der-Waals equation ?
5. What will be the form of Van-der-Waals equation for '*n*' mole gas ?
6. Name some other equation of state for real gas.
7. What is a virial equation ?
8. Can you liquefy CO_2 by applying pressure at 32°C ?
9. What are the units of reduced parameters ?
10. What do you mean by continuity of state ?
11. Explain Amagat's curve from Van-der-Waals equation.

ESSAY TYPE

1. Which postulates of kinetic theory are mainly responsible for deviation of gas from ideal behaviour — how did Van-der-Waals explain it ?
2. Deduce Van-der-Waals equation and explain how both attraction and repulsion among gas molecules have been treated.
3. Define critical parameters. Show that for a Van-der-Waals gas compressibility factor is 3/8.
4. What is reduced equation of state ? Derive it from Van-der-Waals equation. What is law of corresponding state ?
5. Describe the experimental methods for the determination of critical parameters.

PROBLEMS

1. Calculate the constants *a* and *b* if T_c = 31°C and P_c = 72.8 atm and R = 0.082 litre atm.

   [*Hints.* $T_c = \frac{8a}{27Rb}$, $P_c = \frac{a}{27b^2}$. Solve for *a* and *b*.]

2. The critical temperature and critical pressure of oxygen are –118.8°C and 49.7 atm. Calculate its Van-der-Waals constants.

   [*Hints.* See problem '1'.]

3. Van-der-Waals constant in litre atm. per mole for CO_2 are a = 3.6 litre^2 atm mole^{-2} , b = 4.28 × 10^{-12} litre mole^{-1}. Calculate T_c and V_c for the gas.

   [*Hints.* $T_c = \frac{8a}{27Rb}$; $V_c = 3b$. Put R = 0.082 litre atm mole^{-1} deg^{-2}.]

4. Two moles of ammonia are enclosed in a five litre flask at 27°C. Calculate the pressure of NH_3 using Van-der-Waals equation. For NH_3, a = 4.17 atm litre^2 mole^{-2}, b = 0.0371 litre mole^{-1}.

   [*Hints.* Use $\left[P + \frac{n^2a}{v^2}\right][v - nb] = nRT$ equation.]

OBJECTIVE TYPE

1. Choose the correct answer.

(a) Pressure of a real gas is ——— that of ideal gas.

Ans : (i) equal to (ii) less than (iii) more than.

(b) Temperature at which a real gas obeys ideal gas equation is ———.

Ans : (i) reduced temperature (ii) critical temperature (iii) Boyle temperature.

(c) Two Van-der-Waals gases have different '*a*' but same *b*-values. Their molecular radius will be ———.

Ans : (i) same (ii) different.

(d) Compressibility factor of an ideal gas is unity at ———.

Ans : (i) all temperature (ii) Boyle temperature (iii) Critical temperature.

(e) Critical volume of Van-der-Waals gas is ———.

Ans : (i) $\frac{8a}{27Rb}$ (ii) $3b$ (iii) $\frac{a}{27b^2}$.

(f) A gas with $a = 0$, ———.

Ans : (i) is easily liquefiable (ii) can not be liquefied.

(g) At relatively high pressure Van-der-Waals equation reduces to ———.

Ans : (i) $PV = RT\frac{a}{V}$ (ii) $PV = RT$ (iii) $PV = RT + bP$.

(h) *PV* vs *P* isothermals are known as ———.

Ans : (i) Amagat's curve (ii) Andrew's curve (iii) Van-der-Waals curve.

2. Match the terms of column A with that of column B.

A	B
(a) Compressibility of gas.	(a) Collision diameter
(b) Liquefiction of gas.	(b) Ideal gas
(c) Mean free path	(c) Intermolecular attraction
(d) Reduced equation of state	(d) Andrew
(e) $Z = 1$	(e) Real gas.

Answers :

1. (a) (ii); (b) (iii); (c) (i); (d) (i); (e) (ii); (f) (ii); (g) (iii); (h) (i).

2. a – c; b – d; c – a; d– e; e – b.

Chapter 3

MOLAR-MASS-DENSITY-ABNORMAL VAPOUR DENSITY

3.1 MOLAR MASS

The term molar mass means — how many times is the molecule of a substance heavier than a hydrogen atom. It is a dimensionless quantity. However, in chemistry knowledge of molar mass of a substance is essential in order to know its geometry and many other properties. That is why so many methods have been developed for the determination of molar mass. For gaseous substances molar mass is determined from (i) ideal gas equation (ii) density measurement and (iii) vapour density measurements.

3.2 BASIC PRINCIPLES

(a) For 'n' moles ideal gas the gas equation is $PV = nRT$. Now if weight of the gas is g and its molar mass is M then $PV = \frac{g}{M}RT$. Hence $M = \frac{g}{PV}RT$ Thus know all quantities of R. H. S. 'M' can be known.

(b) Density of a gas is its weight per unit volume. Apparently it may be thought to be negligible. The assumption may not always be true. Thus while density of H_2 is only 0.00009 gms/c.c., that of UF_6 is about 0.0158 gms/c.c. Now as $M = \frac{g}{V}\cdot\frac{RT}{P} = \left(\frac{d}{P}\right)RT$ [d = density], so measuring density under a given set of condition, molar mass can be known.

(c) Vapour density is the density of a gas relative to hydrogen; it is also known as relative density. According to Avogadro molar mass of a gas is twice its vapour density. Thus measurement of vapour density can also help us in knowing molar mass.

3.3 DIFFERENT METHODS

Following methods are available for the determination of molar masses of gases. (a) Hofmann's method (b) Victor Meyer's method (c) Regnault's method (d) Duma's method (e) Gas density balance method and (f) Limiting density method.

Out of all those methods only the last one gives accurate result, while the rest give approximate result, as they use ideal gas equation in non-ideal condition.

(a) **Hofmann's method :** Hofmann's apparatus is nothing but a one-metre long graduated barometric tube filled with mercury and then inverted over a mercury-trough. Weighed amount of a volatile liquid, whose molar mass is to be determined is taken in a small capsule, known as Hofmann's bottle. It is forced to dip below the tube. When the force is released the bottle will float in the Toricellian vacuum. The barometric tube is surrounded by a jacket through which

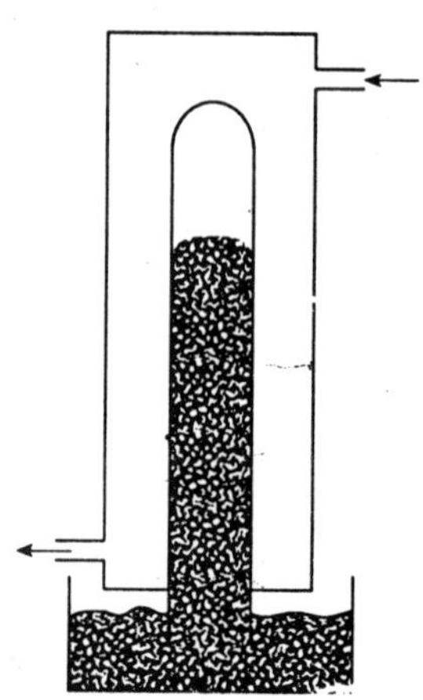

Fig. 3.1

some hot vapour, whose temperature is much higher than boiling point of the liquid in capsule is passed. The liquid at once vaporises and depresses the mercury level at barometer. When steady-state is attained, volume of the vapour is read from barometer; pressure is determined from depression of mercury level and temperature is also read. Since weight of the vapour is equal to weight of liquid taken in the capsule, so all terms of R.H.S of equation $M = \frac{g}{PV} RT$ are known. Hence molar mass is known.

(b) **Victor Meyer's method :** In this method a weighed amount of liquid is vaporised in a container; the vapour displaces an equal volume of air; volume of the displaced air is determined at ordinary temperature and pressure by collecting it in a graduated tube over water. Then molar mass is calculated. Details of the process is as follows.

AB is Victor-Meyer's tube. It is a long tube having a flat bottom. It has a side tube which is kept inside a water-trough. This tube is surrounded by a copper-jacket containing water. A graduated glass tube filled with water is first kept in one side of the water-trough. Now the water of the copper-jacket is boiled. Air inside Victor Meyer's tube expands and comes out in the form of bubble. When temperature becomes steady, no more bubble comes; the graduated tube is now placed over the side tube dipped in water-trough.

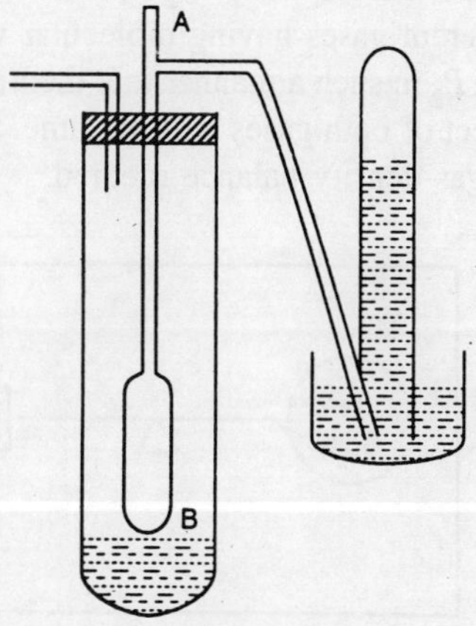

Fig. 3.2

A very small amount of liquid (≤ 0.1 gm) is now weighted in small Hofmann's bottle and dropped inside the tube through the mouth, which is immediately closed. Liquid vapourises and air is displaced. Displaced air comes as bubble and is collected in the graduated tube. When no more bubble comes, the mouth of this tube is closed tightly by thumb and it is dipped in a tall jar containing water. Volume of the air is read after making water level same inside and outside. This is the volume of air at atmospheric pressure.

Calculation : Let volume of air = V; temperature = T°A; barometric pressure = P and vapour pressure of water at this temperature = f. Then $(P - f) V = \frac{g}{M}.RT$; g= weight of liquid.

(c) **Regnault's method :** To determine molar mass of gas Regnault directly weighted a known volume of gas. To eliminate buyoancy he took two metallic bulbs of exactly same volume and hung them in the two arms of the balance. Initially both were empty. Now one of them is filled with the experimental gas at some known pressure. Its weight is now determined. Volume of the vessel is determined by weighing it after filling with water. From weight of water and its density, its volume becomes known. Thus by applying same formula molar mass can be determined.

(d) **Duma's method :** Here the apparatus is a glass bulb with a long drawn-out neck. The bulb is weighted and then few c.c. of volatile liquid is taken inside it by dipping the neck in the liquid and applying alternate heating and cooling technique. The bulb is then heated in a temperature bath much above the boiling point of the liquid. All air is displaced first and then vapour begins to come out. Now the mouth of the tube is sealed and the bulb is weighed after cooling. Pressure (atmospheric) and temperature of the bath are noted. The bulb is then completely filled with water and from weight of this water, its volume is determined. Now by applying same formula molar mass can be determined.

(e) **Gas-density-balance method :** Ramsay used the buoyancy property to determine molar mass of a gas. We know that weight of a substance taken in presence of a gas is equal to its correct weight minus weight of displaced gas. This effect is known as buoyancy. Again weight of displaced gas is dependent on its density (d). But $d = \frac{MP}{RT}$. So at a given temperature density depends on both molecular weight and pressure of the gas. For two different gases having molecular weights M_1 and M_2 we may adjust their pressures at P_1 and P_2 in such a manner that their densities would be same. Under this condition buoyancy effect of both gases will be same. So in this condition $M_1 P_1 = M_2 P_2$. This is the principle of gas-density balance method.

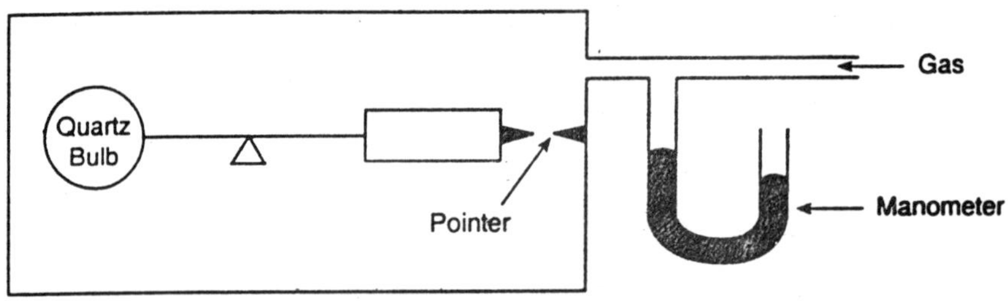

Fig. 3.3

The instrument consists of a large quartz bulb counter poised on a knife edge against a displacable weight. The bulb is kept in a closed container. First the container is filled with the gas, whose molar mass is to be known and pressure is so adjusted that the beam becomes perfectly horizontal, as shown by the pointer. Now the container is filled with a standard gas say O_2 and here also pressure is so adjusted that the beam becomes perfectly horizontal. In both cases pressure is determined by manometer. Now as buoyancy effect is same in each case, so $Po_2\ Mo_2 = P_{gas}.\ M_{gas}$. Thus M_{gas} can be known.

(f) **Limiting-density method :** Methods described so for gives generally approximate results as we are using ideal gas equation in a condition, which is very much non-ideal. We know that all gases behave ideally at very low pressure and (or) at high temperature. The gas-density balance method may give us correct result if we perform our experiment at very low pressure. However, the best method for determining molar mass of a gas is limiting-density method. Its principle is as follows.

As we know that all gases behave ideally at very low pressure, we can write

$$M = \underset{P \to 0}{\text{Lt}} (d/P) \, . \, RT.$$

[$\because PV = \frac{g}{M} RT$ leads to $M = \frac{d}{P} . RT$; but it is valid at low pressure only].

The term $\underset{P \to 0}{\text{Lt}} (d/P)$ is called limiting density. If we can measure this value at a given temperature, then using above equation we can determine its molar mass. To find limiting density of a gas we measure (d/P) values at such low pressures as possible; we then plot

(d/P) vs. P. A straight line is obtained, this is extrapolated to zero pressure; intercept on ordinate gives us limiting density (fig. 3.4.) Thus 'M' can be known exactly.

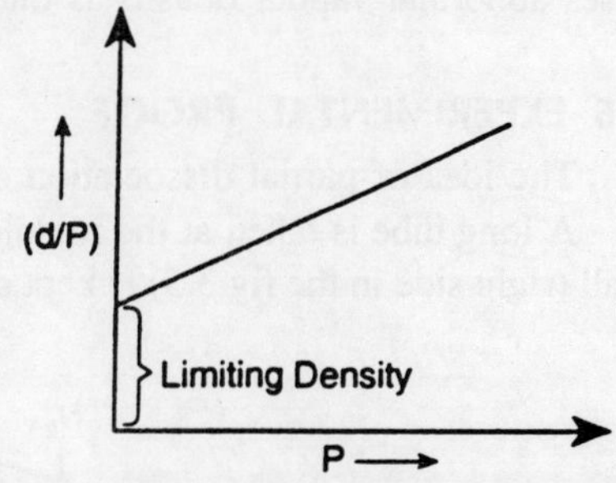

Fig. 3.4

3.4 NORMAL DENSITY AND LIMITING DENSITY

We know for a given gas at a given temperature density depends on pressure $\left[d = \frac{MP}{RT}\right]$. Density of a gas divided by its pressure is called normal density. Thus if 'd' is density measured at 'P' atm. pressure, then (d/P) is normal density. When P approaches zero normal density becomes limiting density. Thus limiting density is always normal density, but normal density is not always limiting density. A relation between two can be established. Let 'g' gm of a gas shows density 'd' when measured at some ordinary pressure P; volume being V. Then,

$$\text{normal density} = \frac{d}{P} = \frac{g}{PV}$$

$$\text{or, } g = \text{normal density} \times PV.$$

If 'g' gm of the same gas shows density d_0, when measured at extremely low pressure P_0 ($\simeq$ zero);

volume being V_0. Then,

$$\text{limiting density} = \frac{d_0}{P_0} = \frac{g}{P_0 V_0}$$

$$\text{or, } g = \text{limiting density} \times P_0V_0.$$

$$\text{So normal density} \times PV = \text{limiting density} \times P_0V_0$$

$$\text{or, limiting density} = \text{normal density} \times \frac{PV}{P_0 V_0}.$$

At low pressure gas becomes ideal; Boyle's law is valid and $PV = P_0V_0$. So limiting density = normal density. But at high pressure, gas is real and $PV \neq P_0V_0$; here two densities are different.

3.5 ABNORMAL VAPOUR DENSITY

Molecular weight of NH_4Cl is 53.5 and so according to Avogadro's hypothesis, its vapour-density is expected to be 26.75. Experimentally found value is however much less e.g. at 350°C the vapour density has been found to be only 14.6. To explain this anomalous observation it was suggested that at such temperature NH_4Cl partially dissociates in to NH_3 and HCl i.e.

$$NH_4Cl\ (g) \rightleftharpoons NH_3\ (g) + HCl\ (g)$$

Such a dissociation will increase number of moles and hence volume. Since mass is remaining same but volume is increasing therefore density and vapour density will decrease.

Similar is the case with many other compounds e.g. PCl_5 ; $COCl_2$; N_2O_4 etc. In all such cases abnormal vapour density is due to partial dissociation.

3.6 EXPERIMENTAL PROOF

The idea of partial dissociation can be verified by a simple experiment.

A long tube is taken at the middle of which there is a porous wall. On one side of the wall (right side in the fig 3.5) is kept some solid NH_4Cl. The mouths of the tube are corked

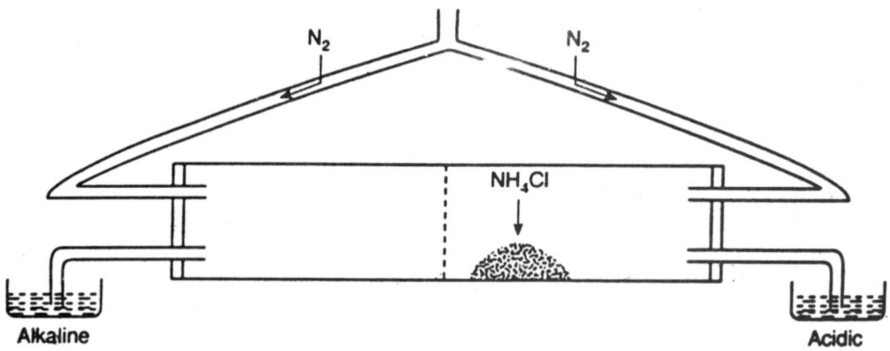

Fig. 3.5

through which two outlets are introduced. These outlets are dipped in water in beakers with some acid-base indicator. Now NH_4Cl is strongly heated; NH_3 and HCl will be produced. NH_3 being lighter will diffuse at a faster rate through porous wall. So very soon concentrate of NH_3 on left side of wall will be high and that of HCl will be high on right side. If now N_2 gas is passed to expell these gases, then gas coming from left outlet will show alkaline character while gas coming from right outlet will show acidic character. The experiment conclusively proves that on heating NH_4Cl dissociates to NH_3 and HCl gas.

3.7 DEGREE OF DISSOCIATION AND VAPOUR DENSITY

We have just now seen that on heating NH_4Cl partially dissociates into NH_3 and HCl gas. Fraction of the total amount undergoing dissociation is called degree of dissociation (α). Its value can be determined from density or vapour density measurement. The principle is as follows.

We consider $NH_4Cl\ (g) \rightleftharpoons NH_3\ (g) + HCl\ (g)$ equilibrium. Let initially we have taken one mole NH_4Cl and α is its degree of dissociation, then after attaining equilibrium number of moles of different substances will be —

$$n_{NH_4Cl} = (1 - \alpha);\ n_{NH_3} = \alpha \text{ and } n_{HCl} = \alpha.$$

So total moles $= (1 - \alpha + \alpha + \alpha) = (1 + \alpha)$.

Had there been no dissociation, no. of moles would have been 1 and volume = V_0 (let). But after dissociation no. of moles have become $(1 + \alpha)$ and volume = V (let). Since volume and number of moles are proportional, so

$$\frac{V_0}{V} = \frac{1}{1 + \alpha}.$$

Again density is inversely proportional to volume. So if d_0 is density assuming no dissociation and d is density if there is dissociation, then

$$\frac{V_0}{V} = \frac{d}{d_0}.$$

So
$$\frac{d}{d_0} = \frac{1}{1+\alpha}.$$

In place of density we can put vapour density. Now V. D. assuming no dissociation is half the molar mass. So if we can determine vapour density experimentally after dissociation, we can calculate degree of dissociation, using the above equation.

3.8 ASSOCIATION

We have just now seen that when dissociation increases number of moles there is a considerable lowering in vapour density value. We can then easily assume that if association occurs some where to decrease number of moles and hence volume, there will be an increase in vapour density. This is the case with water vapour; acetic acid vapour etc. If their vapour densities are measured at respective boiling points we get high values due to association. However, at very high temperature associated molecules separate and it exists as hundred per cent monomer. Then we get normal vapour density value.

POINTS TO REMEMBER

Most of the methods used to determine molar mass of gases use ideal gas equation $M = \frac{g}{V} \cdot \frac{RT}{P}$ in non-ideal condition. That is why we do not get exact molar mass from these methods. Only limiting density method uses this equation in ideal condition (*i.e. when* $P \to 0$); so it gives correct value. While *normal density is* $\left(\frac{d}{P}\right)$; limiting density is $\underset{P \to 0}{\text{Lt}} \left(d/P\right)$. Thus limiting density is normal density but the reverse is not true.

Certain substances, whose vapour state show dissociation or association accompanying with a change in number of moles (increase or decrease) exibit abnormal vapour density. From this value and theoretical value of *vapour density* $\left(i.e.\ \frac{M}{2}\right)$ *degree of dissociation or association* can be calculated.

QUESTIONS

SHORT TYPE

1. Why limiting density method is considered as accurate method for molar mass determination ?
2. "Gas-density balance" method may give accurate or approximate molar mass depending on experimental condition — Explain.
3. Hydrogen iodide dissociates partially on heating. Shall we get any abnormal vapour density here.
4. In Victor-Meyer's experiment what is collected over water ?
5. The amount of volatile liquid to be taken in Victor Meyer's experiment must be very small — why ?
6. Vapour density of acetic acid vapour measured at its boiling point is higher than theoretical value — why ?

ESSAY TYPE

1. Describe Victor Meyer's method for molar mass determination.
2. What are normal and limiting density. Describe limiting density method for determining molar mass.
3. Explain the term abnormal vapour density. Describe an experiment to prove dissociation of NH_4Cl.
4. How can you know degree of dissociation of N_2O_4 from its vapour density measurement ?

PROBLEMS

1. 0.0623 gms. of a substance gave by Victor Meyer's method 31.5 c.c. of air measured at 15°C and 750 m.m. Find the molar mass of the substance. Tension of aqueous vapour at 15°C = 12.7 m.m.

   [*Hints.* $PV = \frac{g}{M}RT$; $P = \frac{750 - 12.7}{760}$ atm.; $V = \frac{31.5}{1000}$ litre.
   $g = 0.0623$; $R = 0.082$; $M = ?$]

2. The following observation were made in a Victor-Meyer determination.

Weight of liquid used	= 0.0572 gm
Volume of dry air displaced	= 15.3 c.c.
Room temperature	= 23.5°C
Barometic pressure	= 763.2 m.m.

   Calculate the molar mass and density of the vapour at N. T. P.

   [*Hints.* See problem '1'.]

3. Ammonium chloride when vaporised at 380°C and 813.8 m.m. pressure has a density of 0.6257 gm. per litre. Calculate the degree of dissociation.

[*Hints.*

	NH_4Cl (g) =	NH_3 (g) +	HCl (g)
Initial	1	0	0
Final	$1 - \alpha$	α	α

Total = $(1 + \alpha)$

$$\frac{1}{1+\alpha} = \frac{d_{theo}}{d_{expt}}; \quad d_{theo} = \text{Vap. density} \times 0.00009$$

$$d_{expt} = 0.6257. \quad = \frac{M}{2} \times 0.00009 \quad \text{Find } \alpha.]$$

4. The vapour density of nitrogen tetraoxide (N_2O_4) at 26°C and one atmospheric pressure is 38. Calculate its degree of dissociation into NO_2 molecules.

[*Hints.* See problem '3'.]

5. 4.5 gms of PCl_5 were completely vaporised at 252°C and the vapour occupied 1700 c.c. at one atmosphere. Calculate the degree of dissociation at this temperature.

[*Hints.* Try as problems '3' and '4'.]

OBJECTIVE TYPE

1. State whether following statements are 'true' or 'false.'

(a) Victor Meyer method gives exact molar mass.

(b) In the ideal gas equation if R is in litre-atm. and P is atm. then density comes as gm/cc.

(c) Limiting density can not be measured directly.

(d) If vapour density is less than theoritical value it means dissociation.

(e) Vapour density of CH_3COOH measured at its boiling point is more than theoritical value.

(f) Limiting density is a type of normal density.

(g) Regnault's method for molar mass determination is for volatile liquids.

(h) In Victor-Meyer's method vapour is collected over water.

(i) In Victor-Mayer's method amount of liquid should be very small.

Answer :

(a) F; (b) F; (c) T; (d) T; (e) T; (f) T; (g) F; (h) F; (i) T.

Chapter 4

THE LIQUID STATE

4.1 WHAT IS A LIQUID ?

Among the three states of matter, abundant in our earth liquid state resides in between solid and gas. Because of its intermediate position between solid and gas it has some properties similar to that of solid and also it possesses some properties similar to gas. Thus liquid is sometimes regarded as a gas with some memory of solid phase.

4.2 SIMILARITIES AND DISSIMILARITIES WITH SOLID AND GAS

We know perfectly crystalline solids are highly ordered substances with strong co-hesive forces. Their densities are high due to compactness and they have definite geometrical shape. X-ray diffraction follows Bragg's equation (vide Chapter-5) and such study can give us distance between different crystalline planes. Because of strong co-hesive force the atoms or molecules can not have any translational motion; they have only vibrational motion.

On the contrary gases are highly disordered substances. Due to absence of any cohesive force among molecules of a perfect gas they move in a most random manner. They possess only translational motion; velocity distribution follows Maxwell's equation. As each molecule is totally free, the question of vibrational motion does not arise.

Now let us peep in to the domain of liquid state. They have certain degree of both order and disorder; of which the former is predominant. On heating a solid disorder gradually increases. According to Eyring some holes begin to appear (fig. 4.1).

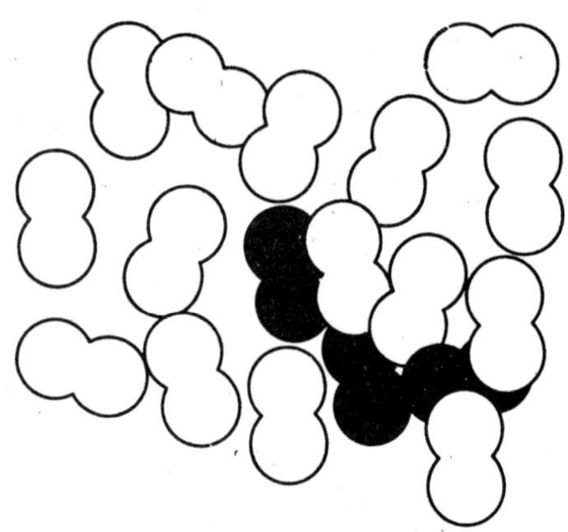

Fig. 4.1

The molecules which had vibrational motion only, now begins to acquire limited translational motion and they can move through these holes. Geometrical shape is thus distorted and liquid state appears. However, cohesive force is still strong enough to confine the molecules within a definite volume; thus liquids have fixed volume. Because of rather high cohesive forces, density of the liquid is much higher than gases and much closer to that of solid. Generally due to appearance of these holes during melting volume of liquid becomes nearly three per cent more than solid. If we go on increasing temperature, number of such holes go on increasing and translational motion also increases. Behaviour of liquid becomes more and more close to gas. Thus it may be said that near melting point a liquid behaves as an imperfect crystal and near critical temperature a liquid behaves as an imperfect gas.

4.3 NATURE OF FORCES IN A LIQUID

We have just now discussed that relatively strong co-hesive forces exist in a liquid. The nature of such forces depends on type of species that constitute the liquid system i.e., atoms;

polar or non polar molecules, ions etc. Brief description of different types of forces are as follows —

(a) **Van-der-Waals force :** When two molecules (similar or not) are quite close then nucleus of one molecule attracts the electron cloud of the other. This type of attraction is Van-der-Waals attraction. At very very close distance nucleus-nucleus and electron-electron repulsion also may be significant. Since in a liquid state molecules are quite close Van-der-Waals force will exist in a liquid.

(b) **Hydrogen bonding :** Strong electronegative elements such as F; O; N; etc. when combine with hydrogen, the resultant compound becomes polar with positive charge at hydrogen end. When such compound exists in liquid state then positive hydrogen end of one molecule is attracted by negative end of the nearest neighbour and a cohesive force appears. This is called hydrogen bonding or dipole-dipole attraction.

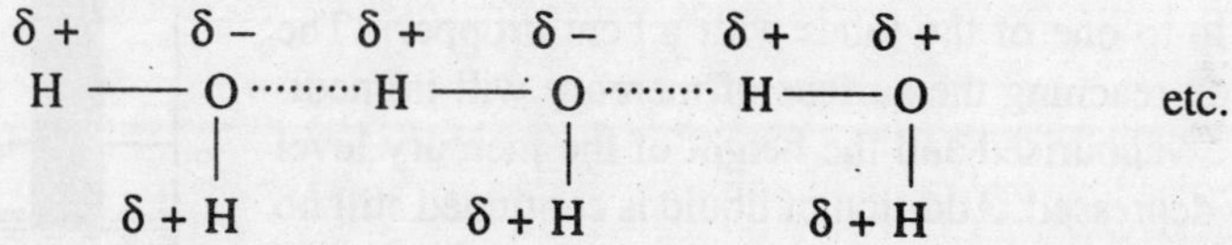

(c) **Coulombic attraction :** Polar molecules carry charge at both ends. So there will be Coulombic attraction between such molecules. Because of this cohesive interaction polar gaseous molecules are easily liquefied.

(d) **Ionic forces :** When ionic substances are present in liquid state e.g., fused NaCl then strong ionic force (i.e., electrostatic force) exists between oppositely charged ions.

4.4 CHARACTERISTIC PROPERTIES OF LIQUIDS

We shall now discuss characteristic properties of liquids. These will be discussed under the following headings. (a) Vapour pressure (b) Surface tension (c) Parachor and (d) Viscosity.

(a) Vapour pressure : It is our common experience that when some liquid is placed in an open vessel it dries up after some time which may be few minutes or even few months. We say that vapourisation took place. Again it is also our experience that if the mouth of the vessel is tightly covered then the liquid is not dried. Does vapourisation stop here ? To understand it let us see why vapourisation occurs ? We have already read that liquids have some holes in their structure where translational motion is found. Distribution of translational velocity follows Maxwell's equation. So there are very few molecules having very high translational velocity. They will overcome cohesive attraction and will leave the surface to be vapourised. Hence number of such high energetic molecules in the liquid state will decrease. To maintain steady-state some of less energetic liquid molecules will become more energetic by collision and they will be vapourised again. The process will be continued in an open vessel till the entire liquid is vapourised. However, if the mouth of the vessel is covered, then vapour can not escape and owing to its own increased pressure, it will begin to condense. When rate of evaporation and rate of condensation becomes same an equilibrium is established. Pressure due to this vapour in equilibrium with the liquid at this temperature is called vapour pressure. It will increase with increase in temperature, because from Maxwell's equation we know that increase in temperature increases the number of molecules moving with high velocity and they are eligible for escaping. Variation of vapour pressure with temperature is given by Clapeyron equation, which will be discussed in the chapter of thermodynamics.

Measurement of vapour pressure

Various method are available for measuring vapour pressure of a liquid. They are broadly of two types : Static and Dynamic. Static method includes barometric method, while Dynamic method includes isoteniscope method, Ramsay and Young's method and transpiration method. In this book we shall discuss only barometric method.

Two identical barometric tubes are filled with mercury and are inverted over a mercury cup. They are surrounded by a jacket maintained at constant temperature. Now small amount of experimental liquid is introduced in to one of the tubes with a bent dropper. The liquid on reaching the surface of mercury will immediately be vapourised and the height of the mercury level will be depressed. Addition of liquid is continued still no more vapourisation occurs and a liquid-vapour equilibrium is established. The difference in mercury levels between two tubes is the vapour pressure of the liquid at this temperature.

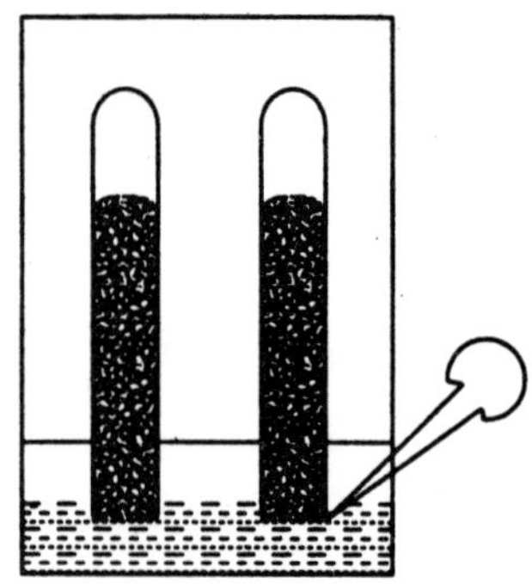

Fig. 4.2

Vapour pressure — Critical temperature — Boiling point

An empirical equation connecting vapour pressure with critical temperature and pressure was proposed by J. D. Van-der-Waals (1881). The equation is

$$\log \frac{P}{P_c} = K\left(1 - \frac{T_c}{T}\right).$$

Here P is the vapour pressure at temperature T and P_c and T_c are critical pressure and temperature. K is a constant with value approximately 3.

Now if T is normal boiling point (T_b) then $P = 1$ atm. So

$$\log \frac{1}{P_c} = K\left(1 - \frac{T_c}{T_b}\right)$$

or,
$$\frac{T_b}{T_c} = \frac{K}{K + \log P_c}.$$

For most substances P_c value is in between 50 to 100 atms. Taking log P_c as approximately 2 and K as 3 the value T_b/T_c comes as 0.6. Experimental value for T_b/T_c for some liquids are given :

Table 4.1

Name	T_b/T_c
He	0.81
H_2	0.61
N_2	0.61
O_2	0.58
CO_2	0.64
NH_3	0.59
H_2O	0.58

(b) Surface tension — Surface energy : Surface of a liquid is always in a state of strain due to unsaturation. This can be understood as follows.

A liquid molecule in the bulk is equally attracted by its neighbours from all sides and so net attractive force on it is zero. But a liquid molecule at the surface experiences an unbalanced backward force. This force tends to decrease the surface area. If we imagine a line of unit length at the surface, then the force acting on it at right angle to decrease the surface area is called *surface tension*. Its traditional unit is dyne cm^{-1} while S.I. unit is Newton $metre^{-1}$.

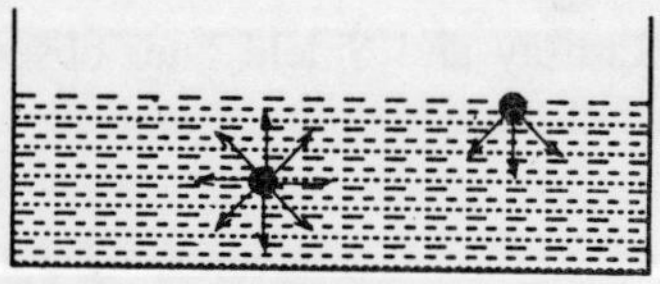

Fig. 4.3

Since surface tension tends to lower the surface area, therefore work must be done against it if we want to increase the surface area. Work done to increase the surface area by unity is called *surface energy*, because this work is stored in the surface as potential energy. If surface tension is equal to γ then work done to increase surface area by unity is also γ. So numerically surface tension and surface energy are same. However, their units are different. Thus while unit of surface tension is dyne cm^{-1} (or Newton $metre^{-1}$) that of surface energy is dyne cm (or Newton metre).

Effects of surface tension

Surface tension of a liquid is responsible for some interesting behaviour of liquid phase. These are described below —

(i) **Spherical drops :** We know that out of different shapes of a substance a sphere has the smallest surface area. Since total surface energy depends on surface area and since every substance has a tendency to minimise its free energy, so liquid will also try to have a spherical shape. This is evident from the fact that falling liquid drops are always spherical in shape.

(ii) **Excess pressure in a bubble :** When air is blown through soap solution bubbles are produced. This is due to lowering of surface tension of water by soap powder. Laplace derived an equation to express the mechanical pressure across an interface and its curvature.

(iii) **Capillarity :** When a capillary is dipped in a liquid, then (if the liquid wets the solid by which capillary is made) the liquid will instantaneously rise in the capillary. These liquids show a concave meniscus. On the contrary liquids which do not wet the solid will show a lowering of level (in comparison to outer level) inside the capillary. Such liquids show convex meniscus. This capillary rise (or fall) is again due to surface tension. Water is transported in plants by this method; plant tissues act as capillary. In a kerosene lamp oil is transported through wick due to this capillary phenomena.

(iv) **Floating of needles :** If a needle is kept over liquid it will float. Here the liquid film under the needle becomes curved and thus an upward force greater than gravitational pull is generated. So the needle floats.

Surface tension and temperature

Surface tension of a liquid decreases with increase in temperature. Eotvos proposed an equation connecting surface tension and temperature. The equation is

$$\gamma\left(\frac{M}{\rho}\right)^{2/3} = K\left(T_c - T\right)$$

γ = surface tension ; M = mol. wt; ρ = density; K = constant; T_c = critical temperature and T = abs. temperature.

The equation was deduced on the assumption that surface tension becomes zero at critical temperature.

Ramsay and Shield later observed that γ becomes zero, somewhat below T_c. They modified the equation as —

$$\gamma\left(\frac{M}{\rho}\right)^{2/3} = K\left(T_c - T - 6\right).$$

This is Ramsay-Shield's equation.

(c) Parachor : Mc Leod gave the following relation between surface tension (γ) and densities of liquid (D) and vapour (d) —

$$\gamma = K(D - d)^4.$$

Neglecting density of vapour in comparison to that of liquid,we get

$$\gamma = K.\, D^4$$

$$\text{or,}\quad \frac{\gamma^{1/4}}{D} = \text{constant.}$$

Multiplying both side by mol. wt. M we get

$$\frac{M}{D}\cdot\gamma^{1/4} = \text{constant}$$

$$\text{or,}\quad V_m \cdot \gamma^{1/4} = \text{constant} = P.\ [V_m = \text{Molar vol.}]$$

This constant P is called *parachor*. So parachor may be defined as the molar volume of the liquid, whose surface tension is unity. Since volume is an additive property, so parachor is mainly additive in nature. However, it has certain amount of constitutive nature also. Parachor values of some substances are given in the following table.

Table 4.2

Name	Parachor
C	4.8
H	17.1
O	20.0
O (in ester)	60.0
=	23.2
≡	46.6
⬡	6.1

Parachor values are often used to select the correct structure from a number of proposed structures. Thus two structures were proposed for benzoquinone. These are

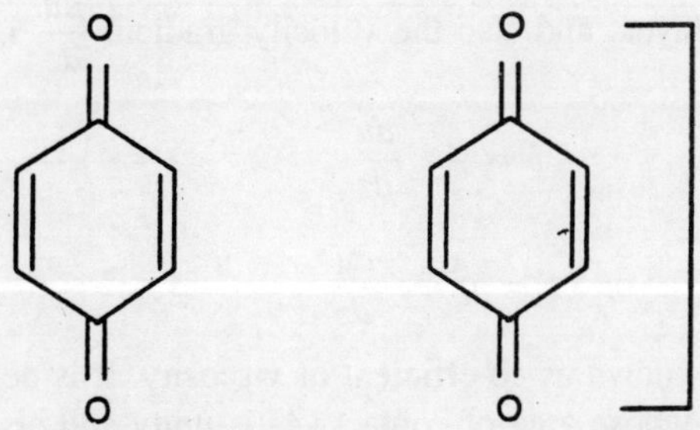

I II

For structure I the parachor value is —

6 C	atoms	6 × 4.8 =	28.8
4 H	,,	4 × 17.1 =	68.4
2 O	,,	4 × 20.0 =	40.0
4 =	bonds	4 × 23.2 =	92.8
1 ⬡	ring	1 × 6.1 =	6.1
		Total =	236.1

For structure II parachor is

6 C	atoms	6 × 4.8 =	28.8
4 H	,,	4 × 17.1 =	68.4
2 O	,,	4 × 20.0 =	40.0
2 ⬡	rings	2 × 6.1 =	12.2
3 =	bonds	3 × 23.2 =	69.6
			219.0

Experimental parachor value is 236.8; hence structure I is accepted. Once upon a time when modern spectroscopic methods for structure determination were not available, this technique was widely used to know the structure.

(d) Viscosity : When a liquid flows through a tube there must exist a difference in pressure between any two parts of the tube. If this pressure difference remains constant with time then the liquid flows smoothly and the motion is called *stream-lined motion*. On the contrary if pressure difference changes with time the motion is turbulent.

Let us consider the stream lined motion of a liquid flowing through a tube. The liquid film, which is in contact with the solid tube does not move at all. If the entire liquid is considered to be a packing of successive liquid films, then there is a difference in velocity between two successive liquid films, the central film moves with highest velocity.

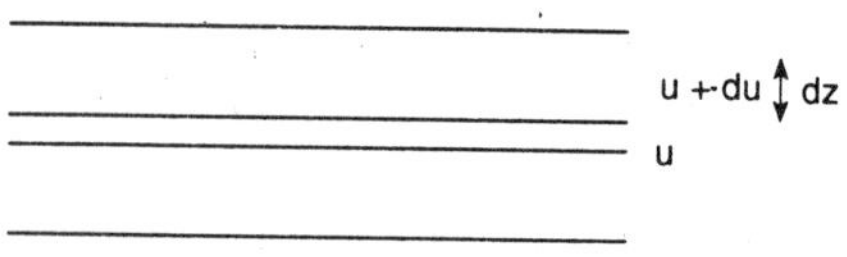

Fig. 4.4

Let u and $u + du$ be velocities of two liquid layers separated by a distance 'dz' then $\frac{du}{dz}$ is velocity gradient. When these two layers are in contact with each other, a frictional force acts between them. This force (F) is called viscous force. According to law of viscosity, this force is directly proportional to area of contact (A) between two layers and also the velocity gradient $\frac{du}{dz}$ i.e.,

$$F \propto A \,.\, \frac{du}{dz}$$

$$\text{or} \quad F = \eta\, A \,.\, \frac{du}{dz}$$

Here η is a constant, known as co-efficient of viscosity. It is defined as the frictional force between two layers whose area of contact (A) is unity and also velocity gradient is unity. Its unit is poise (dyne sec cm^{-2}) in honour of scientist Poiseuille, who did a lot of work on this subject. Reciprocal of viscosity $\frac{1}{\eta}$ is called fluidity and is symbolised φ i.e.,

$$\varphi = \frac{1}{\eta}.$$

Poiseuille's equation

When a liquid flown through a tube in stream-lined motion then according to Poiseuille

$$\eta = \frac{\pi P r^4 t}{8 V l},$$

where V volume of liquid flown through a tube of length l and radius r in time t under a driving pressure of P i.e. difference in pressure at the two ends.

The equation is very important and is used for many purposes. Viscosity of a liquid is determined by Ostwald's viscometer using this formula; details of determination is beyond the scope of this book.

Viscosity and temperature

Like surface tension viscosity of a liquid decreases with temperature. The equation is

$$\eta = A \,.\, e^{-E/RT}.$$

Here A and E are constants. It is similar to Arrhenius equation (see chemical kinetics chapter). Drawing anology 'E' is called activation energy for viscosity.

POINTS TO REMEMBER

Liquid state is something in between gas and solid. It may be regarded as a highly ordered gas or highly disorderd solids. Liquid molecules have limited translational motion. There are many holes in its structure through which molecules can move obeying Maxwell's distribution law.

Nature of forces in a liquid are mainly Van-der-Waals. Besides this certain other forces viz. H–bonding coulombic attraction and ionic forces are found in some cases.

Important properties of liquid are vapour pressure; surface tension, parachor and viscosity. Once parachor values were used to decide molecular geometry.

Eotovos equation connects surface tension (γ) with temperature as

$$\gamma\left(\frac{M}{\rho}\right)^{2/3} = K\left(T_c - T\right).$$

Equation connecting viscosity with temperature is $\eta = A \cdot e^{-E/RT}$

QUESTIONS

SHORT TYPE

1. "Boiling point of a liquid gives an idea of the magnitude of Van-der-Waals force"— explain.
2. What are 'holes' in the structure of liquids ?
3. Why boiling points of H_2O and HF are abnormally high ?
4. Is there any relation between boiling point of a liquid and critical temperature of its vapour ?
5. What are the units of surface tension and surface energy ?
6. Why Hg falls in a capillary but water rises ?
7. A falling liquid drop is always spherical in shape — explain ?
8. Why a needle floats over water ?
9. What is Ramsay-Shield's equation ?
10. 'Parachor is not wholly additive property' — explain.
11. What are traditional and S. I. units of viscosity ?
12. What is Poiseuille's equation ? For what type of motion in liquid is applicable ?

ESSAY TYPE

1. Give a description of liquid state. Is it an intermediate between solid and gaseous states ? — explain.
2. What is meant by vapour pressure of a liquid ? How you can account for it ?
3. Describe a method for measuring vapour pressure of a liquid. How is it related with boiling point.
4. Distinguish between surface tension and surface energy. Describe a method for determining surface tension of a liquid.
5. What is a parachor ? What type of property is it ? Describe its application in elucidating structure of molecules.
6. What is co-efficient of viscosity ? How does it change with temperature ?

PROBLEMS

1. A liquid A has half the surface tension and twice the density of liquid B at a certain temperature. If in a capillary, the rise is 10 cm. for A, what will be the rise of B at the same temperature.

   [*Hints.* $\frac{\gamma_A}{\gamma_B} = \frac{h_A \rho_A}{h_B \rho_B}$ as $\gamma = \frac{1}{2} h\rho gr$. Here $\frac{\gamma_A}{\gamma_B} = \frac{1}{2}$ $\frac{\rho_A}{\rho_B} = 2$ and $h_A = 10$.]

2. A spherical drop of a liquid weighing 0.04 gm. is dispersed into 1000 homogeneous microglobules of radius 0.02 cm each by a suitable experimental device. Find the resultant increase in surface energy. [$\rho_t = 0.8$ gms/c.c.; $\gamma = 27$ dynes/cm]

   [*Hints.* Vol. of drop = $\frac{0.04}{0.8} = 0.05$ c.c. $= \frac{4}{3}\pi r_1^3$. Find surface are $= 4\pi r_1^2$. Find surface area of 1000 droplets. Difference is increase in surface area. Multiply this with surface tension.]

3. At 20°C pure water with an absolute viscocity of 0.01009 dyne sec cm^{-2} requires 102.2 sec to flow through the capillary of an Ostwald viscometer. At 20°C toluene requires 68.9 sec. If the densities of water and toluene be 0.998 and 0.866 gm/cm^3 respectively, calculate the viscosity of toluenc.

[***Hints.*** $\eta_1 = \eta_2 \cdot \dfrac{\rho_1 t_2}{\rho_2 t_1}$.]

4. A liquid is flowing through a capillary tube of length l and radius r under a pressure P at the rate of 1 c.c./min. Calculate the rate of flow of the liquid under a pressure $P/2$ through a capillary of length $4l$ and radius $2r$ at the same temperature.

[***Hints.*** Poiseuille's equation is $\eta = \dfrac{\pi P r^4 t}{8 V l}$; for same liquid flowing.

Under different conditions $\dfrac{\pi P_1 r_1^4 t_1}{8 V_1 l_1} = \dfrac{\pi P_2 r_2^4 t_2}{8 V_2 l_2}$.]

OBJECTIVE TYPE

1. Fill in the blanks.

(a) Liquid molecules are held together mainly due to ——— force.
(b) Association of water molecules is due to ———
(c) Liquid molecules have ——— translational motion.
(d) S. I. unit for surface tension is ———
(e) Liquid drops tend to be spherical because sphere has ———
(f) Viscosity of liquid ——— with rise in temperature.

2. Match terms of column A with those of column B.

A	B
(a) Eyring	(a) Molecular Geometry.
(b) Eotovös	(b) Liquid.
(c) Poiseulle	(c) Holes in a liquid
(d) Parachor	(d) Surface tension.
(e) Disordered Solid	(e) Viscosity.

Answers :

1. (a) Van-der-Waals; (b) H–bonding; (c) limited; (d) Newton $metre^{-1}$; (e) smallest surface area; (f) decreases.

2. a – c; b – d; c – e; d – a; e – b;

Chapter 5

THE SOLID STATE

5.1 WHAT IS A SOLID ?

Of the three common states of matter the solid-state has some characteristic features in comparison to that of liquids and gases. Most important characteristics are —

(a) Solids have definite shape, because of the absence of any translational motion. This makes solids rigid; whereas presence of some sort of translational motion makes gases and liquids fluid.

(b) Solids are practically imcompressible having a very slow diffusion.

5.2 CLASSIFICATION OF SOLIDS

Solid substances can exist in two forms which are (a) crystalline and (b) amorphous. If a strong solution of alum is cooled solid blocks of alum separate out, each of which is found to consist of a number of small units having definite geometric shape and being bound by plane faces. These are crystals. Crystals of various forms are found e.g., cubic, tetrahedral; octahedral; prismatic etc. and can be obtained by slow cooling of molten substance or saturated solution.

On the other hand a precipitate of $Al(OH)_3$ separates out as a jelly like substance having no definite shape. Cooling of molten glass also does not give any solid at a particular temperature but gradually gets solidified to a hard transperent solid. These are called *amorphous.* They are prepared by super cooling a viscous liquid, which prevents crystallisation.

5.3 CLASSIFICATION IS NOT RIGID

The distinction between crystalline and amorphous substance is not sharp. Thus cement which is regarded as amorphous powder appear as minute crystals under microscope. Also many amorphous solids show very slow processes of crystallisation (known as devitrification). This is found in glass panels of certuries old found in windows of many churches.

5.4 ANISORTROPICITY — CHARACTERISTIC OF CRYSTAL

One of the most important property of crystalline substance is its anisotropicity. Liquids, gases and amorphous solids are isotropic i.e. their physical properties, such as refractive index; mechanical strength; thermal conductivity etc. are same in all directions. Crystalline solids, however, are anisotropic. Thus co-efficient of thermal expansion of AgI crystal is positive in one direction but negative in the other. R. I. of Na-D light is different in different directions for $K_2Cr_2O_7$ crystal. Mechanical strength of a mica crystal is abnormally low in horizontal direction but very high in vertical direction. These are found because of arrangement of atoms in a crystal in a symmetrical manner. This is illustrated by the following figure —

Next page figure shows arrangements of two different types of atoms (circle and dot) in a crystal. AB and CD are two light rays passing through the crystals. The two sides of ray $\overrightarrow{AB}$ consists of both circles and dots; but for ray $\overrightarrow{CD}$ one side has only dot and other side circle. So influence of the surrounding on the two rays will be different and they will show different refractive index values.

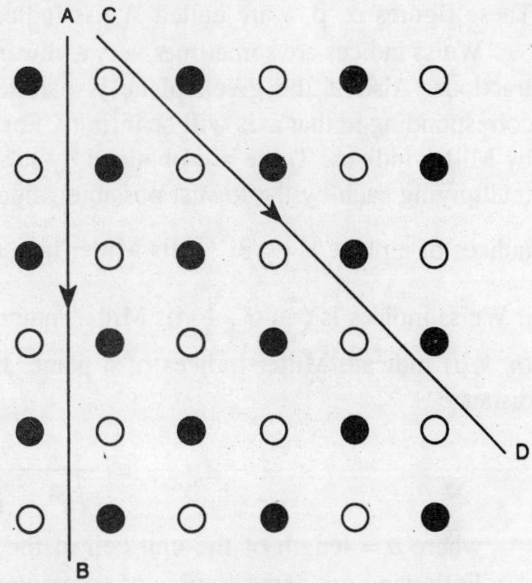

Fig. 5.1

5.5 CRYSTALLOGRAPHY

Crystallography is that branch of science, which deals with both external and internal geometry of the crystal. External geometry is outward spatial arrangement of crystal planes and geometric shape of the crystal. Internal geometry means arrangement of different atoms inside the crystals lattice.

5.6 LAWS OF CRYSTALLOGRAPHY

External geometry of a crystal is studied with the help of three laws, known as laws of crystallography. These are as follows —

(a) Steno's law of constancy of interfacial angle : A crystal may crystallise in different forms depending on external condition. Thus NaCl crystallises as a cube from aqueous solution; but it crystallises as an octahedral from its aqueous solution containing urea also. In each case angle between identical faces is same. Thus it may be stated that the angle between the corresponding faces or planes forming the external surface of a crystal remain constant for a given substance and does not depend on how the face is developed.

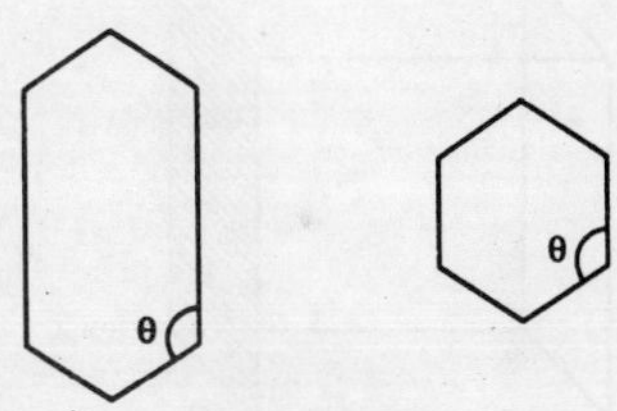

Fig. 5.2

(b) Law of rationality of indices or intercepts (Hauy's law) : A crystal has different planes, some of which are real and some imaginary. In order to name different crystalline planes this law has been proposed. According to this law we choose a suitable point inside the crystal as origin and consider three crystallographic axes. The inter-axial angles may or may not be 90°. However, we first try with 90° angle. If we find that there is at least one plane in the crystal, which makes intercept on all the three axes then 90° angle will serve our purpose. However, if no such plane is found making intercept on all the three axes, then some axis is to be bent to meet the demand of intercept.

In the following figure, let LMN be such a plane in the unit cell of the crystal making intercepts of lengths '*a*', '*b*' and '*c*' on '*X*', '*Y*' and '*Z*' axes. Then *a*, *b*, *c* are unit distances. If the plane which is to be named makes an intercept at

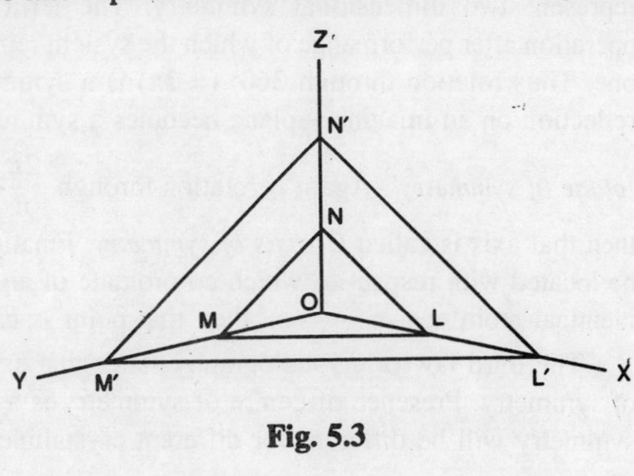

Fig. 5.3

distances α *a*; β *b* and γ *c* on *X*, *Y*, *Z* axes then the plane will be called (α, β, γ) plane. These figures α, β, γ are called Weiss indices.

Weiss indices are sometimes very awkward. Because α, β and γ may sometimes become fractions. Also if the given plane is parallel to either of the axis, then Weiss indices corresponding to that axis will be infinity. For these reasons Weiss indices are now replaced by Miller indices. These are obtained by taking the reciprocal of Weiss indices and then multiplying each by the lowest possible value to make them whole number. Thus if Weiss indices of a plane is (2, 3, 1). Its Miller indices are ($\frac{1}{2}$× 6; $\frac{1}{3}$× 6; 1×6) i.e. (3, 2, 6). Again if Weiss indices is ($\frac{1}{2}$; α ; $\frac{1}{3}$) its Miller indices will be (2, 0, 3) etc. In general the symbols (*h*, *k*, *l*) indicate Miller indices of a plane. For different cubical lattices the interplanner distance

$$d_{h,k,l} = \frac{a}{\sqrt{h^2 + k^2 + l^2}},$$

where *a* = length of the unit cell of the cubical lattice.

Following are some figures of some planes of cubical lattice.

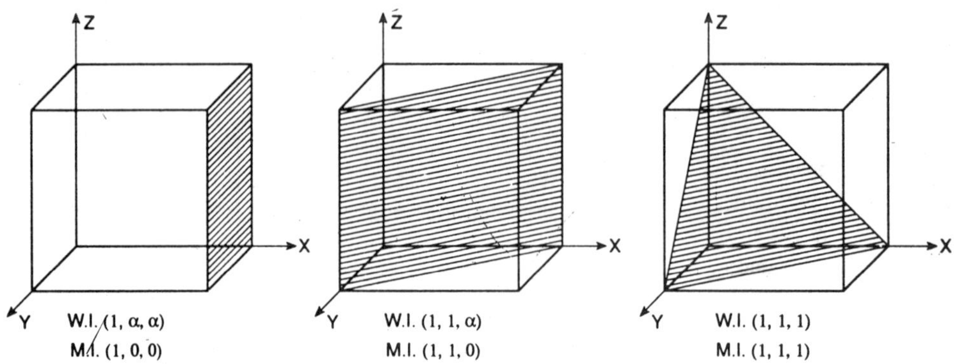

Fig. 5.4

(c) Law of symmetry : The third law of crystallography is the law of symmetry. If a particular pattern is repeated after regular intervals, then we say that there is symmetry. A set of equidistant points is a one-dimensional symmetry. Small squares of a graph paper represent two dimensional symmetry. The term symmetry operation means such an operation after performance of which the system remains indistinguishable from the original one. Thus rotation through 360° (= 2π) is a symmetry operation, for every substance. If reflection on an imaginary plane becomes a symmetry operation, then that plane is called '*plane of symmetry*'. Again if rotation through $\frac{2\pi}{n}$ along an axis is a symmetry operation, then that axis is called C_n *axis of symmetry*. Finally if within the system such a point can be located with respect to which co-ordinate of an atom is (*x*, *y*, *z*) and if then we get an identical atom at (–*x*, –*y*, –*z*), then that point is called *centre of symmetry*.

The third law of crystallography states that every crystal must have certain elements of symmetry. Presence of centre of symmetry is a must and number of other elements of symmetry will be different for different crystalline lattices.

Let us calculate number of symmetry elements present in a cubical lattice.

Plane of Symmetry

Type - I

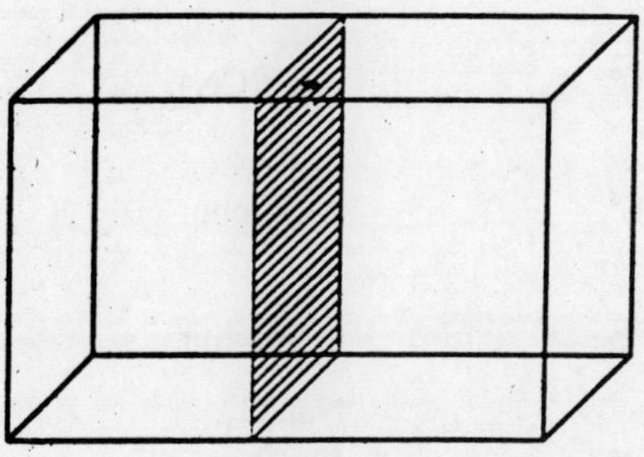

'3' such planes bisecting roof & floor; one wall & the other facing it; & the remainder pair of wall

Type II

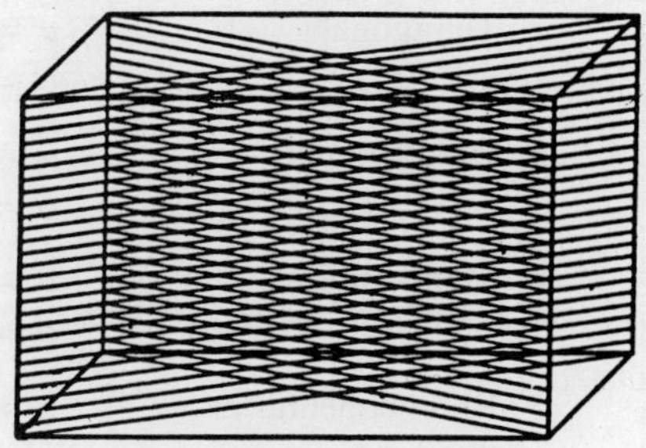

'6' such planes two for each of the 3 mutually opposite pair of walls

Total = 9

Axis of Symmetry

C_2-axis : (Rotation through 180°) :

This is the axis passing through the middle point of two mutually opposite sides. Since there are 12 sides in a cube, six (6) such axis is possible.

C_3-axis : (Rotation through 120°) :

This is the axis passing through mutually opposite corners of the cube. As a cube has eight corners, four such axis is possible.

C_4-axis : (Rotation through 90°) :

This is the axis passing through the certre of two opposite planes. Since a cube has six planes, three such axis is possible.

So total axis of symmetry = 6 + 4 + 3 = 13.

Centre of symmetry

A cubical crystal has also a centre of symmetry, which is the centre of the crystal.

So total number of symmetry elements present in a cubical crystal = 9 + 13 + 1 = 23.

One interesting point to note is that a crystal can have C_1 ; C_2 ; C_3 ; C_4 and C_6 axis of symmetry, but it can not have C_5 or C_n ($n > 6$) axis of symmetry. The reason is that a crystal is a close packed structure of unit cell. The unit cells should be of such shape that during close packing there should not be any void space among them. Such a close packing can be done by objects with C_1 ; C_2 ; C_3 ; C_4 and C_6 axis of symmetry. Just like a floor which can not be close packed [i.e. completely covered] by pentagon tiles (C_5 axis) a space can not be close packed with objects having C_5 or C_n [$n > 6$] axis of symmetry.

5.7 CRYSTAL SYSTEMS

Two hundred thirty different crystalline systems are known. They can be classified in to thirty-two groups. [Bravais did this]. For the sake of simplicity these are again classified in to seven crystalline systems depending on axial intercepts and axial angles between them. If we designate the intercepts on X, Y and Z axes as 'a', 'b', and 'c' and the axial angles as α , β and γ then the seven crystalline systems are —

Name	Characteristic	Example
1. Cubic	$a = b = c$ $\alpha = \beta = \gamma = 90°$	NaCl; Diamond
2. Tetragonal	$a = b \neq c$ $\alpha = \beta = \gamma = 90°$	Sn; $K_4Fe(CN)_6$
3. Orthorhombic	$a \neq b \neq c$ $\alpha = \beta = \gamma = 90°$	KNO_3; α-Sulphur
4. Monoclinic	$a \neq b \neq c$ $\alpha = \gamma = 90° \neq \beta$	β-Sulphur
5. Rhombohedral or Trigonal	$a = b = c$ $\alpha = \beta = \gamma \neq 90°$	$NaNO_3$; Ice
6. Hexagonal	$a = b \neq c$ $\alpha = \beta = 90°$ $\gamma = 120°$	Graphite
7. Triclinic	$a \neq b \neq c$ $\alpha \neq \beta \neq \gamma \neq 90°$	$CuSO_4$; $5H_2O$

Of all the different types of crystal systems, the simplest to visualise, is the cubical system. Such a system is of three types —

(a) Simple cubic, where each corner of the cube is occupied by an atom.

(b) Body-centred-cubic (b – c – c) lattice, where in addition to eight corners, the certre of the unit is occupied by an atom.

(c) Face-centred cubic (f – c – c) lattice, where in addition to eight corners, centres of each face consists of one atom.

5.8 LATTICE AND UNIT CELL

If a particular pattern is repeated after regular intervals we say that there is some symmetry in it. Thus a set of equidistant points in a line is a symmetry in one dimension. The points of a graph paper is also a symmetry in two dimension. Here each point has same environment. Now if parallel lines are drawn through similar points then growth of these lines in Z-direction will give a three-dimensional crystalline structure. The very fact that such structures have regularities in shape indicate regular arrangement of points in space. This regular arrangement is known as space-lattice; each point of lattice has the same neighbours.

Now if parallel lines are drawn through identical points of the space lattice, the entire lattice can be broken down to a number of rhombs, the smallest of which still resembles the shape of the crystal. This smallest rhomb is known as unit cell. The space lattice is made by the translation of this primitive unit cell in all directions. If the crystal be of a monoatomic element e.g. metal, then there will be only one space lattice. However, if it is a compound then the space lattice is made up by interpenetrating lattices of the constituent atoms. Thus space lattice of NaCl is produced by two inter-penetrating lattices of Na^+ and Cl^- each of which is a f – c – c lattice.

5.9 MORE INFORMATIONS ABOUT DIFFERENT CUBICAL LATTICES

From elementary geometrical consideration it can be shown that distance between two crystalline planes with Miller-indices (h, k, l) in cubical system is —

$$d_{h,k,l} = \frac{a}{\sqrt{h^2 + k^2 + l^2}}$$

where a = length of a side of unit cell i.e. smallest crystalline unit.

Thus for a *simple cubic*

$$d_{100} = a; \qquad d_{110} = \frac{a}{\sqrt{2}}; \qquad d_{110} = \frac{a}{\sqrt{3}}.$$

In a body centred lattice the plane passing through centre may be parallel to 100 plane but have Weiss indices $(\frac{1}{2}, 0, 0)$ i.e. Miller indices (2, 0, 0). So d_{100} is actually d_{200} for b – c – c, (1, 1, 0) plane of a simple cube pass through centre, so d_{110} for simple cubic is also d_{110} for b – c – c. But (1, 1, 1) plane of a simple cube does not pass through centre. The central point lies exactly at middle of two such (1, 1, 1) planes i.e. its Weiss indices is $\frac{1}{2}, \frac{1}{2}, \frac{1}{2}$ i.e. Miller indices is (2, 2, 2). So for b – c – c lattice

$$d_{100} \quad (\text{i.e. } d_{200}) = \frac{a}{\sqrt{2^2 + 0 + 0}} = \frac{a}{2}$$

$$d_{110} \quad (\text{i.e. } d_{110}) = \frac{a}{\sqrt{1^2 + 1^2 + 0}} = \frac{a}{\sqrt{2}}$$

$$d_{111} \quad (\text{i.e. } d_{2,2,2}) = \frac{a}{\sqrt{2^2 + 2^2 + 2^2}} = \frac{a}{2\sqrt{3}}.$$

Similar consideration shows that for a f – c – c d_{100} is actually d_{200}. d_{110} is actually d_{220} and d_{111} is same. So in a f – c – c lattice

$$d_{100} \quad (\text{i.e. } d_{200}) = \frac{a}{\sqrt{2^2 + 0 + 0}} = \frac{a}{2}$$

$$d_{110} \quad (\text{i.e. } d_{220}) = \frac{a}{\sqrt{2^2 + 2^2 + 0}} = \frac{a}{2\sqrt{2}}$$

$$d_{111} \quad (\text{i.e. } d_{111}) = \frac{a}{\sqrt{1^2 + 1^2 + 1^2}} = \frac{a}{\sqrt{3}}.$$

Though a crystal is supposed to be most closely packed structure, yet there are some gaps (known as notches) in crystal lattice. The percentages of occupied and also void spaces in a cubical lattice can be calculated as follows —

Simple-cubic system :

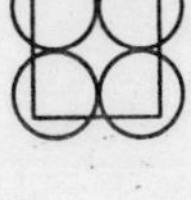

In a simple cubic system, if 'a' is the side length of a unit cell, then $a = 2r$; i.e. $r = \frac{a}{2}$. Now there are eight corners and during primitive translation of this unit cell each corner atom will be shared by eight cubes. So in a given unit cell $\frac{1}{8}$th part of an atom takes shelter. Since there are eight such atoms.

So occupied volume = volume of $(\frac{1}{8} \times 8) = 1$ atom $= \frac{4}{3} \pi r^3$

But volume of unit cell $= a^3 = (2r)^3 = 8r^3$.

$$\text{So \% occupied} = \frac{\frac{4}{3}\pi r^3}{8 r^3} \times 100 = 52\%$$

Body-Centred cubical system :

In the unit cell of a b – c – c lattice, the first plane consists of four atoms; next plane has only one atom placed above this and then again a plane of four atoms sandwiches the central atom. Thus the diagonal of the cube consists of the centres of three atoms. If side length of the cube = a ; then its diagonal $= \sqrt{3}\,a$. And so $\sqrt{3}\,a = 4\,r$; or $r = \frac{\sqrt{3}}{4}\,.\,a$.

To calculate occupied space, we see that central atom belongs to only one unit cell, but each corner atom is shared by eight unit cells.

So occupied space = volume of $(1 + 8 \times \frac{1}{8}) = 2$ atoms $= 2 \times \frac{4}{3} \pi r^3$.

$$\text{So \% of occupied space} = \frac{2 \times \frac{4}{3}\pi r^3}{\left(\frac{4r}{\sqrt{3}}\right)^3} \times 100 = 68\%.$$

Face-centred cubical system :

The first plane of the unit cell of a f – c – c lattice consists of five atoms arranged as —

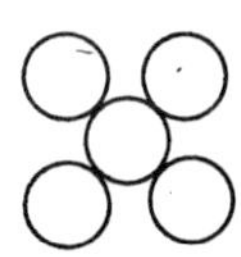

The plane above this consists of four atoms placed over four notches. Then again come five atoms in the next plane. If side length of the unit cell is 'a', then diagonal of the square shaped first plane $= \sqrt{2}\;a = 4\,r$. So $r = \frac{a}{2\sqrt{2}}$. Now in a f – c – c lattice while each corner atom is shared by eight cubes, each facial atom is shared by two cubes. So occupied volume of a unit cell = vol. of $(\frac{1}{8} \times 8 + \frac{1}{2} \times 6) = 4$ atoms $= 4 \times \frac{4}{3} \pi r^3$.

Vol. of unit cell $= a^3 = (2\sqrt{2}\;r)^3$.

$$\text{So \% of occupied space} = \frac{\frac{16}{3}\pi r^3}{\left(2\sqrt{2}\,r\right)^3} \times 100 = 74\% .$$

Hence we see that f – c – c is most closely packed lattice.

The following tables should be remembered.

Table 5.1

Lattice type	Radius	Occupied Volume	d_{100}	d_{110}	d_{111}
Simple cubic	$\frac{a}{2}$	52%	a	$\frac{a}{\sqrt{2}}$	$\frac{a}{\sqrt{3}}$
b – c – c	$\frac{\sqrt{3}a}{4}$	68%	$\frac{a}{2}$	$\frac{a}{\sqrt{2}}$	$\frac{a}{2\sqrt{3}}$
f – c – c	$\frac{\sqrt{2}a}{4}$	74%	$\frac{a}{2}$	$\frac{a}{2\sqrt{2}}$	$\frac{a}{\sqrt{3}}$

Table 5.2

Type	d_{100} : d_{110} : d_{111}
Simple Cubic	1 : 0.71 : 0.58
b – c – c	1 : 1.41 : 0.58
f – c – c	1 : 0.71 : 1.16

5.10 BRAGG'S METHOD OF CRYSTAL ANALYSIS

It can be shown that distance between two consecutive crystal planes is of the order of 10^{-8} cm. Since wavelength of X-rays are of this order, crystals can be used as a natural gratings for X-rays. It was Bragg, who used X-rays to determine the internal arrangement of crystals. His approach was as follows —

Let the lines AB and CD be two parallel crystal planes (passing through identical points) separa-ted by a distance 'd'. Let PQ and QR be the incident and reflected X-ray beam being reflected at Q; θ being the angle of incidence (glancing angle). Let P′Q′ be another X-ray beam being reflected at some identical point Q′ in the next plane. Its reflected part Q′R′ will be in phase with the previous reflected beam QR only when path difference between P′Q′R′ and PQR is an integral multiple of wavelength (λ) of the X-ray used. Now path difference = P′Q′R′ – PQR

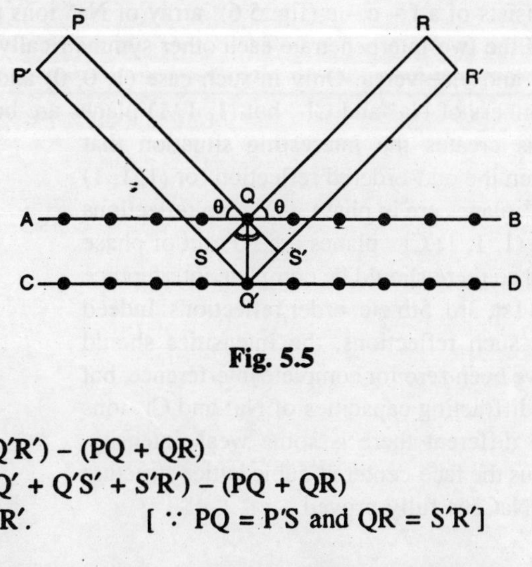

Fig. 5.5

$$= (P'Q' + Q'R') - (PQ + QR)$$
$$= (P'S + SQ' + Q'S' + S'R') - (PQ + QR)$$
$$= SQ' + Q'R'. \qquad [\because PQ = P'S \text{ and } QR = S'R']$$

But $\frac{SQ'}{QQ'} = \sin\theta$ and $\frac{Q'S'}{QQ'} = \sin\theta$ and QQ′ = distance between two planes 'd'.

So SQ′ = d sinθ and Q′S′ = d sinθ.

Hence path difference $= d \sin\theta + d \sin\theta = 2d \sin\theta$.

For two reflected X-ray beams to be in phase, this path difference must be an integral multiple of wave-length λ.

$$\text{i.e. } n\lambda = 2d \sin\theta.$$

This is famous Bragg's equation.

When $n = 1$, it is called first-order reflection; if $n = 2$ it is called second order reflection etc.

Since maximum value of $\sin\theta = 1$, so for 1st order reflection ($n = 1$), the maximum permissible wave length that can be used is $\lambda = 2d$.

5.11 CRYSTAL STRUCTURE OF NaCl

Determination of the structure of NaCl crystal by W. H. Bragg is historically important as it was the first success of X-ray method. The apparatus used by Bragg consists of a single crystal of NaCl mounted at the centre of a revolving table. A monochromatic X-ray beam is allowed to be incident on the crystal and the reflected beam is received in an ionisation chamber connected to an electrometer. The deflection of the electrometer measures the intensity of the reflected beam.

Bragg rotated the crystal and measured the deflection of electrometer for various angles of incidence of the X-rays. It was found that for certain angles of incidence there is strong scattering; evidently these are positions where reflected X-rays are in phase i.e. Bragg's equation holds.

Bragg found that for NaCl the ratio of the spacings of three crystal planes is $d_{100} : d_{110} : d_{111}$ is 1 : 0.707 : 1.154. This matches with the calculated ratio for a f – c – c lattice. Thus NaCl must have a f – c – c lattice. This did not solve the problem of internal arrangement, as the question of relative arrangements of Na and Cl atoms still remains. The problem was solved by noting the intensities of different orders from (1, 1, 1) plane of NaCl. It was found that 1st, 3rd, 5th etc. order reflections are weak, whereas 2nd, 4th, 6th order reflections are strong. This can only occur if the structure of NaCl consists of a f – c – c (fig. 5.6) array of Na^+ ions and another f – c – c array of Cl^- ions and the two interpenetrate each other symmetrically so that each Na^+ is surrounded by six Cl^- and vice-versa. Only in such case (1, 0, 0) and (1, 1, 0) planes will consist of equal numbers of Na^+ and Cl^-, but (1, 1, 1) planes are built by either Na^+ or Cl^- alternatively. This creates the interesting situation that when the odd-ordered reflection for (1, 1, 1) Na^+ planes are in phase, the same reflections for (1, 1, 1) Cl^- planes are 90° out of phase and so there should be complete interference for 1st, 3rd, 5th etc. order reflections. Indeed for such reflections, the intensities should have been zero for complete interference, but as diffracting capacities of Na^+ and Cl^- ions are different there is some weak intensity. Thus the face-centered-cubic lattice structure of NaCl is fully proved.

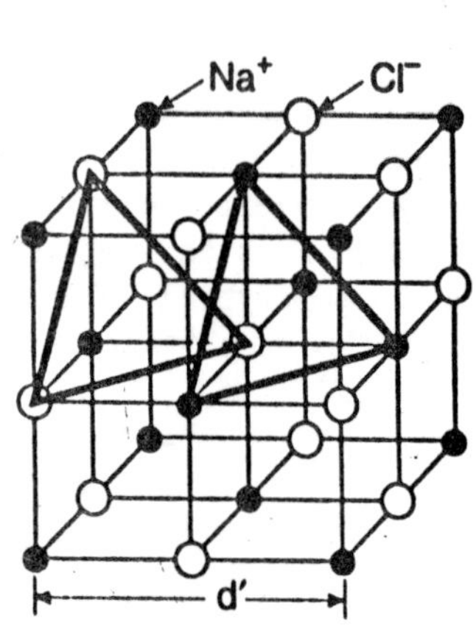

Fig. 5.6 : (f-c-c lattice of NaCl) (1,1,1) plans are alternatively occupied by Na^+ and Cl^- ions.

5.12 STRUCTURE OF KCl

Bragg's method of X-ray analysis indicates that the ratio of $d_{100} : d_{110} : d_{111}$ for KCl crystal is 1 : 0.71 : 0.58. This suggests that KCl should have simple-cubic lattice. But because of great similarity between KCl and NaCl one should expect KCl also to have f – c – c lattice. The answer is given by the fact that K^+ and Cl^- are isoelectronic and so they have same scattering power with respect to X-ray. Like NaCl, KCl has also two interpenetrating f – c – c lattices, one due to K^+ and the other due to Cl^- ; but as K^+ and Cl^- are identical in the eye of X-ray, the overall lattice appear as simple cubic. [Let us look at fig. 5.6, and assume Na^+ and Cl^- as same point; the overall picture then becomes simple cubic]. This is corroborated by the fact that in KCl, 1st, 3rd, 5th etc. order reflection from (1, 1, 1) plane are absent. Because odd-order reflections from (1, 1, 1) K^+ plane and (1, 1, 1) Cl^- plane are out of phase and as K^+ and Cl^- have same scattering capacity they completely cancel each other.

5.13 NATURE OF FORCES IN A CRYSTAL

On the basis of nature of forces acting between particles in a crystal, crystals can be classified into four types. These are (a) Ionic crystal (b) Molecular crystal (c) Co-valent crystal (d) Metallic crystal.

In ionic crystal the lattice consists of cation and anion and they are hold by strong electrostatic forces. Examples being NaCl, KCl etc.

Lattice of molecular crystals consists of repeating units of molecules held together by weak Van-der-Waals force. Ice; Iodine etc. are examples.

In co-valent crystal, the lattice points are occupied by atoms linked by co-valent bond. Classic example is diamond.

Finally the metallic crystal lattices consists of metal atoms held by metallic bond.

5.14 BORN-HABER CYCLE

Because of a close packed structure in a crystal there is a considerable lowering of potential energy of the constituent particles. The total lowering of P. E. of the constituent atoms due to the formation of a lattice is known as lattice energy. For an ionic lattice, this energy is related with ionisation potential; electron affinity etc. Born and Haber devised an imaginary cycle from which the lattice-energy can be calculated. Let us consider an ionic crystal MX. For this following imaginary cycle can be considered.

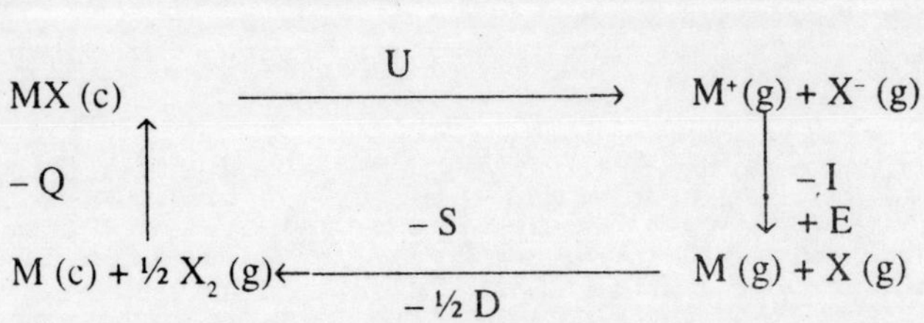

Fig. 5.7

Here U is the lattice energy of the crystal. Actually when M^+ (g) and X^- (g) combine to form a lattice of MX, P. E. is lowered by U. Thus 'U' energy is needed to split MX. I and E are ionisation potential and electron affinity for M and X respectively. Thus M requires I energy for conversion to M^+ ion and X lowers its energy by E when it becomes X^- . S is the sublimation energy for M and D is the dissociation energy of X_2. Q is heat of formation of crystal MX from M(c) and $\frac{1}{2}X_2$ (g).

The thermodynamic condition that total energy change in a cyclic process is zero gives —

$$U - I + E - S - \frac{1}{2}D - Q = 0$$

$$\text{or } U = I - E + S + \frac{1}{2}D + Q.$$

At one time all but 'E' terms of R. H. S was measurable. Then this cycle was used to calculate electron affinity as there is alternative way to know lattice energy U. However later works have devised methods to determine electron-affinity. So Born-Haber cycle can now be used to find out lattice energy. The validity of the cycle can be illustrated with the example of NaF crystal, where lattice energy calculated from Born-Haber cycle is 216.8 K.Cal/mole, whereas the observed value was found to be 216.7 K.Cal/mole.

POINTS TO REMEMBER

Crystallography is that branch of science, which discusses external and internal geometry of crystals.

Crystals are *anisotropic* solids having certain elements of symmetry. There are three laws to study external geometry. These are regarding (i) *constancy of interfacial angle* (ii) *naming different crystalline planes and* (iii) *finding symmetry elements.*

Crystals are classified in to *seven* groups. Cubical system is very important. It is of three types viz. *simple*; *b – c – c* and *f – c – c*. *f – c – c* is most closely packed.

Structure of crystal is determined by X-ray using Bragg's equation which is $n\lambda = 2d \sin\theta$. NaCl and KCl have *f – c – c* laftice.

Born-Hober cycle is used to calculate *lattice-energy.*

QUESTIONS

SHORT TYPE

1. How many (and what types of) symmetry elements are present in a cubical lattice?
2. "A crystal is an anisotropic substance" — explain.
3. A crystal can not have 5-fold axis of symmetry.— why ?
4. What are lattice and unit cell ?
5. X -ray diffraction result shows NaCl in a f – c – c lattice but KCl in a simple cubic lattice though they are isomorphous — why ?
6. Give examples of (a) molecular crystal and (b) Co-valent crystals.

ESSAY TYPE

1. What are laws of crystallography ? Explain Weiss and Miller indices.
2. Draw sketches to show (1, 0, 0); (1, 1, 0) and (1, 1, 1) planes of a simple cubic lattice.
3. Describe the characteristic features of different crystalline systems.
4. Derive Bragg's equation for X-ray diffraction. What is the maximum wavelength permissible for 1st order diffraction in (1, 0, 0) plane of a simple cubic lattice of side length '*a*'.
5. Prove that out of the three different cubical systems, face-centred-cubical lattice is most closely packed.
6. Describe how the structure of Rock Salt (i.e. NaCl) has been established.
7. Classify crystals according to the nature of forces acting in it.
8. Elucidate Born-Haber cycle and state its application.

PROBLEMS

1. A crystal plane intercepts the three crystallographic axes at the multiples of the unit distance $\frac{3}{2}$; 2 and 1. What will be the Miller indices of the plane ?

   [*Hints.* Given W. I = $\frac{3}{2}$, 2, 1

   So M. I. = $\frac{2}{3} \times 6$; $\frac{1}{2} \times 6$; 1×6.]

2. Miller indices of a plane in a crystal are (2, 1, 0). What are the Weiss indices ?

   [*Hints.* $\frac{1}{2}$; 1; ∞.]

3. The first order reflection of a beam of X-ray from a given crystal occurs at 5°15′. At what angle will be the third order reflection ?

   [*Hints.* $\lambda = 2d \sin\theta_1$; $3\lambda = 2d \sin\theta_2$; Take ratio and find θ_2 with θ_1 = 5°15′.]

4. If X-ray of λ = 1.54 Å is used, find the angle at which will be obtained second order reflection from the 111 planes of NaCl. [unit cell length = 5.64 Å]

   [*Hints.* d_{111} for NaCl (f – c – c) = $\frac{5.64}{\sqrt{3}} \times 10^{-8}$ c.m.

   $2d \sin\theta = 2\lambda$; find θ.]

5. The first order reflections from the (1, 0, 0); (1, 1, 0); and (1, 1, 1) planes of a given cubic crystal occurs at angles 7°10′ ; 10°12′ and 12°30′ respectively. To what type of cubic lattice the crystal belong ?

[*Hints.* For (1, 0, 0) $\lambda = d_{100} \sin\theta_1$ For (1, 1, 0) $\lambda = d_{110} \sin\theta_2$
For (1, 1, 1) $\lambda = d_{111} \sin\theta_3$.

Find $d_{100} : d_{110} : d_{111}$ and see that this corresponds to value for simple cubic.]

OBJECTIVE TYPE

1. **State whether the following statements are true or false.**
   (a) Glass is not a solid substance.
   (b) Every crystal maynot have a centre of symmetry
   (c) A crystal cannot have a C_5 axis of symmetry.
   (d) NaCl has a b – c – c crystal lattice.
   (e) Simple cubic is most loosely packed out of all cubical systems.
   (f) X-ray cannot be used to determine geometry of lattice.

2. **Out of the given answers choose the correct one.**
   (a) Classification of crystalline substances was made by ——
   [Ans. (i) Bravais (ii) Bragg (iii) Born]
   (b) The term '*n*' in the equation $n\lambda = 2d \sin\theta$ is ——
   [Ans. (i) Mole (ii) R.I. (iii) simple integer]
   (c) X-ray analysis is of KCl reports it as —— .
   [Ans. (i) simple cubic (ii) Tetragonal (iii) f – c – c]
   (d) Born-Baber cycle can be used to calculate —— .
   [Ans. (i) Internal energy (ii) Lattice energy (iii) P.E.]
   (e) Naming of crystalline planes was done by —— .
   [Ans. (i) Bragg (ii) Miller (iii) Haber]

Answers :

1. (a) T; (b) F; (c) T; (d) F; (e) T; (f) F.
2. (a) (i); (b) (iii); (c) (i); (d) (ii); (e) (ii).

Chapter 6

THERMODYNAMICS — FIRST LAW

6.1 INTRODUCTION

The name *Thermodynamics* indicates that it deals with heat-motion relation. In that sense the name is a misnomer. During discussion we shall see that the subject discusses interconversion of various forms of energies and also conversion of energy to work. So, its correct name should be *energetics*.

There are three main laws of thermodynamics and all these laws are empirical i.e., result of human experience. That is why thermodynamics is sometimes regarded as axiomatic science.

During discussion of thermodynamics we shall very often use some common english words, in some uncommon senses. For this reason, we must first acquaint ourselves with the thermodynamic meanings of these words.

6.2 TERMINOLOGY

System : Any part of the universe, which is the subject matter of our study is called a *system*.

Surrounding : Everything excepting the system is called *surrounding*.

Thus, system + surrounding = universe.

Different types of System : If both mass and energy of a system can be changed, then it is an **open-system.**

When only energy and not mass can be changed, the system is a **closed one.**

And when neither mass, nor energy can be changed, the system is an **isolated system.**

Wall : The partition, which separates a system from surrounding is called *wall.* If heat can pass through it, then the wall is **diathermal** and if passage of heat is prohibitted, the wall is an **adiabatic** wall.

State : States of a system are those variables, on which energy of the system depends. For a gas in open system, n (moles); P; V and T are states; but when it is a closed system P, V, and T are states.

State Function : State functions are those parameters, which depend only on its state and are independent of the path of achieving the given state and previous history. Thus energy is a state function as energy of a gas depends only on P, V & T and not on the method of preparation we write

$$E = f(P, V, T) \text{ for closed system.}$$

Again as out of P,V, and T any two are independent while the rest becomes dependent we can write

$$E = f(P, V) \text{ or } f(P, T) \text{ or } f(V, T).$$

Extensive and Intensive properties : Properties, which are dependent on amount are called *extensive.* They are additive properties e.g. mass; volume; energy etc.

On the contrary properties, which are independent of amount are intensive. They are point properties and same at each point of the system e.g. temperature; pressure; density etc.

Perfect differential: If $y = f(x, z)$ and if $\frac{d^2y}{dzdx} = \frac{d^2y}{dxdz}$, then y is said to be a *perfect differential*. They are path independent and can be integrated between limits i.e., they are state-functions.

Energy : The most common word used in thermodynamics is energy. It is something which when enters a system or leaves a system, the system suffers a change. Thus energy is the root behind any change. Out of different forms of energies, heat energy has some peculiarities. Thus, while changes produced by other energies are purely temporary, changes created by heat energy is permanent. It brings distortion in shape of the system and so a part of heat is wasted in doing unwanted work. Hence conversion of heat to work is never profitable.

Process : When some sort of energy change takes place, we say that a process has occurred. If during the process energy received from surrounding is fully converted to work then energy of the system and hence its temperature remains same. Such a process is called **isothermal.** However some processes do take place without any external inspiration. The system does not receive anything from surroundings. Such processes are called **adiabatic**.

6.3 WORK

If you are asked to hold a five kilo weight few feets above the ground level without moving it; then very soon you will feel strain and a layman will say that you are doing a lot of work. But to a scientist work done by you is zero, as you have not moved anything. Thus when there is any sort of displacement (material or charge) in the direction of force we say that some work has been done. Any work is the product of two terms, viz. intensity factor and capacity factor. The former measures amount of opposition and the latter the extent by which opposition has been overcome. Some examples are given below —

Nature of Work	I.F.	C.F.	I.F. × C.F.
1. Passage of Q coulombs across a p.d. of 'E' volt. (Electrical work).	E	Q	EQ (Volt-coulomb i.e., Joule)
2. Lifting a body of mass 'm' through a height 'h'	mg	h	mgh
3. Expansion of a gas against an external pressure P_{ext} from a vol. V_1 to V_2.	P_{ext}	$(V_2 - V_1)$	$P_{ext}(V_2 - V_1)$

6.4 REVERSIBLE AND IRREVERSIBLE PROCESS

The most important terms used in thermodynamics are 'reversibility' and 'irreversibility' Before defining these terms let us consider the following example.

Here we have considered two identical cylinders, I & II fitted with weightless and frictionless pistons. They consist of same amount of same gas at pressure P_1. Pressure above the pistons is P_2 and let $P_2 << P_1$. So the piston will move upwards. In order to stop this movement let application of 5 kg. weight is necessary. In cylinder I we put a 5 kg. weight on the pan over the piston, but in cylinder II we put 5 kg. sand. Now following thing. done.

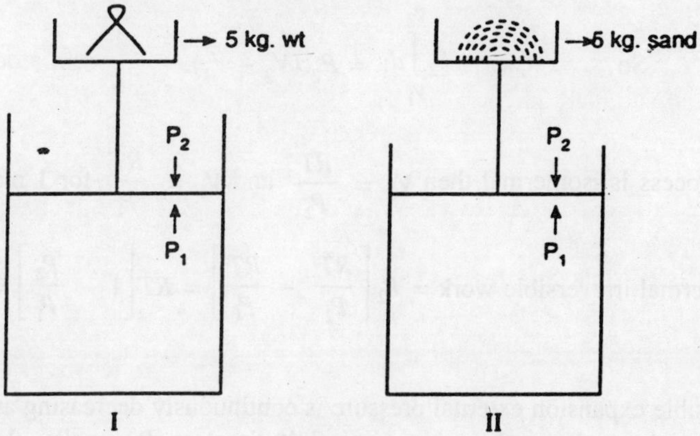

Fig. 6.1

Let from cylinder I, the 5 kg. weight is removed suddenly. Piston will immediately jump upward and reach a position where internal pressure of the gas also becomes P_2. This movement is unidirectional; it is so quick that there was no equilibrium during movement. Also during movement external pressure was constant. Such a process is called irreversible. It is to be noted that during the process the difference between driving and opposing force is very high. [$P_1 \sim P_2 = 5$ kg.]

On the contoary if a man is engaged in removing the sand from the pan over cylinder II, and if he is asked to remove only one grain of sand at a time, then after each removal external pressure will be decrease by *dp* and the gas will expand by *dv*. A time will come when all the sand will be removed and final state will be same as cylinder I. However here the movement was so slow that it appeared as a static process; there was equilibrium always. The process was completely under control. It can be stopped at any moment and can be made to run in either direction by placing or removing a grain of sand. Such a process is called *quasi-static* or *reversible* process.

Now on the basis of above example let us define irreversible and reversible process.

If during a process difference between driving and opposing force in very high, then the process occurs in one direction; there is no equilibrium. Such a process is called *irreversible*. Generally such processes are quick. Expansion of a gas against a constant external pressure is an example of such a process.

On the contrary, if during a process difference between driving and opposing force is infinitesimally small, then the process may occur in either direction depending on which one is becoming more. System is always in equilibrium. Such a process is *reversible*. Generally these are slow process; however slow does not mean reversibility. All natural changes (e.g. growth of a man) are slow but they are irreversible. Expansion of a gas against a continuously decreasing pressurre is an example of reversible process.

6.5 IRREVERSIBLE AND REVERSIBLE WORK CALCULATION

We know that expansion work of a gas from volume V_1 to V_2 is

$$W = \int_{V_1}^{V_2} P_{ext} dv.$$

In irreversible process external pressure is constant during expansion (say it is P_2).

So $$W_{irrev} = P_2 \int_{V_1}^{V_2} dv = P_2 (V_2 - V_1).$$

If the process is isothermal then $V_2 = \frac{RT}{P_2}$ and $V_1 = \frac{RT}{P_1}$ for 1 mole ideal gas.

So isothermal irreversible work $= P_2 \left[\frac{RT}{P_2} - \frac{RT}{P_1}\right] = RT\left[1 - \frac{P_2}{P_1}\right] = RT \cdot \frac{P_1 - P_2}{P_1}$.

....(6.1)

In reversible expansion external pressure is continuously decreasing and as difference between external and internal pressure is infinitesimal so $P_{ext} \approx P_{int}$. Hence reversible work is —

$$W_{rev} = \int_{V_1}^{V_2} P_{int} \, dv = \int_{V_1}^{V_2} \frac{RT}{V} dv \text{ [for 1 mole of ideal gas]}$$

For isothermal reversible process

$$W_{rev} = RT \int_{V_1}^{V_2} \frac{dv}{v} RT \ln \frac{V_2}{V_1} = RT \ln \frac{P_1}{P_2}. \qquad \text{....(6.2)}$$

6.6 COMPARISON OF TWO WORKS

From equation (6.1) isothermal irreversible work

$$W_{irrev} = RT \cdot \frac{P_1 - P_2}{P_1}$$

and from (6.2) isothermal reversible work

$$W_{rev} = RT \ln \frac{P_1}{P_2} = RT \ln\left[1 + \frac{P_1 - P_2}{P_2}\right].$$

If $\frac{P_1 - P_2}{P_2}$ is small then $W_{rev} \approx RT \cdot \frac{P_1 - P_2}{P_2}$.

Now $P_2 < P_1$ so $W_{rev} > W_{irrev}$..

6.7 FIRST LAW OF THERMODYNAMICS

Once upon a time man tried to create energy by constructing perpetual motion machine. But very soon he realised the bitter fact that creation or destruction of energy is beyond his capacity; he can at best convert energy from one form to other. This experience, stated in terms of language, is regarded as first law of thermodynamics. Some popular statements are given below —

(a) Energy can neither be created nor be destroyed.

(b) When one form of energy disappears, an exactly equivalent amount of energy of some other form will make its appearance.

(c) Total energy of the universe is constant.

(d) Energy of an isolated system is constant.

All these statements point toward the same thing which is conservation of energy. Thus first law can be regarded as law of conservation of energy. However Einstein's equation $E = mc^2$ indicate inter- convertibility of mass and energy and so present-day concept is law of conservation of mass and energy.

6.8 INTERNAL ENERGY AND ENTHALPY

If a piece of coal is burnt it gives us heat and light. It means some energy was stored inside it. Every system has certain structure and some work had been done to give it this structure. This work is stored inside it as potential energy; besides this there are internal motion, e.g. rotation, vibration etc. which gives rise to kinetic energy. Summation of these two energies is called internal energy. Thus internal energy is the energy possessed by a system by virtue of its position, structure and internal motion. It can do work at the cost of this energy. It is symbolised as U (or E by few authors). It is a state function perfect differential and extensive property.

Let us consider a solid of internal energy U; let it is sublimed, then the energy of the vapour = Energy of solid + Work done during sublimation = $U + P(V_g - V_s)$. As volume of solid (V_s) << volume of vapour (V_g) so energy of Vap $\simeq U + PV_g$. This total energy is called entharpy or heat content (H). Thus $H = U + PV$. It is the total energy content of a body; for solids and liquids $H = U$. It is also a state function; perfect differential and extensive property.

6.9 MATHEMATICAL FORM OF FIRST LAW

Let a system receives dq heat from surrounding during some process and thereby increases its internal energy by dU and performs work w, then according to law of conservation of energy i.e,. first law —

$$dq = dU + w. \quad \text{....(6.3)}$$

It work is mechanical in nature i.e., expansion by a volume dv against an external pressure P, then. $w = Pdv$, so equation (6.3) becomes

$$dq = dU + Pdv \quad \text{....(6.4)}$$

Equations (6.3) and (6.4) are mathematical form of first law.

Now enthalpy H is

$$H = U + PV$$

Differentiating at constant pressure we get

$$dH = dU + Pdv \quad \text{....(6.5)}$$

So comparing equations (6.4) and (6.5) it can be said, **"increase in enthalpy is equal to heat absorbed at constant pressure when work done is mechanical in nature."**

6.10 CRITICAL COMMENTS ON FIRST LAW'S EQUATION

From equations (6.3) and (6.4) we can arrive at following conclusions —

(1) For isothermal process as $dU = 0$; $dq = w$. Here all the absorbed heat have been spent in doing work and as no heat is stored and temperature has not risen.

(2) For adiabatic process as $dq = 0$, $w = -dU$. Thus during expansion (adiabatic) w is positive and hence internal energy will decrease so the system will be cooled. During adiabatic compression w is negative i.e., internal energy will increase and so the system will be heated.

(3) In a cyclic process $\oint dU = 0$, as U is a state function and in a cyclic process state

does not change. So here also $dq = w$. Though the equation becomes same as in case (1), the two processes are different. In isothermal process P and V change ; in cyclic process none of the states changes. In isothermal process $w \neq 0$. In cyclic process if forward and backward paths are same then total work becomes zero.

6.11 JOULE'S EXPERIMENT — FREE EXPANTION OF GASES

Joule took two bulbs, one of which consists of a gas at low pressure (ideal), the other was evacuated. They are connected by a connecting tube provided with a stop-cock (fig. 6.1). They are immersed in a thermostat containing water and a very sensitive thermometer is placed.

The stop-cock is now opened and the gas expands. As external pressure is zero the gas does not face any opposition during expansion. Such **expansion against zero pressure**

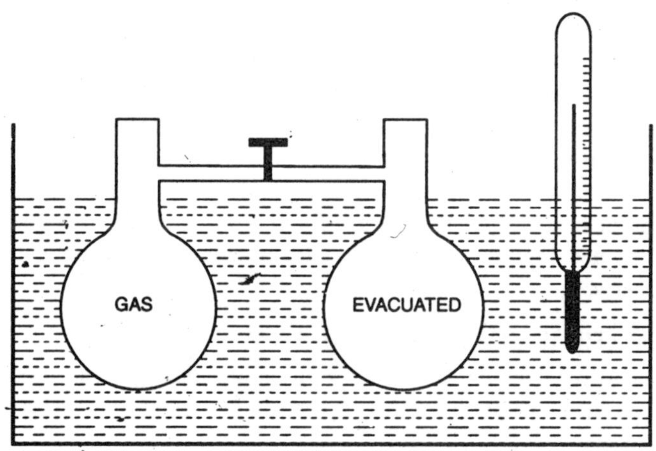

Fig. 6.2

is called free expansion. As here no heat is exchanged with surrounding $dq = 0$, the process is adiabatic. As adiabatic expansions generally produces cooling so it was expected that temperature will fall. But thermometer recorded no drop in temperature. This is because, work done is equal to zero in this case. As $w = P_{ext} \times$ Volume change, so $w = 0$. Now $dq = dU + w$. Here $dq = 0$, $w = 0$, so $dU = 0$. Hence internal energy and that is why temperature remain unchanged. So it can be said that **free expansion of an ideal gas is both isothermal and adiabatic in nature.**

However as it is an expansion, so P and V have changed, so we can write that for ideal gas

$$\left(\frac{dU}{dV}\right)_T = \left(\frac{dU}{dP}\right)_T = 0. \qquad ....(6.6)$$

Now $H = U + PV$.

Hence $$\left(\frac{dH}{dV}\right)_T = \left(\frac{dU}{dV}\right)_T + \left[\frac{d(PV)}{dV}\right] = 0 + 0 = 0.$$

For PV = Constant at constant temperature.

Similarly $\left(\frac{dH}{dP}\right)_T = 0$ for ideal gas.

If the gas becomes real then $\left(\frac{dU}{dV}\right)_T$ will not be zero. Real gases have intermolecular attraction. Hence during expansion they are to work not only against external pressure but also against intermolecular attraction. During free expansion, the former is zero but due to latter internal energy will be spent. It can be proved that for one mole real gas

$$\left(\frac{dU}{dV}\right)_T = \frac{a}{V^2}. \qquad \text{....(6.7)}$$

6.12 HEAT CAPACITIES OF GASES

In chapter-1 we have learnt about two different types of molar heat capacities (C_P and C_V) of gases. Now from thermodynamics we shall establish their relation.

Let one mole of a gas receives dq heat at constant volume and increases its temperature by dT, then

$$C_V = \left(\frac{dq}{dT}\right)_V.$$

Now at constant volume no mechanical work is done, so $dq = dU$ [$P\,dV = 0$].

So $$C_V = \left(\frac{dU}{dT}\right)_V.$$

For n-mole gas $\quad dU = n\,C_V\,dT. \qquad$ ....(6.8)

If however heating is done at constant pressure and if dq heat raises temperature of one mole gas by dT, then

$$C_P = \left(\frac{dq}{dT}\right)_P.$$

At constant pressure $dq = dH$.

So $$C_P = \left(\frac{dH}{dT}\right)_P.$$

For n - mole gas $\quad dH = n\,C_P\,dT \qquad$ ....(6.9)

To establish relation between C_P and C_V we write

$U = f(T, V)$ for a closed system.

So $$dU = \left(\frac{\partial U}{\partial T}\right)_V .dT + \left(\frac{\partial U}{\partial V}\right)_T .dV.$$

At constant pressure $dU = dH - P\,dV$.

So $$dH - PdV = C_V \,.\, dT + \left(\frac{\partial U}{\partial V}\right)_T .dV.$$

Dividing each term by dT and making rearrangement —

$$\left(\frac{dH}{dT}\right)_P - C_V = \left[P + \left(\frac{\partial U}{\partial V}\right)_T\right]\left(\frac{dV}{dT}\right)_P$$

or, $$C_P - C_V = \left[P + \left(\frac{\partial U}{\partial V}\right)_T\right]\left(\frac{dV}{dT}\right)_P. \qquad \text{....(6.10)}$$

Above equation is a general relation and it is valid for any gas ideal or real. For ideal gas we know from Joule's experiment.

$$\left(\frac{\partial U}{\partial V}\right)_T = 0.$$

So $$C_p - C_V = P\left(\frac{dV}{dT}\right)_P.$$

But here $PV = RT$ for 1 mole

So $$P\left(\frac{dV}{dT}\right)_P = R.$$

Hence $C_p - C_V = R$ for one mole ideal gas.

For real gas $$\left(\frac{\partial U}{\partial V}\right)_T = \frac{a}{V^2}.$$

So $$C_p - C_V = \left[P + \frac{a}{V^2}\right]\left(\frac{dV}{dT}\right)_P$$

$$\text{or,}\quad C_P - C_V = \frac{RT}{V-b}\left(\frac{dV}{dT}\right)_P \qquad \text{....(6.11)}$$

[from V. D. Waals equation].

To find $\left(\frac{dV}{dT}\right)_P$ we write V. D. Waals equation as —

$$\left(P + \frac{a}{V^2}\right)(V-b) = RT$$

or, $PV + \frac{a}{V} - Pb = RT.$ (neglecting $\frac{ab}{V^2}$ which is small).

Differentiating w. r. t. T at constant P —

$$P\left(\frac{dV}{dT}\right)_P - \frac{a}{V^2}\left(\frac{dV}{dT}\right)_p = R$$

$$\text{or,}\quad \left(\frac{dV}{dT}\right)_P = \frac{R}{P - \frac{a}{V^2}} = \frac{R}{\left(P + \frac{a}{V^2}\right) - \frac{2a}{V^2}}$$

$$\text{or,}\quad \left(\frac{dV}{dT}\right)_P = \frac{R}{\frac{RT}{V-b} - \frac{2a}{V^2}} = \frac{R}{\frac{RT}{V-b}\left[1 - \frac{2a(V-b)}{RTV^2}\right]}$$

As 'b' is very small so

$$\left(\frac{dV}{dT}\right)_P \simeq \frac{R}{\frac{RT}{V-b}\left[1 - \frac{2a}{RTV}\right]}.$$

Putting this value in equation (6.11) we get —

$$C_p - C_V = \frac{RT'}{V-b} \cdot \frac{R}{\frac{RT}{V-b}\left[1-\frac{2a}{RTV}\right]}$$

$$\text{or,} \quad C_p - C_V = R\left(1-\frac{2a}{RTV}\right)^{-1}$$

$$\text{or,} \quad C_p - C_V \simeq R\left(1+\frac{2a}{RTV}\right) \qquad ....(6.12)$$

(neglecting other terms, which are small).

Thus for real gas $C_p - C_V > R$.

The interesting thing to note is that for real gas $C_p - C_V$ value depends on nature of gases as different gases have different '*a*' and '*b*' values. For ideal gases $C_p - C_V$ is same for each gas and is equal to '*R*'.

6.13 THROTTLING — JOULE THOMSON'S EXPERIMENT

The term 'throttling' means sudden lowering of pressure. This is generally done by releasing the pressure of a compressed gas through some orifice. Joule and Thomson carried out experiment on throttling of gas at room temperature. They found that all gases excepting H_2 and He show a cooling effect during throttiing. Hydrogen and Helium are warmed up.

Later experiment revealed that during throttling any gas can be heated, cooled or even its temperature may remain unchanged. What will happen that depends on temperature during throttling. If temperature is equal to a particular value, we shall call it inversion temperature, (T_i) then nothing will happen.

If experimental temperature is less than inversion temperature then there will be cooling effect. But when experimental temperature is more than inversion temperature then throttling produces heating. For all gases excepting H_2 and He inversion temperatures are much above room temperature and so throttling produces cooling. For H_2, $T_i = 202°$A and for He, $T_i = 40°$A, so throttling at room temperature warms them up.

The experiment of Joule and Thomson is now described.

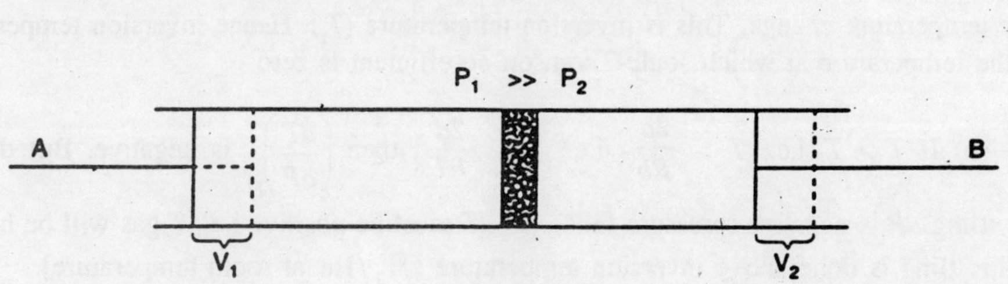

Fig. 6.3

A long tube is taken at the middle of which there is a porous plug packed with cotton wool P (see fig). The plug can be replaced by a silk handkerchief. There are two pistons

A and B at the two ends. Same gas is present in both sides of the porous plug ; but their pressures are different, say P_1 and P_2 $(P_1 >> P_2)$. Now a sudden thrust is given on piston A and it sweeps a volume V_1. So work done on the system $= P_1 V_1$. Certain amount of gas is throttled to the other side and it pushes away piston B by a volume V_2. So work done by the system $= P_2 V_2$. If $P_1 V_1 = P_2 V_2$; internal energy of the system remains unchanged and its temperature will remain same. If $P_1 V_1 > P_2 V_2$; internal energy will increase and gas will be heated and it $P_1 V_1 < P_2 V_2$,the gas looses some of its energy and it will be cooled. So theoretically there are all these three possibilities. Now —

net work done by system $w = (P_2 V_2 - P_1 V_1)$.

If U_1 and U_2 are internal energies before and after throttling then internal energy change $\Delta U = (U_2 - U_1)$. From first law $dq = dU + w$.

Since here no heat exchange with surrounding occurred i.e., $dq = 0$, so –

$$0 = (U_2 - U_1) + (P_2 V_2 - P_1 V_1)$$

$$\text{or,} \quad U_1 + P_1 V_1 = U_2 + P_2 V_2$$

$$\text{or,} \quad H_1 = H_2.$$

Thus enthalpy of the system does not change. Hence Joule-Thomson process is iso-enthalpic in nature.

6.14 JOULE-THOMSON CO-EFFICIENT — INVERSION TEMPERATURE

Let dP be the **drop** in pressure during throttling and dT be **change** in temperature. Then $\left(\frac{dT}{dP}\right)_H$ is called Joule-Thomson co-efficient; it is symbolised as μ_{J-T}.

So $$\mu_{J-T} = \left(\frac{dT}{dP}\right)_H.$$

Its value can be calculated . After some approximations, the value comes as

$$\mu_{J-T} = \left(\frac{dT}{dP}\right)_H = \frac{1}{C_P}\left[\frac{2a}{RT} - b\right]. \quad \text{....(6.13)}$$

Three special cases are very much interesting.

(a) If $T = \frac{2a}{Rb}$ then $\mu_{J-T} = \left(\frac{dT}{dP}\right)_H = 0$ i.e., $dT = 0$. So throttling will not produce any temperature change. This is inversion temperature (T_i). Hence inversion temperature is the temperature at which Joule-Thomson co-efficient is zero.

(b) If $T > T_i$ i.e., $T > \frac{2a}{Rb}$ i.e., $b > \frac{2a}{RT}$; then $\left(\frac{dT}{dP}\right)_H$ is negative. But during throttling dP is negative (pressure falls). So dT must be positive i.e., T gas will be heated if throttling is done above inversion temperature [H_2/He at room temperature].

(c) If $T < T_i$ i.e., $T < \frac{2a}{Rb}$ i.e., $b < \frac{2a}{RT}$; then $\left(\frac{dT}{dP}\right)_H$ is positive. As dP is negative, so dT must also be negative. Hence throttling below inversion temperature will produce cooling effect.

6.15 JOULE'S EXPERIMENT AND JOULE-THOMSON EXPERIMENT — A COMPARISON

Both Joule's experiment and Joule – Thomson's experiment are adiabatic. However, in Joule's experiment no work was done by the gas (free expansion) so its temperature did not change. In Joule Thomson experiment work was done on the gas and in return gas did some work. Nature of thermal change (i.e. heating, cooling or nothing) depends on relative amount of two. However, when gas alone does adiabatic work then its temperature will definitely fall. Expression for adiabatic work is —

$$W_{ad} = -dU \; [dq = 0]$$

$$= C_v\,(T_1 - T_2) \qquad [T_1 ; T_2 \text{ initial and final temperature}]$$

$$= C_v\left[\frac{P_1V_1}{R} - \frac{P_2V_2}{R}\right] = C_v \cdot \frac{P_1V_1 - P_2V_2}{C_P - C_V} \qquad \text{[for ideal gas]}$$

So $$W_{ad} = \frac{P_1V_1 - P_2V_2}{\gamma - 1} = \frac{R\left(T_1 - T_2\right)}{\gamma - 1}$$

When work is done by gas alone W_{ad} is *+ve* and $T_2 < T_1$ i.e. gas will be cooled under all circumstances.

6.16 ADIABATIC RELATIONS

Boyle's law is valid at constant temperature ; Charle's law at constant pressure. So these relations will not be valid for adiabatic conditions where *P*, *V* and *T* all change. In such cases relations between *V* and *T* ; *P* and *V* ; *P* and T can be derived as follows —

From first law $$dq = dU + P_{ext.} dv.$$

During reversible process $P_{ext} \simeq P_{int} = \dfrac{RT}{V}$ for 1 mole gas.

So $dq = dU + \dfrac{RT}{V}\,dV$. Again $dq = 0$ for adiabatic process and $dU = C_v\,dT$ for 1 mole gas.

So $$C_V\,dT + \frac{RT}{V}\,dV = 0$$

or, $$C_V\,dT + (C_P - C_V).\; T.\frac{dV}{V} = 0.$$

Dividing by C_V. T we get

$$\frac{dT}{T} + (\gamma - 1).\,\frac{dV}{V} = 0.$$

On integration

$$\int\frac{dT}{T} + (\gamma - 1).\int\frac{dV}{V} = 0$$

or, $$\ln T + (\gamma - 1)\ln V = \text{Constant}$$

or, $$\ln TV^{\gamma-1} = \text{Constant}$$

Hence $$TV^{\gamma-1} = \text{Constant.} \qquad \text{....(6.14)}$$

Putting $T = \dfrac{PV}{R}$ we get

$$\frac{PV.V^{\gamma-1}}{R} = \text{Constant;} \quad \text{or,} \quad PV^{\gamma} = \text{Constant.} \qquad \text{....(6.15)}$$

Again putting $V = \frac{RT}{R}$

$$P\left(\frac{RT}{P}\right)^{\gamma} = \text{Constant; or, } P^{1-\gamma}.T^{\gamma} = \text{Constant;}$$

$$\text{or, } TP^{\frac{1-\gamma}{\gamma}} = \text{Constant.} \qquad \text{....(6.16)}$$

Equations (6.14) ; (6.15) and (6.16) are adiabatic relations ; these are valid for reversible processes only.

6.17 ISOTHERMALS AND ADIABATICS

When pressure of a gas is dropped isothermally its volume will increase due to pressure effect. However, when pressure is decreased adiabatically its temperature also falls. Decrease of pressure tries to increase the volume but drop in temperature tries to decrease the volume. As a result of these two opposing effects extent of increment in volume due to pressure drop will be some what less than that observed in similar isothermal pressure drop. So adiabatic curve will always run below the isothermal i.e. adiabatics are steeper.

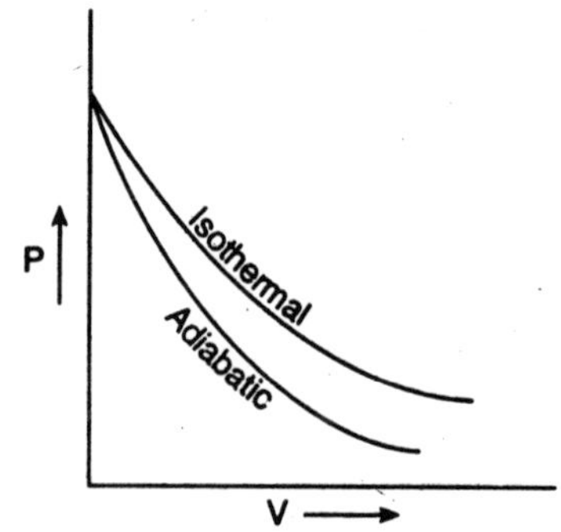

Fig. 6.4

This is also proved mathematically. For isothermals PV = Constant.

So $\quad PdV + VdP = 0.$

Hence slope $\quad \frac{dP}{dV} = -\frac{P}{V}.$

For adiabatics $\quad PV^{\gamma}$ = Constant.

So $\quad P\gamma V^{\gamma-1}dV + V^{\gamma}dP = 0$

or, $\quad \gamma PdV + VdP = 0$

$$\text{Slope} = \frac{dP}{dV} = -\frac{\gamma P}{V}$$

as $\gamma > 1$ so $\left|\frac{\gamma P}{V}\right| > \left|\frac{P}{V}\right|$. So slopes of adiabatics are more; hence they will be steeper.

POINTS TO REMEMBER

Thermodynamics, whose correct name is *energetics*, discusses about interconversion of various forms of energies.

System is some matter, which is being studied. When matter and energy interacts '*process*' is said to be occurred. In '*isothermal process*' temperature of system does not change but in '*adiabatic process*' as nothing is taken (or given) from (or to) surrounding; energy of system changes and hence temperature also changes.

If effect of energy causes displacement, then '*work*' has been done. When difference between driving and opposing force is infinitesimally small it is *reversible*, but when difference is high it is irreversible.

Expansion of a gas against zero pressure is *free expansion.* For *ideal* gases it is both *isothermal* and *adiabatic.*

Sudden lowering of pressure (i.e. *throttling*) is called *Joule-Thomson effect.* This process, when done below inversion temperature produces cooling. For H_2 and He this temperature is much below room temperature but for other gases it is much above room temperature. Plot of P against V under adiabatic condition is *steeper* than same type of plot under isothermal condition.

QUESTIONS

SHORT TYPE

1. "Thermodynamics in an axiomatic science"– explain.
2. The name thermodynamics is a misnomer; what is its correct name ?
3. 'State functions are perfect differentials' – explain.
4. What are intensive and extensive properties ?
   To which class P, V, and T belong ?
5. Failure of perpetual machine of 1st kind gave first law — explain.
6. What will be the changes of U and H in a cyclic process ?
7. Name a process which is isothermal and at the same time adiabatic.
8. What is the value of $\left(\frac{dU}{dV}\right)_T$ for (i) ideal gas (ii) real gas.
9. What is throttling ? What is inversion temperature ?
10. Throttling of N_2 at room temperature produces cooling but for H_2 it is a heating effect – explain?
11. Adiabatic relations are valid for reversible processes only – why ?
12. Does adiabatic expansion always produce cooling ?

ESSAY TYPE

1. What are reversible and irreversible processes ? Prove that work done in a reversible process is always more than that in an irreversible process.
2. State first law of thermodynamics in three different forms and show that they have same meaning.
3. Write down mathematical form of first law. Critically examine (a) isothermal (b) adiabatic and (c) cyclic processes.
4. What was Joule's experiment ? What conclusions were drawn from it ?
5. Define molar heat capacities of gases. Establish their relations from first law.
6. Distinguish between Joule-Thomson and adiabatic expansion.
7. Derive adiabatic relationships between P and V ; P and T ; V and T. Why they are valid for reversible processes only ?
8. Prove that in a P–V plot adiabatics are steeper than isothermals.

PROBLEMS

1. One mole of a perfect gas expand isothermally and reversibly from 2 atm. to 1 atm. at 25°C. Calculate the maximum work done in ergs.

   [*Hints.* $W = nRT \ln \frac{P_1}{P_2}$; $n = 1$; $R = 8.313 \times 10^7$ $T = 298$, $P_1 = 2$, $P_2 = 1$.]

2. 2 moles of a perfect gas expand reversibly and isothermally at 27°C from 1 litre to 10 litres. Calculate the work done in litre atm.

   [*Hints.* $W = nRT \ln \frac{V_2}{V_1}$; put $R = 0.082$.]

3. A mole of an ideal gas at 0°C is compressed reversibly and adiabatically from 1atm. to 2 atm. What is the final temperature if $C_p = 5$ cal / °C ?

   [*Hints.* $C_v = C_p - R \simeq 3$; $\gamma = \frac{C_p}{C_v} = \frac{5}{3} = 1.66.$ $T_1 P_1^{\frac{1-\gamma}{\gamma}} = T_2 P_2^{\frac{1-\gamma}{\gamma}}$.]

4. Calculate the amount of heat absorbed in calorries when 15 gm mole of a perfect gas expand reversibly and isothermally from 105 litres to 1050 litres at 27°C.

[*Hints.* Same as before.]

5. Calculate the rise in temperature of a piece of copper ($C_p = 6.4$) when dropped from a tower (height 270 ft.) assuming that no heat is lost.

[*Hints.* Work = Mgh ergs heat = $M\ C_p \times (t_2 - t_1)$ cal.

$= \dfrac{Mgh}{4.2 \times 10^7}$ Cal. $= M\ C_p\ (t_2 - t_1)$. Find $(t_2 - t_1)$.]

6. Calculate ΔE and ΔH if 10 litres of helium at N.T. P. are heated inside a gas cylinder from 25°C to 100°C assuming it to behave ideally and having $C_v = \dfrac{3}{2} R$.

[*Hints.* $\Delta E = n\ C_v dT$; $\Delta H = n\ C_p dT$; $n = \dfrac{10}{22.4}$; $C_V = \dfrac{3}{2} R$; $C_p = \dfrac{5}{2} R$.]

OBJECTIVE TYPE

1. Fill in the blank.

(a) Properties dependent on amount are ——— .
(b) Temperature is an ——— property.
(c) System whose energy and not mass can change is ——— .
(d) Throttling at room temperature produces ——— for N_2 but ——— for He.
(e) Adiabatics are ——— than isothermals.
(f) Work obtained during irreversible expansion is ——— than similar expansion done reversibly.
(g) ——— is a process, which is both isothermal and adiabatic.
(h) In isochoric process ——— remains constant.
(i) The relation PV^r is valid for adiabatic ——— process.
(j) Both internal energy and enthalpy are ——— function.

2. State whether the following statements are 'true' or 'false'.

(a) During reversible expansion external pressure is constant.
(b) In the formula $W = Pdv$; 'P' is really internal pressure.
(c) During adiabatic change the system can be regarded as isolated.
(d) Volume is an extensive property.
(e) All slow processes are reversible.

3. Match the terms of column 'A' with those of column 'B'

A	B
(a) Thermoflask.	(a) Joule-Thomson.
(b) Water in a beaker.	(b) Isobaric.
(c) Mass.	(c) Open system.
(d) Constant pressure.	(d) Extensive property.
(e) Throttling.	(e) Isolated system.

Answers :

1. (a) Extensive; (b) intensive; (c) closed; (d) cooling, heating; (e) steeper; (f) less; (g) Free expansion; (h) volume; (i) reversible; (j) state.

2. (a) F; (b) F; (c) T; (d) T; (e) F.

3. a – e; b – c; c – d; d – b; e – a.

Chapter 7

THERMOCHEMISTRY

7.1 WHAT IS THERMOCHEMISTRY ?

All chemical changes and some physical changes are associated with energy change. This energy appears in the form of heat in most cases (it comes as light during reaction in the body of fire-fly; as electricity in electrochemical cells). If we can know the amount of heat change, we can proceed to determine many other things e.g. bond energy, flame temperature; electron affinity, calorific values of food etc. The subject which deals in all these, is the subject of thermochemistry.

7.2 THERMOCHEMICAL EQUATION

Let us consider a simple chemical equation $C + O_2 = CO_2$. One of the limitations of this equation is, it does not indicate the amount of energy change associated with this reaction. If someone writes as $C + O_2 = CO_2 + 94.4$ k.cal. — then even many question regarding energy change remain unanswered. For, energy is a state function and here states are not mentioned. If we write

$$C\ (s) + O_2\ (g) = CO_2\ (g) + 94.4 \text{ k.cal at } 25°\text{C and 1 atm}$$

then all questions regarding energy change are answered. Such an equation is called a *thermochemical equation.* Thus a *thermochemical equation* is a chemical equation, where not only energy change but also complete states are mentioned.

7.3 HEAT OF REACTION

According to old nomenclature, heat of reaction was defined as the amount of heat evolved or absorbed during a reaction written in a balanced equation (stoichiometric). Thus for exothermic reaction it was written with positive sign (heat evolved) and for endothermic reaction it was written as negative (heat absorbed). Examples are —

$$C\ (s) + O_2\ (g) = CO_2\ (g) + 94.4 \text{ k.cal}$$

$$\frac{1}{2}N_2\ (g) + \frac{1}{2}O_2\ (g) = NO\ (g) - 21.6 \text{ k.cal}$$

At present we use thermodynamic parameters. Since during formation of CO_2 energy is liberated, so energy of one mole CO_2 is less than total energy of one mole carbon and one mole oxygen. And 94.4 k.cal is the difference. So heat of reaction is nothing but the difference in total energy between products and reactants in a balanced equation. When the reaction is carried out at constant volume, no PV-work is done and internal energy (U) is total energy. So

$$\text{heat of reaction at constant volume} = \Sigma U_{\text{products}} - \Sigma U_{\text{rectants}} = \Delta U .$$

But when reaction is carried out at constant pressure, PV-work can be done and enthalpy (H) becomes total energy. So

$$\text{heat of reaction at constant pressure} = \Sigma H_{\text{products}} - \Sigma H_{\text{rectants}} = \Delta H .$$

Difference between two becomes evident for reactions involving gases.

7.4 INTER-RELATION

Let us consider a reaction $r \rightarrow p$ involving gaseous substances. Its heat of reaction at constant pressure is —

$$\Delta H = H_p - H_r$$

But $H = U + PV.$

So $\Delta H = (U_p - U_r) + (PV_p - PV_r).$

Assuming the gas to be ideal we put $PV_p = n_p\,RT$ and $PV_r = n_r\,RT$. So

$$\Delta H = \Delta U + (n_p - n_r)\,RT$$

or, $\Delta H = \Delta U + \Delta\,nRT$

where $\Delta\,n$ is the difference in number of moles of gaseous reactants and products. For the reaction C (s) + O_2 (g) = CO_2 (g); $\Delta n = 0$ because as carbon is present in the solid state, its number of moles is not taken in to account. So this is the relation between two different types of heats of reactions i.e. at constant pressure and constant volume. When $\Delta n = 0$; e.g. 2HI (g) $\rightleftharpoons$ H_2 (g) + I_2 (g) ; $\Delta H = \Delta U$.

7.5 EXPERIMENTAL MEASUREMENT OF HEAT OF REACTION

Heat of reaction of different reactions are known either (a) by direct measurement or (b) by indirect calculation.

Direct measurement involves use of (i) ordinary calorimeter and (ii) bomb calorimeter. Reactions, which take place rapidly in solution e.g. neutralisation, solution, dilution, hydration, other ionic reactions etc. can be carried out in an ordinary calorimeter and by applying simple calorimetric principle i.e. Heat lost = Heat gained, heat of reaction can be calculated. A self-explanatory figure is drawn. (fig. 7.1).

From water equivalent of calorimetric material and temperature before and after

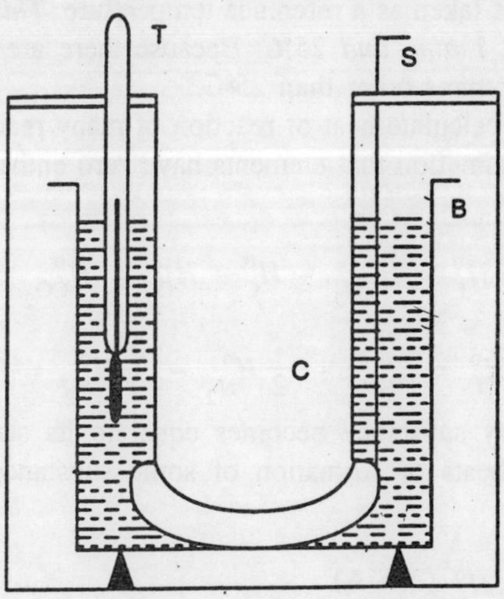

S — Stirrer
T — Thermometer
A — Calorimeter
B — Inner beakers having water
C — Thin-walled test tube where reaction occurs

Fig. 7.1

reaction, heat of reaction can be measured.

Combustion type reactions are carried out in a special calorimeter, known as bomb calorimeter (fig. 7.2). Here the substance is ignited electrically inside the calorimeter. The calorimeter is made of thick steel, sufficient to withstand high pressure evolved during combustion. The inside of the calorimeter is lined with platinum. Two electrodes are dipped; at the end of one there is a spoon where combustible material is taken. The other end is a spiral Pt-wire which just touches the spoon. The bomb is filled with O_2-gas; at high pressure (20 – 25 atm.); the reactant is taken in the spoon and it is electrically ignited. The bomb is generally dipped in water taken in a large bath. From the temperature rise of water bath and known water-equivalent of calorimeter heat of reaction can be measured.

Indirect measurement of ΔH can be done also. According to Vant Hoff, plot of $\ln K$ [K= Eqm. constant for a reversible reaction] Vs $\frac{1}{T}$ is a str. line having slope $-\frac{\Delta H}{R}$. Thus by measuring K at three or more different temperatures, ΔH can be known graphically. Most modern method of determining 'ΔH' is spectroscopic method, which is beyond the scope of this book.

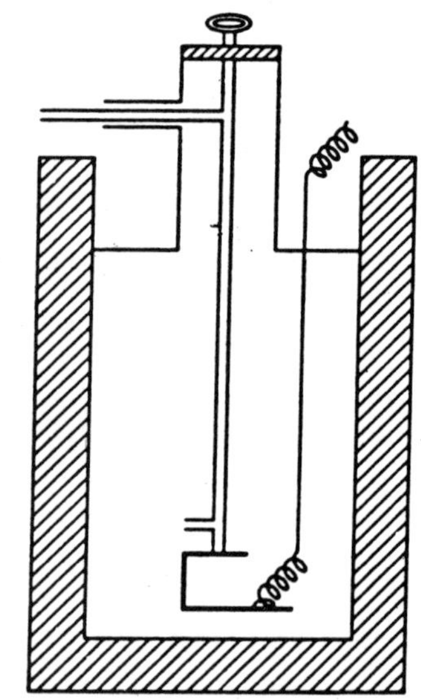

Fig. 7.2

7.6 HEAT OF FORMATION

Heat of formation of any substance is the enthalpy change when **one mole** of it is produced from constituent elements. We sometimes define a term standard heat of formation (ΔH_f^o). It is the heat of formation when reactants and products are in their standard state. By standard state we mean each constituent is at a pressure of 1 atm; **temperature may be anything - else**. However, usually 25°C is taken as a reference temperature. *This does not mean* that standard state is the state at *1 atm. and 25°C.* Because there are many examples of standard heat change at temperatures other than 25°C.

Heat of formation values are utilised to calculate heat of reaction of many reactions. During all such calculation we make the assumption that elements have zero enthalpy in the standard state. Thus

$$C\ (s) + O_2\ (g) = CO_2\ (g); \qquad \Delta H_f^o = H^o_{CO_2} - H^o_C - H^o_{O_2} = H^o_{CO_2}$$

$$\frac{1}{2} N_2\ (g) + \frac{3}{2} H_2\ (g) = NH_3\ (g);\ \Delta H_f^o = H^o_{NH_3} - \frac{1}{2} H^o_{N_2} - \frac{3}{2} H^o_{H_2} = H^o_{NH_3}$$

Thus standard heat of formation of any substance becomes equal to its standard enthalpy. In the following table standard heats of formation of some substances are given.

Table 7.1 ΔH_f^o (298°A)

Substance	ΔH_f^o (K.Cal)	Substance	ΔH_f^o (K.Cal)
CO (g)	– 26.42	HI (g)	5.9
CO_2 (g)	– 94.05	C_2H_4 (g)	12.56
NH_3 (g)	– 11.03	C_2H_2 (g)	54.23
H_2O (l)	– 68.32	C_6H_6 (l)	11.70
HCl (g)	– 22.06	CH_4 (g)	– 17.89
H (g)	– 52.09	C (g)	– 172

Let us consider the following reaction

$$CH_4\ (s) + 2O_2\ (g) = CO\ (g) + 2H_2O\ (l)$$

Its heat of reaction —

$$\Delta H = \left(H^o_{CO} + 2H^o_{H_2O}\right) - \left(H^o_{CH_4} - 2H^o_{O_2}\right) \text{ at } 298°A$$

$$= [-26.42 + 2(-68.32)] - (-17.89 - 0) \text{ k.cal}$$

$$= -212.8 \text{ K.cal.}$$

Pauling used heat of formation data to calculate bond energy. As heat of formation = – heat of decomposition and decomposition of a compound AB_x means breaking of 'x' number of A–B bonds. So A–B bond energy can be calculated. Thus from heat of formation of CH_4, the average bond dissociation energy for a C–H bond can be calculated as follows. Bond energy of a C–H bond is actually one-fourth of the heat of reaction for the following reaction.

$$CH_4\ (g) = C\ (g) + 4H\ (g)$$

ΔH for this reaction is —

$$\Delta H = H^o_{C\,(g)} + 4H^o_{H\,(g)} - H^o_{CH_4\,(g)}$$

$$\text{or,} \quad \Delta H = [-172 + 4(-52.09) - (-17.89)] \text{ K.cal}$$

$$\text{or,} \quad \Delta H = [-172 - 208.36 + 17.89] \text{ K.cal}$$

$$\text{or,} \quad \Delta H = -362.47 \text{ K.cal.}$$

So C–H bond energy (or bond enthalpy) is $\frac{+362.47}{4} = 90.62$ K.cal. Empirical (experimental) value is ~ 99 K.cal/mole at 25°C. So the results are very much approximate. But once. when spectroscopic methods for accurate determination were not available, such methods were extensively used.

7.7 LAWS OF THERMOCHEMISTRY

There are two laws in thermochemistry. These are (a) Law of Lavoisier and Laplace and (b) Hess's law of constant heat summation. We shall see that these laws are also telling us about the law of conservation of energy, i.e. first law. Hence they are sometimes regarded as 'corollary to the first law of thermodynamics.'

(a) Law of Lavoisier and Laplace : This law states that heat of formation of a substance must be equal in magnitude with the heat of dissociation of the same substance. Or in other words heats of reaction for forward and backward paths of a reversible reaction must be same in magnitude.

Explanation :

We know that $C\ (s) + O_2\ (g) = CO_2\ (g) + 94.4$ K.cal. Thus generation of CO_2 at standard state is associated with evolution of 94.4 K.cal heat/mole. If decomposition of CO_2 can be done by spending less heat, say 90 K.cal then we can carry out a profitable cyclic process in which CO_2 will be produced; 94.4 K.cal heat will be obtained; 90 K.cal will be spent; CO_2 will be decomposed and again the cycle will be repeated. We shall get 4.4 K.cal heat per cycle at the expense of nothing. Thus energy will be created. This is against first law. Hence decomposition of CO_2 will also need 94.4 K.cal. So heat of formation = – heat of decomposition.

(b) Hess's law of constant heat summation : According to Hess's law *if a reaction takes place in a number of paths then for the same balanced equation total heat change is independent of path.*

Explanation :

Let us consider preperation of an aqueous solution of ammonium chloride. We can do it via two paths.

Path-I :

NH_3 (g) + HCl (g) = NH_4Cl (s) ; ΔH_1
NH_4Cl (s) + large H_2O = NH_4Cl (aq.) ; ΔH_2
Total energy change = $(\Delta H_1 + \Delta H_2)$.

Path-II :

NH_3 (g) + large H_2O = NH_3 (aq.) ; ΔH_3
HCl (g) + large H_2O = HCl (aq.) ; ΔH_4
NH_3 (aq.) + HCl (aq.) = NH_4Cl (aq.) ; ΔH_5
Total energy change = $(\Delta H_3 + \Delta H_4 + \Delta H_5)$.

According to Hess's law $(\Delta H_1 + \Delta H_2) = (\Delta H_3 + \Delta H_4 + \Delta H_5)$.

This law defines energy as a state function and independent of path; thus it is also corollary to first law. This law is often used in many thermochemical calculation e.g. computation of bond enthalpy (see section 7.6), calculation of lattice energy and electron affinity in Born Haber cycle (see section 5.14 chapter 5) etc.

7.8 HEAT OF COMBUSTION

It is the energy change when one mole of a substance is completely burnt in presence of oxygen. Since combustion processes are always exothermic, heats of combustion (ΔH) are always negative. These values are utilised in

(a) Calculating heat of formation of the substance
(b) Calculating flame temperature and
(c) Assigning calorific values towards food.

Case (a) :

Let us take the following example —

$$C_3H_8\ (g) + 5O_2\ (g) = 3CO_2\ (g) + 4\ H_2O\ (l)$$

Heat of combustion = – 530610 cal.

Here $\Delta H^\circ = 3H^\circ_{CO_2} + 4H^\circ_{H_2O} - H^\circ_{C_3H_8}$. $[\because H^\circ_{O_2} = 0]$

Now we know $H^\circ_{CO_2} = -94050$ cal and $H^\circ_{H_2O} = -68320$.

So

$$-530610 = 3(-94050) + 4(-68320) - H^\circ_{C_3H_8}.$$

So $H^\circ_{C_3H_8} = -24{,}820$ cal.

Case (b) :

Flame is nothing but very very hot gas emitting light. If only one gas burns and if its heat capacity (molar) is known then heat of combustion = $C_p (t_f - t_i)$; t_i and t_f are temperature before and during burning. Thus from heat of combustion flame temperature can be calculated.

Case (c) :

Carbohydrate type foods are hydrolysed to glucose in our stomach. This glucose slowly burns and heat of combustion is absorbed by ATP and retained as our internal energy. If we can know how much glucose is produced from a definite amount of a certain food (say 250 gm rice) then from heat of combustion value of glucose we can calculate heat available from that food. This is the calorific value of that food.

7.9 HEAT OF NEUTRALISATION

Neutralisation process is exothermic in nature. Heat of neutralisation is the energy change when one equivalent of an acid is exactly neutralised by one equivalent of a base. It is very interesting to note that during reaction between any strong acid and strong base, heat of neutralisation is almost same and is ~ 13.7 K.cal. Examples are —

$$HCl\ (aq.) + NaOH\ (aq.) = NaCl\ (aq.) + H_2O\ (l) \qquad \Delta H = -\ 13.68\ \text{K.cal}$$

$$HCl\ (aq.) + LiOH\ (aq.) = LiCl\ (aq.) + H_2O\ (l) \qquad \Delta H = -\ 13.70\ \text{K.cal}$$

$$HNO_3\ (aq.) + KOH\ (aq.) = KNO_3\ (aq.) + H_2O\ (l) \qquad \Delta H = -\ 13.87\ \text{K.cal}$$

The explanation for near constancy of ΔH-values can be given on the basis of ionic reaction. In dil. solution strong acids and bases are completely ionised and above reactions occur as

$$H^+\ (aq.) + Cl^-\ (aq.) + Na^+\ (aq.) + OH^-\ (aq.) = Na^+\ (aq.) + Cl^-\ (aq.) + H_2O$$

$$\text{i.e. } H^+\ (aq.) + OH^-\ (aq.) = H_2O$$

Thus every strong acid, strong base reaction is nothing but combination of H^+ and OH^- to give water. For this reason enthalpy change is same in each case. However, when either of the acid and base or both is (are) weak, then much less heat is liberated. Because, here some energy is spent to cause ionisation of the weak acid (or base). In fact heat of ionisation can be determined from heat of neutralisation value. An example is given. HCN is a weak acid. Its reaction with NaOH occurs in two steps.

$$HCN = H^+\ (aq.) + CN^-\ (aq.) \qquad \Delta H_1$$

$$H^+\ (aq.) + Na^+\ (aq.) + OH^-\ (aq.) = Na^+\ (aq.) + H_2O \qquad \Delta H_2$$

$$\text{Adding} \quad HCN + OH^- = H_2O + CN^- \qquad \Delta H_3$$

Experimental value of heat of neutralisation (ΔH_3) is – 2.46 K.cal; ΔH_2 is heat of neutralisation when ions are already present; it is – 13.68 K.cal for HCl and NaOH. If ΔH_1 is heat of dissociation (ionisation) then from Hess's law

$$\Delta H_3 = \Delta H_1 + \Delta H_2$$

$$-\ 2.46 = \Delta H_1 + (-\ 13.68)$$

$$\text{or,} \quad \Delta H_1 = 13.68 - 2.46 = +\ 11.22\ \text{K.cal}$$

7.10 HEAT OF SOLUTION AND DILUTION

It is our common experience that when a solute is dissolved in a solvent heat is liberated or absorbed. Extent of heat change at a particular pressure and temperature depends on both amount of solute and solvent. Let us see the following table —

Table 7.2

Mole of water added to 1 mole H_2SO_4	ΔH (Cals.)
0	0
1	– 6820
2	– 9960
6	– 14,740
10	– 16,240
3200	– 20,050
∞	– 22,990

The term integral heat of solution is the enthalpy change when one mole of solute is dissolved in a definite amount of solvent. Thus integral heat of solution for producing H_2SO_4, 10 H_2O is –16,240 cals. Limiting value of integral heat of solution is called *heat of solution* (or *dilution*). It is the enuıalpy change when one mole solute is dissolved in such a large volume of solvent, that on further dilution no heat change will take place. Such solutions are aqueous solution and written as H_2SO_4 (aq). Difference between any two integral heat of solution is integral heat of dilution. Thus

$$H_2SO_4\,,\ 2\ H_2O\ +8H_2O = H_2SO_4\,,\ 10\ H_2O$$

Here integral heat of dilution = – 16,240 – (– 9960) = – 6280 cal.

7.11 THERMONEUTRALITY OF SALT SOLUTION

Since salts are strong electrolytes and strong electrolytes are completely dissociated; so it is expected that when dilute solutions of two different salts are mixed there should not be any thermal change provided there is no precipitation. Thus mixing of NaCl (aq.) and KNO_3 (aq.) will not produce any heat change. This principle is known as *principle of thermoneutrality*. When however some precipitation occurs then some heat change takes place, which is equal to heat of solution of the solid ppt.

7.12 VARIATION OF HEAT OF REACTION WITH TEMPERATURE — KIRCHOFF'S EQUATION

Let us consider a reaction A + B $\rightleftharpoons$ C + D

Its heat of reaction $\Delta H = (H_C + H_D) - (H_A + H_B)$ ....(7.1)

Now enthalpy of a substance depends on temperature and increase in enthalpy for same rise in temperature is different for different substances because they have different heat-capacities ($dH = n\ C_p\ dT$). So if a reaction is carried out at different temperature, heat of reaction values will be different. Variation of heat of reaction with temperature is shown by equations, known as Kirchoff 's equations. To derive these we differentiate equation (7.1) with respect to temperature and get

$$\frac{d}{dT}(\Delta H) = \left(\frac{dH_C}{dT} + \frac{dH_D}{dT}\right) - \left(\frac{dH_A}{dT} + \frac{dH_B}{dT}\right)$$

$$\text{or,}\quad \frac{d}{dT}(\Delta H) = \left(C_{P_C} + C_{P_D}\right) - \left(C_{P_A} + C_{P_B}\right)$$

$$\text{or,}\quad \frac{d}{dT}(\Delta H) = \Sigma C_{P_{\text{product}}} - \Sigma C_{P_{\text{reactant}}} = \Delta C_P$$

$$\text{or,}\quad \int_{\Delta H_1}^{\Delta H_2} d(\Delta H) = \int_{T_1}^{T_2} \Delta C_P dT$$

$$\text{or,}\quad \Delta H_2 - \Delta H_1 = \int_{T_1}^{T_2} \Delta C_P dT. \qquad ....(7.2)$$

For some gases C_P's dependence on T is negligible, but for some other substances

$$C_P = a + bT + cT^2 + ...$$

Similarly it can be proved that

$$\Delta U_2 - \Delta U_1 = \int_{T_1}^{T_2} \Delta C_V dT. \qquad ....(7.3)$$

Equations (7.2) and (7.3) are Kirchoff's equations; they tell us how energy change during a process (chemical or physical change) depends on temperature. An example is —

Latent heat of fusion of ice (ΔH) = 1440 cal/mole at 273°A and

We want to find it at – 10°C i.e. 263°A

$$\Delta H_{273} - \Delta H_{263} = \int_{263}^{273} \left(C_{P_{\text{water}}} - C_{P_{\text{ice}}} \right) dT$$

$$1440 - \Delta H_{263} = \int_{263}^{273} (18 - 8.7)\, dT = 9.3\ (273 - 263).$$

$$\text{So} \quad \Delta H_{263} = 1440 - 93 = 1347 \text{ cal/mole.}$$

POINTS TO REMEMBER

Thermochemistry deals in heat changes associated with physical and chemical changes. Depending on the nature of change these heat changes are named. Thus heat changes may be called — *heat of reaction; heat of formation; heat of combustion; heat of neutralisation; heat of solution* etc. As 'H' is heat content, heat changes are symbolised as ΔH. For *exothermic processes product has less energy and so* ΔH is *negative* for *endothermic* processes ΔH *is positive.* For reactions at constant volume *heat of reaction is* ΔU, because $\Delta H = \Delta U$ when no P-V work is done. They are related as $\Delta H = \Delta U + \Delta n\, RT$ where 'Δn' is the difference in number of moles between *gaseous* products and *gaseous* reactants. These 'ΔH' and ΔU values change with temperature following *Kirchoff's equation*, which is

$$\Delta H_2 - \Delta H_1 = \int_{T_1}^{T_2} \Delta C_P dT \quad \text{and} \quad \Delta U_2 - \Delta U_1 = \int_{T_1}^{T_2} \Delta C_V dT.$$

QUESTIONS

SHORT TYPE

1. What is a thermochemical equation ?
2. For the following reactions state whether ΔH will be > ; < or = ΔU :
   (i) C (s) + O_2 (g) = CO_2 (g)
   (ii) NH_4Cl (s) = NH_3 (g) + HCl (g)
   (iii) 2HI (g) = H_2 (g) + I_2 (g)
   (iv) CO (g) + Cl_2 (g) = $COCl_2$ (g).
3. For what purpose heat of formation data are utilised ?
4. What is the utility of heat of combustion data ?
5. For any strong acid - strong base reaction heat of neutralisation value is same — why ?
6. From heat of neutralisation value of a weak acid, strong base reaction what information regarding weak acid can be known ?
7. What is principle of thermoneutrality ?
8. Distinguish between integral heat of dilution and heat of dilution ?
9. Define standard state. What is standard heat of formation.
10. Name the equation which shows variation of heat of reaction with temperature.

ESSAY TYPE

1. Define two different types of heats of reaction and establish a relation between them when their values will be same ?
2. State the two laws thermochemistry and show that they are corollary to the first law.
3. Name two instruments used to determine heat of reaction. Describe one of them in detail.
4. What is standard heat of formation. Explain the application of such data.
5. 'Heat of neutralisation for reaction between HCl and NaOH is the same as that between HNO_3 and KOH but is different for CH_3COOH and NaOH' — explain in details.
6. Derive that equation which shows variation of heat of reaction with temperature and show some of its applications.

PROBLEMS

1. The heat of combustion of carbon is 97,000 cals. and that of carbon monoxide 68,000 cals. Calculate the heat of formation of carbon monoxide.

   [*Hints.* Given 1. $C + O_2 = CO_2$; $\Delta H_1 = -97000$

   $$2.\ CO + \frac{1}{2} O_2 = CO_2 ; \Delta H_2 = -68000$$

   Eq. (1) – Eq. (2) is $C + \frac{1}{2} O_2 = CO$; $\Delta H = -29000$ cal.]

2. Calculate the heat change accompanying the following reaction :

   $$CaCl_2 + 2Na = 2NaCl + Ca$$

   Given the heat of formation of $CaCl_2$ = 1,91,000 cals and of NaCl = 97,700 cals

   [*Hints.* Given 1. $Ca + Cl_2 = CaCl_2$; $\Delta H_1 = -1{,}91{,}000$ cals

   $$2.\ Na + \frac{1}{2} Cl_2 = NaCl ; \Delta H_2 = -97{,}700 \text{ cal}$$

   Eq. (2) × 2 – Eq. (1) = $CaCl_2 + 2Na = 2NaCl + Ca$ $\quad \Delta H = 2(-97{,}700) - (-1{,}91{,}000)$.

3. The heats of combustion of carbon, hydrogen and methane are 96,000 cals 68,400 cals and 2,12,000 cals respectively. Calculate the heat of formation of methane.

[*Hints.* 1. $C + O_2 = CO_2$; $\Delta H_1 = -96{,}000$ cal

2. $H_2 + \frac{1}{2} O_2 = H_2O$; $\Delta H_2 = -68{,}400$

3. $CH_4 + 2O_2 = CO_2 + 2H_2O$; $\Delta H_3 = -2{,}12{,}000$

Eq. (1) + 2 × Eq. (2) – Eq. (3) $C + 2H_2 = CH_4$; $\Delta H = \Delta H_1 + 2\Delta H_2 - \Delta H_3$.]

4. The heat of formation of methane at constant pressure is 18,500 cals. at 25°C. Calculate the heat of reaction at constant volume.

[*Hints.* $C\ (s) + 2H_2\ (g) = CH_4\ (g)$; $\Delta n = 1 - 2 = -1$

$\Delta H = \Delta U + \Delta nRT$; Given $\Delta H = 18{,}500$; $\Delta n = -1$; R = 1 987.]

5. The heat of combustion of carbon monoxide at constant volume and at 17°C is 67,710 Cals. Calculate its heat of combustion at constant pressure.

[*Hints.* $CO + \frac{1}{2} O_2 = CO_2$; $\Delta n = -\frac{1}{2}$; Now follow problem '4'.]

6. Calculate the heat of formation of ethylalcohol from the following data :

$$C_2H_5OH + 3O_2 = 2CO_2 + 3H_2O + 3{,}26{,}695 \text{ cals}$$

$$C + O_2 = CO_2 + 94{,}052 \text{ cals}$$

$$H_2 + O = H_2O + 68{,}317 \text{ cals}$$

[*Hints.* Eq. (2) × 2 + Eq. (3) × 3 – Eq. (1) is desired equation.]

7. Calculate the heat of formation of carbon disulphide. The heat of combustion of CS_2, sulphur and carbon are 2,65,100 cals and 71,080 cals and 94,300 cals respectively.

[*Hints.* Try as before.]

OBJECTIVE TYPE

1. State whether following statements are 'true' or 'false'.

(a) The equation $C + O_2 = CO_2 + 94.4$ K.cal; is a thermo chemical equation.
(b) For exothermic reaction ΔH is negative.
(c) For the reaction $H_2\ (g) + I_2\ (g) \rightleftharpoons 2HI\ (g)$; $\Delta H = \Delta U$.
(d) Standard enthalpy of an element is zero.
(e) Hess's law and First law are basically same.
(f) Heats of neutralisation for (NaOH + HCl) reaction and (NH_4OH + HCl) reaction are same.
(g) Flame temperature can be calculated from heat of combustion.
(h) Calorific value of food is calculated from heat of solution.

2. Match terms of column A with those of column B.

A	B
(a). Kirchoff.	(a) Law of thermochemistry.
(b) Heat of combustion.	(b) Effect of T on ΔH.
(c) Born-Haber Cycle.	(c) Salt solution.
(d) Principle of thermoneutrality.	(d) Calorific-value of food.
(e) Lovoisier.	(e) Hess's law.

Answers :

1. (a) F; (b) T; (c) T; (d) T; (e) F; (f) F; (g) T; (h) F;

 2. a – b; b – d; c – e; d – c; e – a;

Chapter 8

THERMODYNAMICS — SECOND LAW

8.1 THE NEED

The very fact that thermodynamics has a second law, indicates that first law is incomplete. In fact limitations of first law are the need for the second law. So let us see the limitations first. First law tells that energy can neither be created nor be destroyed but its different forms are interconvertible and there is nothing to object if two bodies exchange their energies. In fact if a ship sailing in the sea collects energy from sea-water and convert it to energy; first law will not object. But experience tells that such a conversion is not possible since temperature of outer surface of ship and sea-water is same. Experience also teaches us that spontaneous changes are always unidirectional and during such changes energy flows down the hill. Thus limitation of first law is its silence regarding direction of spontaneous energy flow. That is why we needed another law, which would tell us about the direction of flow of energy. And this law is second law. Like first law it is also the result of human experience.

8.2 THE LAW

Second law of thermodynamics can be stated in many ways. Some popular statements are written below.

(a) Heat spontaneously flows from high to low temperature

[**Note.** A heat pumping device e.g. refrigerator can extract heat from low temperature region and reject it at high temperature region; but that is non-spontaneous.]

(b) According to Clausius, *It is impossible for any self-acting machine, unaided by any external agency to convey heat from low to high temperature.*

[**Note.** This statement also tells about the impossibility of spontaneous flow of energy from lower to higher potential.]

(c) Kelvin stated, *It is impossible for a cyclic process to convert heat in to work without the simultaneous transfer of heat from a body at high temperature to a body at low temperature*

[**Note.** This statement means an engine which converts heat to work must work between two reserviors. If it isothermally takes heat from a reservior its shape will be distorted. For further work it must be given original shape i.e. process should be cyclic. If this *bringing back to original state* operation is done on the same reservior then total work will be zero. So in an isothermal cyclic process total work is zero. Hence to have some net work, it must be taken to some other reservior at low temperature, where the engine can spend its heat to do some work. Thus to have some net work in a cyclic process, an engine must work between two reserviors.]

(d) Planck stated second law as — *It is impossible for a self-acting machine to work in a complete cycle and to produce no effect except raising of a weight and cooling a reservior.*

[**Note.** Raising a weight means doing some work. This statement means that when a machine takes heat from source it will not do work alone. Some other effect e.g. expansion of the machine, distortion of its shape etc. will also definitely occur. For this reason we are to do some work to bring the machine back to its original state. The process becomes cyclic then. And here forward and backward paths must be different otherwise net work will be zero. Backward path is so chosen (see - Carnot's cycle) that we get some net work and of course this work is less than heat absorbed.]

(e) *It is impossible to cool a body spontaneously below the coldest of its surrounding.* [**Note.** This statement means, so long as temperature of the body is higher than surrounding, it will reject heat.]

Apparently though the statements differ in the mode of expression, basically their theme is same. And these are —

(i) Impossibility of flow of heat spontaneously from low to high temperature and

(ii) Impossibility of converting heat to work fully in a cyclic process.

8.3 ENGINE; CYCLIC PROCESS; EFFICIENCY

Engine is a device, which converts heat in to work. Now since after absorbing heat shape of the engine also changes, so to regain its original shape some work is to be done. The net result is — 'the process becomes cyclic.' In a cyclic process there are forward and backward paths. If both paths become same i.e. isothermal cyclic process, total work will be zero. Therefore to have some net work the two paths must be different. In any case, since engine gives work (equivalent to heat absorbed) in the forward path and we are to do work (equivalent to heat rejected) in the backward path, the net work obtained is always less than heat absorbed per cycle. If W is the net work and Q is heat absorbed per cycle, then efficiency of the engine $(\eta) = \dfrac{W}{Q}$. Thus efficiency of an engine is the fraction of absorbed heat converted to work. It is also expressed in terms of percentage.

8.4 CARNOT'S CYCLE

Sadi Carnot. a French engineer theoretically calculated maximum efficiency of an engine working in a cyclic process. His imaginary cycle is known as Carnot's cycle. Carnot's engine is nothing but one mole of an ideal gas taken in a cylinder, fitted with a weightless and frictionless piston. There are two heat reserviors, one at high temperature T_1, known as source and another at low temperature T_2, known as sink. Now the following reversible cyclic process consisting of the following four steps are carried out.

STEP–I

Initial state of the gas i.e. engine material is P_l, V_1 and T_l. It is placed over the source at T_l. External pressure is now decreased infinitesimally, gas will expand infinitesimally and work will be done by the gas. But its temperature will not fall as it is kept on the heat reservior from which it will absorb heat equivalent to the work done. The process is repeated till volume increases to V_2. This is a case of isothermal reversible expansion and is shown by the AB-portion of the adjacent PV diagram (fig. 8.1). If during this expansion (from P_l, V_l, T_1 to P_2, V_2, T_1) Q_1 is heat absorbed and W_1 is work done, then

$$W_1 = Q_1 = RT_1 \ln \frac{V_2}{V_1}.$$

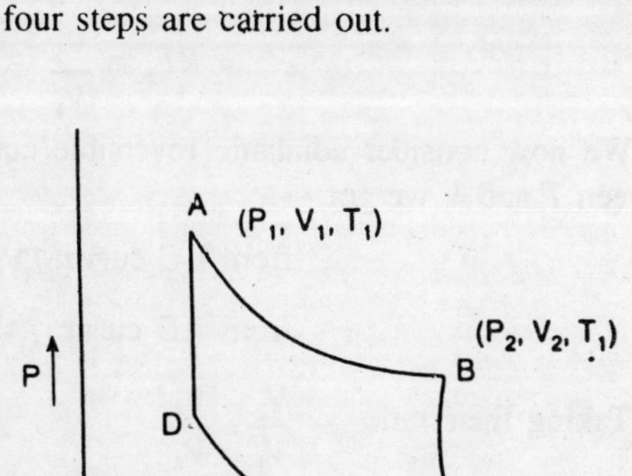

Fig. 8.1

STEP–II

The engine is now covered by adiabatic wall to prevent any radiation loss; and is taken to sink at T_2, where it is allowed to do reversible expansion work. Its temperature will fall and when it becomes T_2 it stops. So here state changes from (P_2, V_2, T_1) to (P_3, V_3, T_2). This is shown by BC curve. Here no heat ex-change with surrounding occurs and work done $W_2 = C_v (T_1 - T_2)$.

STEP–III

After step-II temperature of engine becomes same with the sink so it is unable to do any more work by loosing heat. To make it workable we must take it back to source by doing work on it (negative sign; W_1 and W_2 are + *Ve*). We can not do it by adiabatic compression and subsequent isothermal compression i.e. paths $\overrightarrow{CB}$ and $\overrightarrow{BA}$; because net work will be zero then. That is why we remove the adiabatic wall and compress the gas isothermally and reversibly to such a state (P_4, V_4, T_2) that on subsequent adiabatic reversible compression we can reach initial state. This is shown by CD portion of the diagram. During this compression if Q_2 heat is rejected, then

$$Q_2 = W_3 = RT_2 \ln \frac{V_4}{V_3}.$$

[**Note.** $V_4 < V_3$; so W_3 is $-ve$; it is work done on the system; Negative Q_2 means heat is rejected.]

STEP–IV

The engine is again covered by adiabatic wall and by isothermal reversible compression it is taken back to the initial state (AD curve). Thus the cycle is completed. Here work done = $W_4 = C_v (T_2 - T_1)$. It is also negative work as this has been done on the system.

Hence net work obtained per cycle

$$\begin{aligned} W &= W_1 + W_2 + W_3 + W_4 \\ &= RT_1 \ln \frac{V_2}{V_1} + C_V (T_1 - T_2) + RT_2 \ln \frac{V_4}{V_3} + C_V (T_2 - T_1) \\ &= RT_1 \ln \frac{V_2}{V_1} - RT_2 \ln \frac{V_3}{V_4}. \qquad \text{....(8.1)} \end{aligned}$$

We now consider adiabatic reversible curves BC and AD. Using adiabatic relation between T and V we get —

from BC curve $T_1V_2^{\gamma-1} = T_2V_3^{\gamma-1}$ and

from AD curve $T_1V_1^{\gamma-1} = T_2V_4^{\gamma-1}$.

Taking their ratio $\frac{V_2}{V_1} = \frac{V_3}{V_4}$.

So equation (8.1) becomes —

$$W = RT_1 \ln \frac{V_2}{V_1} - RT_2 \ln \frac{V_2}{V_1} = R \ln \frac{V_2}{V_1} (T_1 - T_2). \qquad \text{....(8.2)}$$

Heat absorbed from source $Q_1 = RT_1 \ln \frac{V_2}{V_1}$. (vide : step-I) ....(8.3)

From equations (8.2) and (8.3),

efficiency of engine $\eta = \frac{W}{Q_1} = \frac{T_1 - T_2}{T_1}$.

In the entire cycle Q_1 heat was absorbed W work was done and Q_2 heat was rejected so from law of conservation of energy

$Q_1 = W + Q_2$

i.e. $W = Q_1 - Q_2$.

Hence efficiency $$\eta = \frac{W}{Q_1} = \frac{Q_1 - Q_2}{Q_1} = \frac{T_1 - T_2}{T_1}. \qquad ....(8.4)$$

This is the expression for efficiency of a Carnot's engine.

8.5 CRITICAL COMMENTS

From the expression for efficiency (eq. 8.4) we see that efficiency of Carnot's engine does not depend on nature of engine material i.e. which gas and what type (ideal or real) has been used. It simply depends on temperature difference between source and sink. The greater this difference, the greater is the efficiency.

If $T_2 = 0°A$ then $\eta = 1$; so

an engine will be 100% efficient only when temperature of sink is 0°A. But as 0°A is unattainable so 100% efficient engine can never be constructed. The idea of Kelvin's scale of temperature also came from this. Zero degree absolute can be defined as the temperature at which when an engine works, its efficiency becomes 100%.

8.6 CARNOT'S THEOREMS

As a corollary to equation (8.4) two theorems can be constructed. These are popularly known as Carnot's theorem.

Theorem–I :

All reversible engines working between same source and sink will have same efficiency.

Theorem–II :

Efficiency of a reversible engine is always greater than that of an irreversible engine working between same source and sink.

Though both the theorems are understood from efficiency calculation in Carnot's cycle, yet they can be proved in a separate way by considering an impossible cyclic process. We discuss proof of theorem –I below :

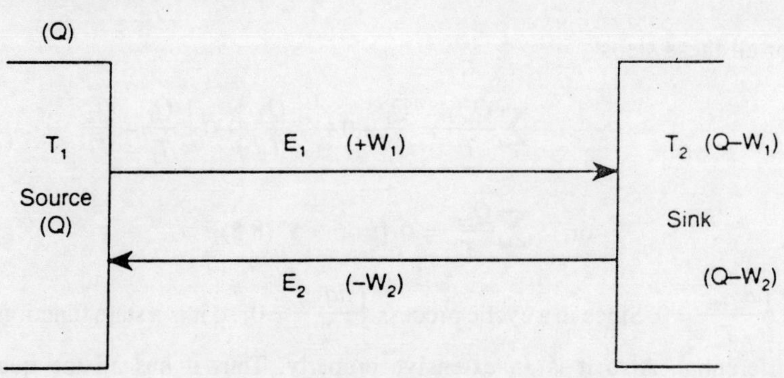

Fig. 8.2

Let us consider two engines E_1 and E_2 working between same source (T_1) and sink (T_2). Let engine E_1 gives W_1 work after absorbing Q heat; but engine E_2 gives W_2 work after absorbing same heat i.e. their efficiencies are different. Now let engine E_1 acts between source and sink in forward direction and E_2 in backward direction. E_1 absorbs Q heat from source performs W_1 work and rejects $(Q - W_1)$ heat at sink. If engine E_2 also works in forward direction it would reject $(Q - W_2)$ heat at sink after absorbing Q heat from source and doing W_2 work. But now it is working in backward direction as a refrigerater. So to absorb $(Q - W_2)$ heat from sink we are to do W_2 (negative work) on it, so that it can reject Q heat at source. Now net work obtained from this cycle $= (W_1 - W_2)$ and

net heat **loss** by the sink $= (Q - W_2) - (Q - W_1)$

$= (W_1 - W_2)$.

So in a cyclic process heat lost by a reservior becomes equal to work done in a cycle. This is against second law. Hence W_1 must be equal to W_2 i.e. efficiency of two reversible cycles must be same.

Theorem-II can also be proved in a similar manner.

8.7. ENTROPY – ROOT OF ALL EVILS ?

Second law of thermodynamics gave us idea of a very important quality of matter, which is entropy. The question — 'what is entropy' is really ridiculous — because entropy is an abstract idea. It can be felt; it can be understood but it is difficult to define it in the usual way of giving definition. To have an idea about entropy let us do some calculation first and analyse the results.

(i) $\oint \frac{dq_{rev}}{T} = 0$.

Proof :

From Carnot's cycle we have seen that $\frac{Q_1 - Q_2}{Q_1} = \frac{T_1 - T_2}{T_1}$ [eq. (8.4)]

$$\text{So} \quad 1 - \frac{Q_2}{Q_1} = 1 - \frac{T_2}{T_1}$$

$$\text{or,} \quad \frac{Q_2}{T_2} = \frac{Q_1}{T_1}. \qquad \text{....(8.5)}$$

Now first step of Carnot's cycle involves absorption of Q_1 heat at T_1 and 3rd step involves rejection of Q_2 heat at T_2 Second and fourth steps are adiabatic involving no heat change.

So for all these steps

$$\sum \frac{Q_{rev}}{T} = \frac{Q_1}{T_1} + 0 + \left(-\frac{Q_2}{T_2}\right) + 0 = \frac{Q_1}{T_1} - \frac{Q_2}{T_2}$$

$$\text{or,} \quad \sum \frac{Q_{rev}}{T} = 0 \text{ [from eq. (8.5)].} \qquad \text{....(8.6)}$$

Thus $\oint \frac{dq_{rev}}{T} = 0$. Since in a cyclic process $\oint \frac{dq_{rev}}{T} = 0$; so it is a state function and hence perfect differential. Also it is an extensive property. Thus it has all the qualities of a thermodynamic parameter like U and H. Let us call it another thermodynamic parameter, entropy (S). So entropy (S) is mathematically defined as

$$S = \int ds = \int \frac{dq_{rev}}{T}. \qquad \text{....(8.7)}$$

8.8 CALCULATION OF ENTROPY CHANGE DURING SOME PROCESS

During calculation of entropy change we should remember that entropy is a state function and hence we should ignore the path. As $\int ds = \int \frac{dq_{rev}}{T}$ we are the to calculate R. H. S. assuming the change as reversible(actually it may or may not be reversible). Some examples are given below :

(a) Heating of a solid (or liquid) : Let 'm gm of a solid (or liquid) of specific heat c is heated from T_1 and T_2. If it is a reversible engine then in each step dq heat is absorbed and $dq_{rev} = m\, c\, dT$. So entropy change

$$\int_{S_1}^{S_2} ds = \int_{T_1}^{T_2} mc \frac{dT}{T}$$

$$\text{or, } (S_2 - S_1) = \Delta S = m\, c \ln \frac{T_2}{T_1}. \qquad \text{....(8.8)}$$

(b) Isothermal expansion of an ideal gas : Let one mole of an ideal gas expands from (P_1, V_1, T_1) state to (P_2, V_2, T_1) state isothermally. Entropy change can be calculated using the same formula $\int ds = \int \frac{dq_{rev}}{T}$.

From 1st law $\quad dq_{rev} = dU + P_{int} dV$.

For isothermal case $\quad dU = 0 \quad [\because dU = C_V\, dT]$

Also internal pressure $\quad P_{int} = \frac{RT}{V}$.

$$\text{So } dq_{rev} = RT.\frac{dV}{V}.$$

$$\text{Hence} \quad \int_{S_1}^{S_2} ds = R \int_{V_1}^{V_2} \frac{dV}{V}$$

$$\text{or, } (S_2 - S_1) = \Delta S = R \ln \frac{V_2}{V_1}. \qquad \text{....(8.9)}$$

For isothermal change $\quad \frac{V_2}{V_1} = \frac{P_1}{P_2}$.

$$\text{Hence} \quad \Delta S = R \ln \frac{P_1}{P_2}. \qquad \text{....(8.10)}$$

(c) Non-isothermal expansion of an ideal gas : Let one mole of an ideal gas expands non-isothermally from (P_1, V_1, T_1) state to (P_2, V_2, T_2) state. Entropy change

$$\int_{S_1}^{S_2} ds = \int \frac{dq_{rev}}{T} = \int \frac{C_V dt + PdV}{T}.$$

$$= C_V \int_{T_1}^{T_2} \frac{dT}{T} + R \int_{V_1}^{V_2} \frac{dV}{V}. \qquad \left[\because P = \frac{RT}{V}\right]$$

$$\text{So } (S_2 - S_1) = \Delta S = C_V \ \ln \frac{T_2}{T_1} + R \ln \frac{V_2}{V_1}. \qquad ....(8.11)$$

Putting $V_2 = \frac{RT_2}{P_2}$ and $V_1 = \frac{RT_1}{P_1}$

$$\Delta S = (C_V + R) \ln \frac{T_2}{T_1} + R \ln \frac{P_1}{P_2}$$

$$= C_P \ln \frac{T_2}{T_1} - R \ln \frac{P_2}{P_1}. \qquad ....(8.12)$$

If above integration are done without limit we get

$$S = C_V \ln T + R \ln V + \text{Const.} \qquad ....(8.13)$$

$$S = C_P \ln T - R \ln P + \text{Const.} \qquad ....(8.14)$$

These are expressions for absolute entropy.

(d) Entropy change during transition : Transition of state e.g. melting, boiling, sublimation etc. at normal transition temperature are reversible processes. Hence tatent heats of transitions are q_{rev}. So entropy change /mole during transition $\Delta S = \frac{L_t}{T}$

[L_t = latent heat of transition/mole.]

8.9 MORE RESULTS OF ENTROPY CALCULATIONS

It can be shown by calculation that during any irreversible change, entropy of the universe (i.e. system + surrounding) increases. Now all spontaneous processes are irreversible and all natural processese. e.g. flow of water from high level to low level; flow of heat from high temperature to low temperature etc. are spontaneous. So entropy of universe is continuously increasing. Sometimes first and second law are jointly stated as – 'while total energy of universe is constant, total entropy of universe is increasing.' The first part of this statement is first law; while latter part is second law.

Let we have two Carnot's engine. The first works between T_1 and T and the second between T_2 and T. We have learnt that if $T_2 < T_1$ less work will be done by second engine even if it also absorbs same heat Q from source. Entropy increase for 1st engine = $\frac{Q}{T_1}$ and that for second = $\frac{Q}{T_2}$. As $T_2 < T_1$ entropy of second engine is more and more entropy makes it less efficient.

8.10 CONCEPT OF ENTROPY

We have already said that idea of entropy is abstract. Let us now try to have a picture of it on the basis of calculations and discussions of sections 8.6 to 8.8. We have seen that entropy is something, which increases during heating, during expansion and during melting and vapourisation. We have also seen that entropy of universe is increasing and finally we found that increase in entropy decreases work-performing ability. Now heating increases random kinetic motion of molecules, expansion brings disorder and fusion and vapourisation brings irregularity. So entropy is a measurement of randomness, disorder and irregularity present in a system. Disorder of universe is always increasing. If a perfectly cleaned room is kept under lock and key it becomes automatically unclean and dirty; thus entropy increase is a spontaneous process for the universe. Finally the more the indiscipline, the less will be work-performing-ability. That is why increase in entropy decreases efficiency of an engine. If temperature of a system is lowered disorder and so also entropy decreases. At absolute zero, there is no disorder and so entropy of a pure crystal absolute zero is zero. At some other temperature every system has some irregularity in it. Entropy (S) measures this. If 'S' is entropy at T°A then it can be shown that unavailable energy = $T \times S$. Thus entropy (S) is capacity factor for unavailable energy.

8.11 WORK FUNCTION — FREE ENERGY

We have just seen that if 'S' is the entropy of a substance at T°A, then TS amount of energy is unavailable for work. If U is its internal energy then energy available for work = $(U - TS)$. This is called *work function* or *Helmholtz free energy* (F). So

$$F = U - TS$$

It is a state function, perfect differential and extensive property.

However, when enthalpy (H) is considered as total energy, then energy available for work is ($H - TS$). This is called *Gibbs free energy* or *Gibbs potential* or *thermodynamic potential* (G). So

$$G = H - TS$$

It is also a state function, perfect differential and extensive property.

8.12 SOME RELATIONS

From first law $dU = dq_{rev} - PdV.$

But $dq_{rev} = TdS.$

So $dU = TdS - PdV.$ ....(8.15)

Again $H = U + PV$

or, $dH = dU + PdV + VdP$

from eq. (8.15) $dH = TdS + VdP.$ ....(8.16)

Also $F = U - TS$

or, $dF = dU - TdS - SdT.$

From eq. (8.15) $dF = -PdV - SdT.$ ....(8.17)

Finally $G = H - TS$

or, $dG = dH - TdS - SdT.$

From eq. (8.16) $dG = VdP - SdT.$ ....(8.18)

These equations are very much general and applicable to all substances under every condition.

8.13 GIBBS-HELMHOLTZ EQUATION

We have learnt that no substance is capable of utilising whole of its energy content to work, because of presence of disorder i.e. entropy in it. If H is the total energy, then available energy 'G' is

$$G = H - TS.$$

But from eq. (8.18) $S = -\left(\frac{dG}{dT}\right)_P$.

$$\text{So} \quad G = H + T\left(\frac{dG}{dT}\right)_P \qquad \text{....(8.19)}$$

$$\text{or,} \quad \Delta G = \Delta H + T.\ \frac{d}{dT}(\Delta G)_P\ . \qquad \text{....(8.20)}$$

Similarly $F = U - TS.$

From eq. (8.17) $S = -\left(\frac{dF}{dT}\right)_V$

$$\text{So} \quad F = U + T\ \frac{d}{dT}(F)_V \qquad \text{....(8.21)}$$

$$\text{or,} \quad \Delta F = \Delta U + T\ \frac{d}{dT}(\Delta F)_V\ . \qquad \text{....(8.22)}$$

Equations (8.19) to (8.22) are known as Gibbs-Helmholtz equation. **Another form of this equation can be obtained from Ist law.** Let a Carnot's engine works between T and $T - dT$. If it absorbs dq heat from source and performs dw amount of work, then

$$\frac{dw}{dq} = \frac{T-(T-dT)}{T} = \frac{dT}{T} \qquad \text{[Vide Carnot's cycle]}$$

$$\text{or,} \quad dq = T.\ \frac{d}{dT}(w).$$

Again from first law $dq = dU + dw$

$$\text{So} \quad T.\ \frac{d}{dT}(w) = dU + dw$$

$$\text{or,} \quad d(-w) = dU + T.\ \frac{d}{dT}(-w). \qquad \text{....(8.23)}$$

This form is similar to eq. (8.22). It is also Gibbs-Helmholtz equation.

8.14 CLAPEYRON EQUATION

Transition of state e.g. melting, boiling etc. takes place at a particular temperature. This transition temperature depends on atmospheric pressure. Variation of transition temperature with pressure is given by Clapeyron equation. It can be derived by considering an imaginary cyclic process similar to Carnot's cycle. We consider vapourisation as an example of transition.

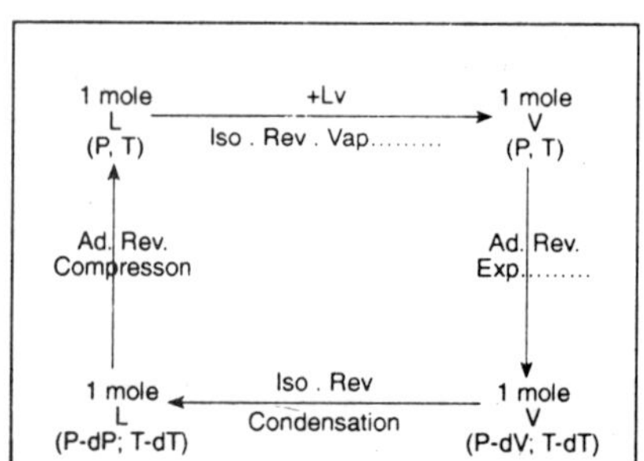

Fig. 8.3

Let one mole of a liquid be at its boiling point T at pressure P. It is now supplied latent heat of vapourisation so that the liquid is isothermally reversibly vaporised. Heat absorbed = L_v (latent heat) and work done = $P(V_g - V_e)$. The vapour is now adiabatically reversibly expanded so that pressure becomes $(P - dP)$ and temperature $(T - dT)$. Then it is isothermally reversibly condensed. Work $= (P - dP)(V_l - V_g)$. Finally the liquid is adjabatically reversibly compressed to the orginal state. This is a cyclic process similar to Carnot's cycle. In such cycles total work is the summation of works in isothermal steps since adiabatic works cancel each other. So total work

$$W = P(V_g - V_l) + (P - dP)(V_l - V_g)$$
$$= dP(V_g - V_l).$$

Heat absorbed $= L_V$.

So from Carnot's engine $\dfrac{dP\left(V_g - V_l\right)}{dT} = \dfrac{dT}{T}$ or, $\dfrac{dP}{dT} = \dfrac{L_V}{T\left(V_g - V_l\right)}$ .....(8.24)

This is **Clapeyron equation.**

8.15 APPLICATION

For any type of transition Clapeyron equation is

$$\frac{dP}{dT} = \frac{L_t}{T(V_2 - V_1)}. \qquad ....(8.25)$$

L_t = Latent heat of transition /mole

Putting $V_2 = \dfrac{M}{\rho_2}$ and $V_1 = \dfrac{M}{\rho_1}$ where M = mol. wt and ρ = density, we get

$$\frac{dP}{dT} = \frac{L_t/M}{T\left[\dfrac{1}{\rho_2} - \dfrac{1}{\rho_1}\right]} = \frac{l_t}{T\left[\dfrac{1}{\rho_2} - \dfrac{1}{\rho_1}\right]}. \qquad ....(8.26)$$

l_t = latent heat of transition/gm.

Now for ice → water transition

$$\frac{dP}{dT} = \frac{L_t}{T\left(V_{water} - V_{ice}\right)}.$$

As $V_{ice} > V_{water}$, so $\dfrac{dP}{dT}$ is negative. If we increase pressure dP becomes positive i.e. dT is negative. Hence melting point of ice will decrease with increase in pressure.

8.16 CLAUSIUS-CLAPEYRON EQUATION

Clausius modified Clapeyron equation for those cases, where final phase (after transition) is vapour i.e. cases of vapourisation and sublimations. Modified equation is called Clausius-Clapeyron equation. Let us consider the case of vapourisation. Here Clapeyron equation is —

$$\frac{dP}{dT} = \frac{L_V}{T\left(V_g - V_l\right)}.$$

As $V_g >> V_l$ so V_l can be neglected in comparison to V_l; though it is an approximation.

Hence

$$\frac{dP}{dT} = \frac{L_V}{TV_g}.$$

Clausius further assumed that the vapour is an ideal gas and he put $V_g = \frac{RT}{P}$ this is another approximation.

$$\text{So} \quad \frac{dP}{dT} = \frac{L_V P}{RT^2}$$

$$\text{or,} \quad \frac{dP}{P} = \frac{L_V}{R} \cdot \frac{dT}{T^2}.$$

If T_1 and T_2 are boiling points at pressure P_1 and P_2 then an integration —

$$\int_{P_1}^{P_2} \frac{dP}{P} = \frac{L_V}{R} \int_{T_1}^{T_2} \frac{dT}{T^2}.$$

[Hence L_V is taken as independent of temperature ; this is another approximation.]

$$\text{or,} \quad \ln \frac{P_2}{P_1} = \frac{L_V}{R}\left[-\frac{1}{T_2} + \frac{1}{T_1}\right] = \frac{L_V}{R} \cdot \frac{T_2 - T_1}{T_1 T_2}. \qquad \text{....(8.27)}$$

This is Clausius -Clapeyron equation; from this equation we can calculate boiling point of a liquid at any pressure, provided we know its boiling point at some other pressure.

8.17 TROUTON'S RULE

Differential form of Clausius-Clapeyron equation is $\frac{dP}{P} = \frac{L_V}{R} \cdot \frac{dT}{T^2}$.

If it is integrated without limit, then we get

$$\int \frac{dP}{P} = \frac{L_V}{R} \cdot \int \frac{dT}{T^2}$$

$$\text{i.e.} \quad \ln P = -\frac{L_V}{RT} + \text{constant.}$$

If T = normal boiling point (T_b) then P = 1 atm.

$$\text{Then} \quad 0 = -\frac{L_V}{RT_b} + \text{constant.}$$

$$\text{So} \quad \frac{L_V}{T_b} = \text{constant.}$$

$$\text{But} \quad \frac{L_V}{T_b} = \text{Molar entropy change during vapourisation.}$$

Hence molar entropy change during vapourisation of any liquid is constant. This statement is called **Trouton's rule**. For simple non-associated liquids, the value of this constant is ~ 21 cal. deg^{-1} mole^{-1}. For associated liquid with H – bonding e.g. water the value is high. For water it is ~ 26 cal.deg^{-1} mole^{-1}.

POINTS TO REMEMBER

The silence of first law regarding direction of low of energy led to the prediction of second Law — whose central theme is — '*Energy flows down the hill*' and '*in a cyclic process energy can never be fully converted to work.*' Sadi Carnot calculated efficiency of an engine $\left[\eta = \frac{W}{Q} = \frac{T_1 - T_2}{T_1}\right]$. It was found that efficiency depends only on temperature difference between source and sink and on nothing else.

The term $\int \frac{dq_{rev}}{T}$ is found to be a state function; perfect differential and extensive property. It was named as *entropy* 'S'. It increases during expansion; heating; vaporisation; fusion. For universe it is always increasing. And its increase decreases efficiency. So it was concluded that 'S' measures *indiscipline; irregularity; disorder; chaos* etc. Which are inherently possessed by every system at a temperature other than 0°A. Only at 0°A, a system becomes perfectly ordered and 'S' becomes zero. At T°A if 'S' is entropy then TS energy is *unavailable* for work. Available energies are $(U - TS)$ and $(H - TS)$. These are called *work function* and *free energy.*

QUESTIONS

SHORT TYPE

1. What is the need for the second law ?
2. In a refrigerator heat is flown from low temperature to high temperature – does it contradict second law ?
3. What is an engine?
4. Define efficiency of an engine. On what factors does it depend ?
5. When an engine will be hundred per cent efficient ?
6. Entropy of universe always increases – explain.
7. Entropy becomes zero at absolute zero – explain.
8. What is your idea about entropy of a substance ?
9. What is free energy ? Why it is so called ?
10. Why melting point of ice decreases with increase in pressure ?
11. What is Trouton's rule ?
12. Name the equation, which shows variation of boiling point of a liquid with atmospheric pressure.

ESSAY TYPE

1. State second law of thermodynamics in three different ways and show that each way bears the same meaning.
2. Show that efficiency of an engine depends only on temperature difference between source and sink.
3. Write down the expression for efficiency of a Carnot's engine and hence arrive at idea of absolute zero.
4. Prove that all reversible engines working between same source and sink will have same efficiency.
5. From the expression for efficiency of a Carnot's engine arive at the idea of entropy.
6. Derive an expression for absolute entropy. Is it determinable ?
7. Elucidate the concept of entropy.
8. Derive Gibbs-Helmholtz equation.
9. From first principle derive Clapeyron equation.
10. What is Clapeyron equation ? How was it modified by Clausius ? Arrive at Trouton's rule.

PROBLEMS

1. An ideal heat engine operates between T_2 and T_1 per cycle, q_1 is -4.5×10^6 cal and ΔS_1 is $-15{,}000$ cal K^{-1}; T_2 is 300°C. Calculate (a) efficiency of the engine and (b) work done per cycle.

   [*Hints.*

   $$\Delta S_1 = \frac{q_1}{T_1} \text{ ; find } T_1 \text{ in °A ; } \quad T_2 = 573\text{°A}$$

   $$\eta = \frac{T_1 - T_2}{T_1}, \quad \text{also } w = \eta \times Q_1 .]$$

2. One mole of an ideal monoatomic gas initially at STP expands isothermally and irreversibly to 44.8 litres under conditions such that $W' = 100$ cal. Calculate (i) ΔS for gas and surroundings (ii) ΔG for the gas.

[*Hints*.

$$\Delta S = R \ln \frac{V_2}{V_1} = 1.987 \ln \frac{44.8}{22.4} \text{ cal. k}^{-1}$$

$$\Delta G = RT \ln \frac{V_1}{V_2}.]$$

3. One mole of an ideal gas in contact with a heat reservoir at 25°C expands isothermally from 100 atm. to 1 atm. pressure. Make a table showing ΔS for the gas; for the heat reservoir and for the two systems combined, if in the expansion (a) 2730 cal work is done (b) 1000 cal work is done and (c) no work is done.

[*Hints*.

Path	ΔS gas	$\Delta S_{\text{Reservior}}$
(a)	$R \ln \frac{P_1}{P_2}$	$-\frac{2730}{298}$
(b)	$R \ln \frac{P_1}{P_2}$	$-\frac{1000}{298}$
(c)	$R \ln \frac{P_1}{P_2}$	0.]

4. Calculate ΔS when 10 gm. of ice at 0°C are added to 50 gm of water at 40°C in an isolated system. The latent heat of fusion of ice is 79.7 cal. gm^{-1}; the specific heat of water 1 cal. gm^{-1} deg^{-1}.

[*Hints*.

Find final temperature calorimetrically. Let it is T°A.

For ice $\qquad \Delta S_1 = \frac{10 \times 79.7}{273} + 10 \times 1 \ln \frac{T}{273}$

For water $\qquad \Delta S_2 = 50 \times 1 \ln \frac{T}{273}$

Total = $\Delta S_1 + \Delta S_2$.]

OBJECTIVE TYPE

1. Choose the correct answer.

(a) Second law gives us concept of ——— .

[Ans. (a) Enthalpy (b) Free engery (c) Internal energy]

(b) Device which converts heat to work is called ——— .

[Ans. (a) Calorimeter (b) Thermometer (c) Engine]

(c) In an irreversible process entropy of ——— increases.

[Ans. (a) System (b) Universe (c) Surrounding]

(d) Increase of P will ——— meeting point of ice.

[Ans. (a) decrease (b) not affect (c) increase]

(e) During sublimation entropy will ——— .

[Ans. (a) not change (b) decrease (c) increase]

2. Match the equation for quantities in column A with column B.

A	B
(a) ds	(a) $= C_v \ln t + R \ln V +$ constant
(b) ΔS_{fusion}	(b) $\frac{T_1 - T_2}{T_1}$
(c) $\frac{dP}{dT}$	(c) $= -\frac{dq_{\text{rev}}}{T}$
(d) S	(d) $= VdP - SdT$
(e) dG	(e) $= \frac{L}{T\left(V_2 - V_1\right)}$
(f) η	(f) $= \frac{L_f}{T}$

3. Write true or false against the following statements.

(a) Refrigerator works against second law.

(b) An engine is 100% efficient at 0°A.

(c) For a cyclic process $\oint ds = 0$.

(d) When water solidifies entropy increases.

(e) $\frac{L_v}{T_b}$ = constant — is known as Trouton's rule.

(f) Relation among ΔG, ΔH & T is Gibbs-Helmholtz equation.

Answers :

1. (a) b; (b) c; (c) b; (d) a; (e) c.
2. a = c; b = f; c = e; d = c; e = d; f = b.
3. (a) F; (b) T; (c) T; (d) F; (e) T; (f) T.

Chapter 9

CHEMICAL EQUILIBRIUM

9.1 REVERSIBLE REACTION

We know that a chemical reaction means breaking of some old bonds and forming of some new bonds. The word 'yield' is very important during any chemical reaction. It means extent of product formation. It was once thought that chemical reaction occurs due to chemical affinity. Thus for a reaction A + BC = AC + B; it was said that chemical affinity of 'C' for A is much greater than its affinity for B and so when A and BC comes in contact with each other C leaves B and joins with A. Though chemical affinity has a role regarding chemical reaction, explanation of the above type is an oversimplified version of a complicated thing. Many other factors are to be considered. One such factor is concentration. The noted scientist Berthelot, (minister of Nepolean) found huge deposits of trona (Na_2CO_3) in the shores of Egyptian lakes. Lake water was saline and there were ores of marble stone nearby. So he assumed that following reaction had been occurring for a long long time.

$$2NaCl + CaCO_3 = Na_2CO_3 + CaCl_2$$

Such an assumption was exactly opposite to their laboratory observation. In the laboratory above reaction does not occur, but reverse reaction occurs. Berthelot concluded that in the laboratory of nature above reaction occurred in the forward direction because both the reactants were present in large excess. Thus, the revolutionary conclusion, that was drawn was — 'concentration has a role on the direction of a reaction'. Next person, who did some important work on this subject was St. Gills. He carried out following esterification reaction at 25°C.

$$C_2H_5OH + CH_3COOH = CH_3COOC_2H_5 + H_2O$$

His first observation was that it is an incomplete reaction. When the reaction does not proceed further, amount of ester was found to be 66.6%. If now pure ethylacetate is allowed to hydrolyse at the same temperature, then also incomplete hydrolysis occurs and 66.6% ester remained unhydrolysed. Thus this reaction was found to occur in both ways and hence it is a reversible reaction. Also at the time of cessation amount of each constituent gets some fixed values. This discovery was also another important mile-stone.

9.2 LAW OF MASS ACTION

Discoveries of Berthelot and St. Gills inspired Guldberg and Waage to put forward a very important law, popularly known as law of mass action. This law states that *at a particular temperature rate of any chemical reaction is directly proportional to 'active-masses' of reactants.* The term rate means amount of reactant (or product) disappearing (or appearing) per unit time. We shall see in the chapter of chemical-kinetics that rate is expressed in the differential form $-\frac{dc}{dt}$ (with respect to reactant) or $\frac{dx}{dt}$ (with respect to product). The term active mass is something closely related with concentration. Sometimes it is taken as molar concentration, Modern thermodynamics however taught as that active mass is actually activity. For gases activity is equal to its partial pressure in atmosphere unit. Activity is actually a dimensionless quantity. For SOLIDS and PURE LIQUIDS activity is arbitrarily taken as unity.

9.3 MATHEMATICAL FORM OF THE LAW—EQUILIBRIUM CONSTANT

Let us consider a reversible reaction A + B $\rightleftharpoons$ C + D. At the beginning there were only A and B and rate of forward reaction was highest. As time passes concentration of A and B begin to fall and rate of forward reaction also decreases; C and D begin to be accumulated and backward reaction starts. Its rate will increase with time (Fig 9.1). A time will come when the two rates will be same. At this time concentration of none of the constituent will change with time. It will appear that the reaction has stopped. But actually it has not stopped but occurring in both direction with equal speed. Such a state is called an equilibrium state. But it is not static. The dynamic nature of equilibrium has been proved by radio isotope.

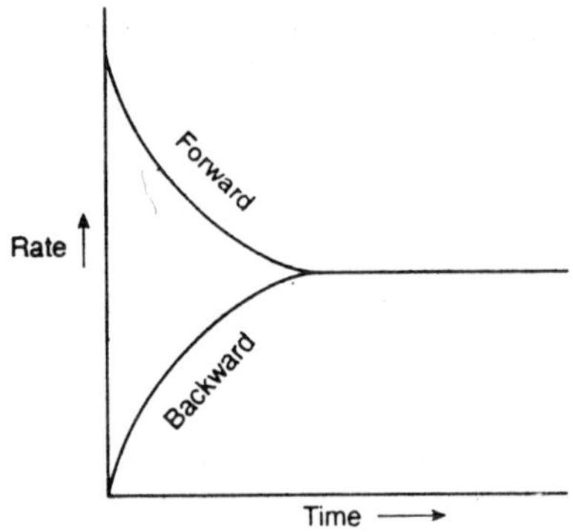

Fig. 9.1

To explain it let us consider N_2 + $3H_2 \rightleftharpoons 2NH_3$ equilibrium. Let we found that at eqm. 'a' mole N_2; 'b' mole H_2 and 'c' mole NH_3 co-exist without undergoing any change in concentration. We then prepare another equilibrium mixture of N_2 ; H_2 and NH_3 with same amounts but this time some radioisotope of N_2 were taken we shall see that concentration are remaining same but radioactivity is being detected in ammonia; it means reaction is still occurring and equilibrium is dynamic in nature.

Now rate of forward reaction at any moment $\propto a_A.a_B$ [a = active mass at that time]

or, rate of forward reaction = $k_1\ a_A.a_B$ [k_1 = forward rate constant]

Rate of backward reaction $\propto a_C a_D$

or, rate of backward reaction = $k_2\ a_C a_D$ [k_2 = backward rate constant]

At equilibrium two rates are same. So

$$k_1\ a_A.a_B = k_2\ a_C a_D$$

$$\text{or, } \frac{a_C \cdot a_D}{a_A \cdot a_B} = \frac{k_1}{k_2} = K$$

Here K is a new constant, known as equilibrium constant. For reaction 2A $\rightarrow$ Product i.e. A + A $\rightarrow$ Product; forward rate $\propto a_A.a_B$ i.e. forward rate $\propto a_A^2$. So co-efficient in the stoichiometric equation comes as power in rate equation. Hence for reaction $aA + bB \rightleftharpoons cC + dD$; equilibrium constant is

$$K = \frac{a_C^{\,c} \cdot a_D^{\,d}}{a_A^{\,a} \cdot a_B^{\,b}}. \qquad ....(9.1)$$

9.4 TYPES OF EQUILIBRIUM CONSTANTS—THEIR DIMENSIONS

We have seen just now that for a reversible reaction $aA + bB \rightleftharpoons cC + dD$; equilibrium constant K is —

$$K = \frac{a_C{}^c \cdot a_D{}^d}{a_A{}^a \cdot a_B{}^b}$$

where a_A ; a_B ; a_C and a_D are active masses of A, B, C and D at equilibrium. We have also learnt that truely speaking active mass is 'activity' which is a dimensionless quantity and so **true eqm. constant (thermodynamic eqm. constant) is a dimensionless quantity**. However, active mass is sometimes approximately taken as concentration and in such cases depending on concentration unit '*K*' may be of three types and may have unit. These are explained in the following table.

Table 9.I

Reaction : $aA + bB \rightleftharpoons cC + dD$		
Conc. Unit	$K = \frac{[C]^c[D]^d}{[A]^a[B]^b}$	unit of *K*
Molar (*C*) i.e 'C' Moles/litre	$K_C = \frac{[C_C]^c\ [C_D]^d}{[C_A]^a\ [C_B]^b}$	$\left(\frac{M}{L}\right)^{\Delta n}$
Partial pressure (*p*) i.e. '*p*' atm.	$K_P = \frac{[p_C]^c\ [p_D]^d}{[p_A]^a\ [p_B]^b}$	$(\text{Atm.})^{\Delta n}$
Mole fraction (*X*)	$K_X = \frac{[X_C]^c\ [X_D]^d}{[X_A]^a\ [X_B]^b}$	NIL

Here $\Delta n = (c + d) - (a + b)$.

Thus we see that there are three varieties of equilibrium constants K_c ; K_p and K_x and all become unit less when $\Delta n = 0$. But if $\Delta n \neq 0$, then K_c and K_p will have unit.

9.5 INTER-RELATIONS

For the reaction $aA + bB \rightleftharpoons cC + dD$; equilibrium constant ($K_p$) is —

$$K_p = \frac{p_C{}^c \cdot p_D{}^d}{p_A{}^a \cdot p_B{}^b}.$$

But partial pressure (*p*) of any component (say *C*) is $p'_c = C_c\ RT$.

So $p_c = \frac{n_c}{V} RT = C_c\ RT$;

[$\because p_c\ V = n_c\ RT$; C_c = Conc. in *M*/*L*]

$$\text{So} \quad K_P = \frac{[C_C]^c\ [C_D]^d}{[C_A]^a\ [C_B]^b} \cdot (RT)^{(c+d)-(a+b)}$$

$$\text{or,} \quad K_p = K_c.(RT)^{\Delta n}. \qquad \text{....(9.2)}$$

Again $p_c = X_c . P$.

[Partial pressure = Mole-fraction × Total pressure]

So $$K_p = \frac{X_C{}^c . X_D{}^d}{X_A{}^a . X_B{}^b} . (P)^{(c+d)-(a+b)}$$

or, $$K_p = K_X . P^{\Delta n} . \qquad ....(9.3)$$

From equation (9.2) and (9.3)

$$K_X = K_c . \left(\frac{RT}{P}\right)^{\Delta n} .$$

But $PV = \sum n \, RT.$

So $$K_X = K_c . \left(\frac{V}{\sum n}\right)^{\Delta n} . \qquad ....(9.4)$$

The interesting thing to note is that if $\Delta n = 0$, $K_p = K_c = K_X$.

9.6 MORE ABOUT *K*

Synthesis of ammonia from nitrogen and hydrogen can be written in either of the following ways —

$$\text{(a)} \ \frac{1}{2} N_2 + \frac{3}{2} H_2 \rightleftharpoons NH_3$$

and $$\text{(b)} \ N_2 + 3H_2 \rightleftharpoons 2NH_3$$

For equation (a), equilibrium constant $$K = \frac{a_{NH_3}}{\left(a_{N_2}\right)^{1/2} . \left(a_{H_2}\right)^{3/2}}$$

and for equation (b), equilibrium constant $$K' = \frac{\left(a_{NH_3}\right)^2}{a_{N_2} . \left(a_{H_2}\right)^3} .$$

Obviously $K' = K^2$.

So we see that value of eqm. constant depends on type of stoichiometric equation written. Units of K_p and K_c also depend on equation. Thus for eq. (a) units of K_p and K_c are $(\text{Atm})^{-1}$ and $(M/L)^{-1}$ but for eq. (b) their units are $(\text{Atm})^{-2}$ and $(M/L)^{-2}$ respectively.

9.7 CALCULATIONS OF K_C ; K_P AND K_X

We shall now learn how to calculate K_c , K_p and K_x for any reaction. As typical examples we will choose three reactions with zero; positive and negative Δn - values.

(a) Reaction : H_2 (g) + I_2 (g) $\rightleftharpoons$ 2HI (g) ; $\Delta n = 0$

Let 1 mole of hydrogen and 1 mole of oxygen are reacting in a vessel of '*V*' litre capacity. Let at equilibrium '*x*' mole H_2 reacted. So at equilibrium number of moles of different constituents are —

$n_{H_2} = (1-x)$; $n_{I_2} = (1-x)$ and $n_{HI} = 2x$.

Total moles $= (1 - x) + (1 - x) + 2x = 2$. Let total pressure $= P$. We now make the following table. Here the symbol [ ] means concentration.

Table 9.2

Conc. Unit	$[H_2]$	$[I_2]$	$[HI]$	$K = \frac{[HI]^2}{[H_2][I_2]}$
Moles/litre (M/L)	$\frac{1-x}{V}$	$\frac{1-x}{V}$	$\frac{2x}{V}$	$K_c = \frac{4x^2}{(1-x)^2}$
Mole-fraction	$\frac{1-x}{2}$	$\frac{1-x}{2}$	x	$K_x = \frac{4x^2}{(1-x)^2}$
Partial pressure	$\frac{1-x}{2}.P$	$\frac{1-x}{2}.P$	xP	$K_p = \frac{4x^2}{(1-x)^2}$

As $\Delta n = 0$; so $K_c = K_x = K_p$.

(b) Reaction : PCl_5 (g) $\rightleftharpoons$ PCl_3 (g) + Cl_2 (g) ; $\Delta n = +1$

Let one mole of PCl_5 is allowed to decompose in a vessel of V litre capacity. At equilibrium 'x' mole PCl_5 reacted and so at eqm. $n_{PCl_5} = (1 - x)$; $n_{PCl_3} = n_{Cl_2} = x$ and total moles $= (1 - x) + x + x = (1 + x)$. Let eqm. pressure $= P$. We make another table —

Table 9.3

Conc. Unit	$[PCl_5]$	$[PCl_3]$	$[Cl_2]$	$K = \frac{[PCl_3][Cl_2]}{[PCl_5]}$
Molar	$\frac{1-x}{V}$	$\frac{x}{V}$	$\frac{x}{V}$	$K_c = \frac{x^2}{(1-x)V}$
Mole fraction	$\frac{1-x}{1+x}$	$\frac{x}{1+x}$	$\frac{x}{1+x}$	$K_x = \frac{x^2}{(1-x^2)}$
Partial pressure	$\frac{1-x}{1+x}.P$	$\frac{x}{1+x}.P$	$\frac{x}{1+x}.P$	$K_p = \frac{x^2P}{(1-x^2)}$

(c) Reaction : N_2 (g) + $3H_2$ (g) $\rightleftharpoons$ $2NH_3$ (g) ; $\Delta n = -2$

Let one mole of nitrogen and three moles of hydrogen are reacting in a vessel of volume V-litre. Let at equilibrium 'α' moles of N_2 has reacted, so according to stoichiometric equation 3α moles of hydrogen has reacted. Hence at equilibrium number of moles are — $n_{N_2} = (1-\alpha)$; $n_{H_2} = 3(1-\alpha)$; and $n_{NH_3} = 2\alpha$ and total moles $= 4 - 2\alpha$. Let total pressure $= P$, we now make the following table.

Table 9.4

Conc. Unit	$[N_2]$	$[H_2]$	$[NH_3]$	$K = \frac{[NH_3]^2}{[N_2][H_2]^3}$
Molar	$\frac{1-\alpha}{V}$	$\frac{3(1-\alpha)}{V}$	$\frac{2\alpha}{V}$	$K_c = \frac{4\alpha^2 V^2}{27(1-\alpha)^4}$
Mole fraction	$\frac{1-\alpha}{4-2\alpha}$	$\frac{3(1-\alpha)}{4-2\alpha}$	$\frac{2\alpha}{4-2\alpha}$	$K_X = \frac{16\alpha^2(2-\alpha)^2}{27(1-\alpha)^4}$
Partial pressure	$\frac{1-\alpha}{4-2\alpha} \cdot P$	$\frac{3(1-\alpha)}{4-2\alpha} \cdot P$	$\frac{2\alpha}{4-2\alpha} \cdot P$	$K_P = \frac{16\alpha^2(2-\alpha)^2}{27(1-\alpha)^4 P^2}$

9.8 DERIVATION OF LAW OF MASS ACTION

Law of mass action can be derived both from kinetic theory of gas and thermodynamics. Here we are considering the kinetic derivation. Let us consider a homogeneous gaseous reaction

$$A + B \rightleftharpoons C + D$$

which is reversible. According to collision theory forward reaction occurs due to collision between A and B molecules and backward reaction will occur due to collision between C and D. But collision frequency between A and B is proportional to respective partial pressures.

Hence

$$\text{forward reaction rate} \propto p_A \cdot p_B \qquad [p = \text{partial pressure}]$$

or, $\text{forward reaction rate} = k_1 \cdot p_A \cdot p_B$

and $\text{backward reaction rate} = k_2 \cdot p_C \cdot p_D$

At equilibrium $k_1 \cdot p_A \cdot p_B = k_2 \cdot p_C\, p_D$

$$\text{or,} \quad \frac{p_C \cdot p_D}{p_A \cdot p_B} = \frac{k_1}{k_2} = K. \qquad [K = \text{Eqm. constant}]$$

This is law of mass action.

9.9 EFFECT OF INERT GAS ON EQUILIBRIUM

Let us consider a homogeneous gaseous equilibrium $aA + bB \rightleftharpoons cC + dD$. Its equilibrium constant

$$K_p = \frac{p_C^{\,c} \cdot p_D^{\,d}}{p_A^{\,a} \cdot p_B^{\,b}}.$$

Now $p_c = X_c \,.\, P = n_c\,(P / \Sigma n)$.

$$\text{So} \quad K_p = \frac{n_C^{\,c} \,.\, n_D^{\,d}}{n_A^{\,a} \,.\, n_B^{\,b}} \cdot \left(\frac{P}{\Sigma n}\right)^{\Delta n}. \qquad [\Delta n = (c + d) - (a + b)]$$

Now when some inert gas is introduced total moles of gas i.e. Σn is increase. The effect on equilibrium depends on several factors. These are —

(a) Addition of inert gas at constant volume : If inert gas is introduced at constant volume, then both P and Σn will increase proportionately and $(P/\Sigma n)$ will remain same. So equilibrium is unaffected.

(b) Addition at constant pressure : If inert gas is introduced at constant pressure, three cases may arise depending on value of Δn.

(i) $\Delta n = 0$; In such case $\left(\frac{P}{\Sigma n}\right)^{\Delta n}$ is always unity and eqm. is unaffected.

(ii) $\Delta n = +$ ve ; In such case increase of Σn will decrease $\left(\frac{P}{\Sigma n}\right)^{\Delta n}$. So to keep K_p constant, n_C, n_D should increase, n_A, n_B should decrease. Hence eqm. will shift towards right.

(iii) $\Delta n = -$ ve ; In such case increase of Σn will increase $\left(\frac{P}{\Sigma n}\right)^{\Delta n}$. Hence to keep K_p constant, n_C, n_D should decrease, n_A, n_B should increase. Hence equilibrium will shift in the backward direction (left).

9.10 EFFECT OF CONCENTRATION (OR PRESSURE) — VANT HOFF'S ISOTHERM

Let us consider a homogeneous chemical reaction $aA + bB \rightleftharpoons cC + dD$. Two important question can be asked regarding this reaction. These are —

(i) What will be the effect if some quantity of any one of the constituents is added at equilibrium?

(ii) If some one collects A, B, C and D and mix them up in any arbitrary proportion, in which direction the reaction will proceed i.e. whether A and B will be consumed or produced ?

Answer to these questions is given by an equation known as Vant Hoff's reaction isotherm. Derivation of the equation is beyond the scope of this book, but the equation can be written and discussed. It is —

$$\Delta G = -RT \ln K + RT \ln \frac{a_C{}^c . a_D{}^d}{a_A{}^a . a_B{}^b} \quad \text{....(9.5)}$$

or,
$$\Delta G = -RT \ln K + RT\Sigma\upsilon \ln a_i \quad \text{....(9.6)}$$

Here K is equilibrium constant. ΔG is free energy difference between products and reactants when their activities are a_C , a_D , a_A and a_B respectively.

At equilibrium $\frac{a_C{}^c . a_D{}^d}{a_A{}^a . a_B{}^b} = K$ and $\Delta G = 0$.

During reaction $\frac{a_C{}^c . a_D{}^d}{a_A{}^a . a_B{}^b} \neq K$ and $\Delta G \neq 0$.

Now let at equilibrium some C or D are added ΔG becomes +ve then. This means reverse process will occur. Similarly if some A or B are added ΔG becomes –ve and forward process will occur. Again if we mix up A, B, C and D in any arbitrary proportion, we are to find out ΔG value from their activity values. If ΔG is found to be –ve forward process will occur, if ΔG is +ve backward process will occur and if $\Delta G = 0$ no change in concentration will occur.

9.11 EFFECT OF TEMPERATURE ON EQUILIBRIUM — VANT HOFF'S EQUATION

Rate of reaction is very much sensitive towards temperature. Rate constant of any reaction increases with temperature. Now for a reversible reaction, both forward and backward rate constant (k_1 and k_2) will increase with temperature. But rate of increment is not same. So equilibrium constant K, which is (k_1 / k_2) (see eqn. 9.1) may increase or decrease with increase in T. Variation of equilibrium constant (K) with temperature is given by two equations, known as Vant Hoff's equation. These are —

$$\frac{d \ln K_p}{dT} = \frac{\Delta H}{RT^2} \quad \text{....(9.7)}$$

$$\frac{d \ln K_c}{dT} = \frac{\Delta U}{RT^2}. \quad \text{....(9.8)}$$

Here ΔH and ΔU are heat of reaction at constant pressure and constant volume. That is why these equations are also known as Vant Hoff's isobar and isochore. Integrating equation (9.7) we get

$$\ln \frac{K_{P_2}}{K_{P_1}} = \frac{\Delta H}{R} \cdot \frac{T_2 - T_1}{T_1 T_2}. \quad \text{...(9.9)}$$

For exothermic reaction ΔH is –ve. If we further assume that $T_2 > T_1$ then R.H.S. is –ve. So $Kp_2 < Kp_1$. Hence for exothermic reaction equilibrium constant decreases with rise in temperature.

For endothermic reaction ΔH is +ve. Here also if $T_2 > T_1$ then R.H.S. is +ve. So $Kp_2 > Kp_1$. Hence for endothermic reaction equilibrium constant increases with rise in temperature.

9.12 LE CHATELIER PRINCIPLE

In the previous sections we have seen effect of pressure, temperature etc. on equilibrium. Vant Hoff gave quantitative relation between yield (eqm. constant K is a measurement of yield) and pressure (conc.); temperature etc. But before Vant Hoff Le-Chatelier and Von Braun explained qualitatively the effects of pressure, temperature and concentration on equilibrium. Let us discuss them one by one.

(a) Effect of pressure : If in a reaction Δn is –ve e.g. $N_2 + 3H_2 = 2NH_3$, then forward reaction is associated with decrease in number of moles and hence volume. So if we increase pressure forward reaction will be favoured and yield will increase. But if for a reaction Δn = +ve e.g. $PCL_5 \rightleftharpoons PCl_3 + Cl_2$ then backward reaction is associated with decrease in volume and hence increase of pressure will favour backward reaction i.e., decrease the yield. For reaction with $\Delta n = 0$, pressure will have no effect.

(b) Effect of temperature : Let us consider an endothermic reaction e.g. $N_2 + O_2 = 2NO - Q$ cal. Tendency of forward reaction is to absorb heat; so if heat is supplied forward reaction will be favoured. Conversely in an exothermic reaction e.g. $2SO_2 + O_2 = 2SO_3 + Q$ cal. forward reaction releases heat. So if heat is taken away the reaction will be helped. Thus endothermic reactions are favoured by raising temperature and exothermic reactions are favoured by lowering temperature. However in industrial methods exothermic reactions are not always done at low temperature. Because at low temperature equilibrium yield is high no doubt but it takes a long time to reach equilibrium. Thus a moderate temperature is selected in such a manner that yield is not negligible but time of reaching equilibrium is shoot. This temperature is called *optimum temperature.*

(c) Effect of concentration : Let us consider a homogeneous chemical equilibrium $aA + bB \rightleftharpoons cC + dD$. Its eqm. constant $K = \frac{a_C^c . a_D^d}{a_A^a . a_B^b}$. At eqm. if we add either C (or D) then to keep 'K' constant, some of added 'C' (or 'D') will combine with the other reactant to produce A and B i.e. eqm. will shift in backward direction. Similarly addition of A (or B) will shift the equilibrium in forward direction.

All these three effects have been predicted by the following statement, which is popularly known as LeChatelier's principle. It states — *if any one of the factors viz. pressure, temperature or concentration, conditioning an equilibrium is changed, the equilibrium will shift in a direction, so as to neutralise the effect of such a change. The principle is the eternal principle, that every incident opposes the cause behing it.*

9.13 EFFECT OF CATALYST

Catalysts are those which influence rate of a reaction. Thus a question which may arise in our mind, is whether catalysts influence yield or not. The answer is **NO**. If a reaction $R \rightleftharpoons P$ takes place first in absence of and then in presence of catalyst 'X' then in first case $\Delta G = G_P - G_R$ and in the second case $\Delta G = (G_P + G_X) - (G_R + G_X) = (G_P - G_R)$. Thus free energy change is same in each case. From Vant Hoff's reaction isotherm (eq. 9.7) standard free energy change (activity = 1) $\Delta G° = -RT \ln K$. So 'K' is same both in presence and absence of catalyst. Hence catalyst does not affect equilibrium i.e. yield.

9.14 ARE ALL REACTIONS REVERSIBLE ?

We have seen many reaction proceed in such a manner that almost no reactant is left. We call them irreversible. However, theoretically all reactions are reversible. During so called irreversible reactions either products are constantly being removed (e.g. evolving out as gas etc.) so that equilibrium is not attained or equilibrium constant is so high that conc. of reactant remaining is negligibly small (e.g. in the oxidation of $C_2O_4^{2-}$ by MnO_4^- eqm. constant $K = 10^{238}$). That all reactions are reversible can be proved mathematically. For irreversible reaction, conc. of reactant at eqm. = 0; so eqm. constant becomes infinity. We have seen $\Delta G° = -RT \ln K$. So $\Delta G°$ also becomes infinity; but it is the difference in free energy between product and reactant at standard state. So it must be finite. Hence 'K' must be finite. So some reactant must be left at eqm. i.e. reaction should be reversible. Thus all chemical changes are reversible! Hence we may hope that biological changes inside our body are also reversible and by suitable cultivation we can shift the equilibrium in backward direction and thus go back to our childhood !

POINTS TO REMEMBER

Chemical equilibrium is generally studied by '*law of mass action.*' According to this law ratio of product of activities of products and product of activities of reactants (all with appropriate power dependent on stoichiometry) is constant at a given temperature. Thus for reaction $aA + bB \rightleftharpoons cC + dD$, equilibrium constant $K = \frac{a_C{}^c . a_D{}^d}{a_A{}^a . a_B{}^b}$ where a_i = *activity* of ith substance. Activity is dimensionless, so truely *K has no dimension.* But approximately we put $a_i = C_i$ [C_i = Conc. of ith substance]. Then depending on its unit some units for K are written when $(c + d) \neq (c + d)$. Three types of K viz. K_p, K_c and K_X are used where concentrate are expressed in partial pressure, molar conc. and mole-fraction respectively. When $\Delta n = 0$; $K_p = K_c = K_X$. The relations are $K_p = K_c (RT)^{\Delta n}$; $K_p = K_X . P^{\Delta n}$.

Equilibrium of a reaction is affected by temperature; pressure; concentration and presence of inert gas. Catalysts alter the rate but not equilibrium. *LeChatelier's principle* explains qualitatively role of P; T and C. *Vant Hoff's isotherm and equation* explains quantitatively role of C and T.

QUESTIONS

SHORT TYPE

1. 'The reaction $CaCO_3 + 2NaCl = CaCl_2 + Na_2CO_3$ does not occur in laboratory but occurs near some Egyptian lakes' — why ?
2. Clearly explain the word 'active-mass'.
3. What is rate of a reaction ? What is its correct mathematical expression ?
4. How can it be proved that chemical equilibrium is a dynamic equilibrium ?
5. For a reaction $pP + qQ = rR + sS$. What are dimensions of K_c ; K_p and K_X ?
6. Will equilibrium constants for $2H_2$ (g) + O_2 (g) = $2H_2O$ (l) and H_2 (g) + $\frac{1}{2}O_2$ (g) = H_2O (l) be same or different ?
7. Name an equation which explains effects of concentration on reaction rate.
8. How does a catalyst influence equilibrium ?
9. For what type of reaction K_c ; K_p and K_X values will be same ?
10. Are all reaction reversible ?

ESSAY TYPE

1. State and explain law of mass action and arrive at the idea of equilibrium and equilibrium constant.
2. How many types of equilibrium constants are known to you ? Establish a relation between each possible pairs.
3. In a closed vessel of V litre capacity degree of dissociation of PCl_5 (g) is α ; total pressure is P. Derive expressions for K_c ; K_p and K_X for the reaction PCl_5 (g) $\rightleftharpoons$ PCl_3 (g) + Cl_2 (g).
4. Consider the equilibrium C(s) + O_2 (g) = CO_2 (g) ; $\Delta H = -$ 99.4 K. cal. Predict the effect upon equilibrium, when —
   (i) Amount of C (s) is increased without changing any other.
   (ii) Temperature is raised.
   (iii) Pressure is increased.
   (iv) He gas is introduced (a) at constant pressure and (b) at constant volume.
5. State and explain LeChatelier's principle. Predict on the basis of this principle what will happen when
   (i) Temperature of the equilibriated system H_2O (l) $\rightleftharpoons$ H_2O (v) is raised.
   (ii) Pressure of the equilibriated system N_2 (g) + $3H_2$ (g) = $2NH_3$ (g) is reduced.
   (iii) Catalyst in the equilibriated system SO_2 (g) + $\frac{1}{2}O_2$ (g) = SO_3 (g) is removed and
   (iv) Ester is added in the equilibriated system. Acid + Alcohol = Ester + Water.
6. "Theoretically all reactions are reversible" — explain.

PROBLEMS

1. When ethyl alcohol and acetic acid are mixed together in equimolecular proportion 66.5% are convered into ester. Calculate how much ester will be formed if 1 mole acid is treated with 0.5 moles of alcohol.

[*Hints.*

$$EtOH + AcOH = EtOAc + H_2O$$

$$K_c = \frac{665 \times 665}{(100-665)\,(100-665)}\text{; Case - 1}$$

$$K_c = \frac{x \times x}{(1-x)\,(05-x)}\text{; Case - 2}$$

Solve for x.]

2. Calculate K_p for the reaction $N_2O_4 \rightleftharpoons 2NO_2$ at 49.7°C from value of $K_c = 8.55 \times 10^{-3}$ gm. moles per litre and R = 0.0821 litre atm./degree.

[*Hints.* $K_p = K_c\,(RT)^{\Delta n}$
Here $R = 0.0821$, $T =$ 312.7°A, $\Delta n = (2 - 1) = 1$.]

3. Calculate the value of K_c for the reaction $\frac{3}{2}H_2 + \frac{1}{2}N_2 \rightleftharpoons NH_3$ at 400°C from the value of $K_p = 0.0129$ atm.

[*Hints.*

Same as before ; put $\Delta n = [\,1 - (\frac{3}{2} + \frac{1}{2})] = -\,1$.]

4. The equilibrium constant in the reaction between ethyl alcohol and acetic acid to form ethyl acetate and water is 4 at 25°C. What weight of ethyl acetate will be obtained from 50gms of alcohol and 50 gms of acid, the reaction being carried out at 25°c in a closed vessel ?

[*Hints.* See problem '1'.]

5. Calculate the percentage dissociation of PCl_5 in an experiment in which 2 gms of the substance were heated to 200°C in a closed litre vessel which contained only PCl_5 and the product of the dissociation. The equilibrium constant (K_c) for the reaction $PCl_5 \rightleftharpoons PCl_3 + Cl_2$ is 0.00793 at 200°C.

[*Hints.*
If 'a' is initial concentrate of PCl_5 and x amount reacts at equilibrium, then

$$K_c = \frac{x^2}{(a-x)\,V}.$$

Here $K_c = 0.00793$; $V = 1$ litre; $a = \frac{2}{M_{PCl_5}} = \frac{2}{208.2}$ moles; find x .]

6. If a mixture of 3 moles H_2 to 1 mole N_2 react under a constant pressure of 100 atm. to form at equilibrium 0.5 mole ammonia, calculate K_p for the reaction $\frac{3}{2}H_2 + \frac{1}{2}N_2 \rightleftharpoons NH_3$ at this temperature.

[*Hints.*

If x mole NH_3 is produced at eqm. then $\frac{x}{2}$ mole N_2 and $\frac{3x}{2}$ mole H_2 reacted.

So $$n_{N_2} = 1 - \frac{x}{2} = 1 - 0.25$$

$$n_{H_2} = 3 - \frac{3x}{2} = 3 - 0.75$$

$$n_{NH_3} = x = 0.5.]$$

$$K_p = \frac{n_{NH_3}}{\left(n_{N_2}\right)^{1/2}\left(n_{H_2}\right)^{3/2}} \cdot \left(\frac{P}{\Sigma n}\right)^{-1} .]$$

7. When 2.94 moles of iodine and 8.1 moles of hydrogen are heated at constant volume at 444°C until equilibrium is established, 5.64 moles of hydroiodic acid are formed. If we start with 5.3 moles of iodine and 7.9 moles of hydrogen, how much hydroiodic acid is present at equilibrium at the same temperature ?

[*Hints.*

Find eq. constant from the first set of data; then from this value solve the latter part.]

8. When 1 gm of PCl_5 crystals is vaporised at 182°C and 1 atm, the density of the vapour referred to air is 5.08 when equilibrium is reached. Calculate α, K_c and K_p. (Given air is 14.4 times heavier than H_2).

[*Hints.*

Vapour density = 5.08 × 14.4 = d_{expt}.

Theo. Vap. density = $\frac{(31 + 5 \times 35.5)}{2} = d_{theo}$.

For $PCl_5 \rightleftharpoons PCl_3 + Cl_2$; if α is degree of dissociation then $\frac{d_{expt}}{d_{theo}} = (1 + \alpha)$

$$K_c = \frac{\alpha^2}{(1 - \alpha) V} \; ; \; K_p = \frac{\alpha^2 P}{\left(1 - \alpha^2\right)} .]$$

OBJECTIVE TYPE

1. **Write √ mark in the box beside the correct answer.**

(i) Role of concentration on reaction was first predicted by

Ans. (a) St. Gills ☐ (b) Vant Hoff ☐ (c) Berthelot ☐

(ii) Dimension of activity is ———.

Ans. (a) Atm. ☐ (b) Moles/litre ☐ (c) Nil ☐

(iii) $K_p = K_c = K_X$ when 'Δn' is ———.

Ans. (a) Positive ☐ (b) Zero ☐ (c) Negative ☐

(iv) Reaction isotherm was derived by ———.

Ans. (a) Guld berg ☐ (b) LeChateller ☐ (c) Vant Hoff ☐

(v) For a reaction K will increase with T if ΔH is ———.

Ans. (a) Zero ☐ (b) Negative ☐ (c) Positive ☐

(vi) The reaction $Na_2CO_3 + CaCl_2 = CaCO_3 + 2NaCl$ is found to occur in reverse direction near, some Egyptian lakes because of ———.

Ans. (a) climatic condition ☐ (b) abundance of product ☐ (c) high temperature ☐

2. Write T (true) or F (false) against the following statement.

(a) Equilibrium constant depends on type of equation writing,

(b) For the reaction H_2 (g) + I_2 (g) $\rightleftharpoons$ 2HI (g); $K_P > K_C$.

(c) In the reactions PCl_5 (g) $\rightleftharpoons$ PCl_3 (g) + Cl_2 (g); addition of N_2 at constant pressure increase yield.

(d) All reactions are reversible.

(e) Catalyst can increase yield of a reaction.

3. Match terms of column 'A' with that of column 'B'.

A	B
(a) Berthelot.	(a) $K_p = K_c = K_X$.
(b) Active mass.	(b) Esterification.
(c) $\Delta n = 0$.	(c) Isotherm.
(d) Vant Hoff.	(d) Trona.
(e) St. Gills.	(e) Equilibrium constant.

Answers :

1. (i) c; (ii) c; (iii) b; (iv) c; (v) c; (vi) b.

2. (a) T (b) F (c) T (d) T (e) F.

3. a –d; b – e; c – a; d – c; e – b;.

Chapter 10

SOLUTIONS — GENERAL

10.1 INTRODUCTION

Mixture of two or more substances may be homogeneous or heterogeneous. Homogeneous mixtures of two or more substances are called solutions. Though the terms solvent and solute are relative, but yet substance, present in large excess is called solvent and substance present in relatively small amount is called solute.

10.2 DIFFERENT TYPES OF SOLUTIONS

Depending on nature of solute and solvent different types of solutions have been recognised. This is shown in the following table.

Table 10.1

Solute	Solvent	Example
Gas	Gas	Air
Gas	Liquid	O_2 in water CO_2 in soft-drinks
Gas	Solid	H_2 in Palladium
Liquid	Liquid	Spirit in water
Liquid	Solid	Hg in Ag
Solid	Liquid	Sugar solution
Solid	Solid	Alloys

10.3 CONCENTRATION — DIFFERENT UNITS

The term 'concentration' arises when more than one substances are present together. Relative amount of one with respect to the other is called concentration. It can be expressed in different units. Some of the very popular units are discussed below —

(a) Percentage unit : If we express how many parts of solute are present in hundred parts of solvent then the unit is called percentage unit. It may be volume per cent or weight per cent. Thus a 2% (Vol.) NaCl solution in water means 2gm NaCl is present in 100 c.c. aq. solution; but a 2% (weight) NaCl solution in water means 2 gm NaCl is present in 100 gm. aq. solution.

(b) Normality unit : The term 'normality' means how much equivalent of solute is present in one litre solution. Thus if we dissolve 1.575 gm oxalic acid ($H_2C_2O_4$, $2H_2O$) in 250 c.c. water, then in one litre amount of oxalic acid = 1.575 × 4 = 6.3 gm. Equiv. wt. of oxalic acid = $\frac{126}{2}$ = 63. So 6.3 gm means $\frac{1}{10}$ equiv. As $\frac{1}{10}$ equiv. is present in one litre; the solution has a strength of $\left(\frac{N}{10}\right)$.

(c) Molarity unit : Molarity of a solution is equal to number of moles of solute present per litre of solution. Let 53 gms of Na_2CO_3 is dissolved in 1 litre water. Mol. wt. of Na_2CO_3 = 106. So 53 gms ≡ $\frac{1}{2}$ mole. Hence concentration of this solution is $\frac{M}{2}$ i.e. 0.5 molar.

(d) Molality unit : In normality and molarity units amount of solvent is expressed in volume unit (litre). Since volume of a liquid is dependent on temperature these units become temperature dependent. To make them temperature independent we use molal unit. Here we express how many moles have been dissolved in 1 kg. of solvent. Let 'a' gm solute (mol. wt = m) is dissolved in 'b' gm solvent. Then in 'b' gm solvent we have a/m mole solute. So in 1000 gm solvent we have $\frac{a \times 1000}{b \times m}$ mole. This is molal concentration. As it is temperature independent, it is widely used now.

(e) Mole-fraction Unit : This is yet another unit. Let n_1 mole of solvent and n_2 mole of solute are present in a solution. Then mole-fraction of each of them is —

$$\text{Mole-fraction of solvent } (x_1) = \frac{n_1}{n_1 + n_2};$$

$$\text{and mole-fraction of solute } (x_2) = \frac{n_2}{n_1 + n_2}.$$

Obviously $x_1 + x_2 = 1$.

Mole-fraction is ratio and so it is dimensionless.

10.4 SOLUTION OF GAS IN GAS

All gases are miscible in all proportion. Because of their random motion and isotropic behaviour they form a homogeneous mixture during mixing. Total pressure of the gas mixture is the summation of gas pressures exerted by each gas when it alone occupies the whole vessel. These individual pressures are known as partial pressures and the above statement is known as Dalton's law of partial pressure. Let we have a vessel of V litre capacity; at T°A, n_A moles of gas A is kept in it and the observed pressure is let P_A. Then assuming ideal nature

$$P_A V = n_A RT. \quad \text{....(10.1)}$$

If now gas A is taken out of the vessel and n_B moles of gas B is introduced and if this gas shows a pressure P_B, then

$$P_B V = n_B RT. \quad \text{....(10.2)}$$

Adding equation (10.1) and (10.2) we get

$$(P_A + P_B) V = (n_A + n_B) RT. \quad \text{....(10.3)}$$

Let now n_A moles A and n_B moles B are introduced in the same vessel, and the total pressure of gas mixture is P, then

$$PV = (n_A + n_B) RT. \quad \text{....(10.4)}$$

Comparing (10.3) and (10.4) we get

$$P = P_A + P_B.$$

These P_A, P_B are partial pressures.

Also from (10.1); (10.2) and (10.4) we get

$$\frac{P_A}{P} = \frac{n_A}{n_A + n_B} = X_A \quad \text{and} \quad \frac{P_B}{P} = \frac{n_B}{n_A + n_B} = X_B,$$

Where X = mole - fraction.

Thus **Partial pressure = Mole fraction × Total pressure.**

10.5 SOLUTION OF GAS IN LIQUID

Gases are soluble in many liquids. Generally easily liquefiable gases e.g. CO_2; NH_3; HCl etc. have high solubility values. On the contrary, so called permanent gases e.g. N_2, O_2, H_2

etc. have very little solubility. Solubility of a gas in a liquid depends on pressure of the gas above it. According to Henry's law —

weight of a gas (W) dissolved in a given volume of a liquid at a particular temperature is directly proportional to its pressure (P) above the liquid. Thus

$$W \propto P \quad \text{or} \quad \frac{W}{P} = \text{constant.} \qquad ....(10.5)$$

Now let molar mass of the gas be M then if 'W' gm is dissolved in V litre liquid, then molar conc. of gas in solution phase is $\frac{(W/M)}{V}$ = let C_1. Hence $W = C_1 MV$. Also if C_2 is the molar conc. of gas in the gaseous phase then $P = C_2 RT$. Hence equation (10.5) becomes

$$\frac{C_1 MV}{C_2 RT} = \text{constant.} \qquad ....(10.6)$$

For a given gas dissolved in a given vol. of liquid at a given temperature, M, V, and T are constants. Hence eq. (10.6) becomes —

$$\frac{C_1}{C_2} = \text{constant.} \qquad ....(10.7)$$

Thus Henry's law can be stated as *ratio of conc. of a gas in gaseous state and in solution state (remaining side by side in eqm.) is constant.*

Henry's law states that weight of the dissolved gas is proportional to pressure of the gas over the liquid. If this pressure is doubled, weight of dissolved gas will also be doubled. So volume should also be doubled; but doubling the pressure means halving the volume. So due to two equal and opposite effects volume of dissolved gas will remain same. That is why the apparently paradoxical statement of Henry's law is *volume of a gas dissolved in a given volume of liquid at a given temperature is constant and independent of pressure.* This can also be proved mathematically. Assuming ideal behaviour

$$PV = \frac{W}{M} \, . \, RT.$$

$$\text{or,} \quad V = \frac{W}{P} \cdot \left(\frac{RT}{M}\right). \qquad ....(10.8)$$

From Henry's law $\left(\frac{W}{P}\right)$ is constant, so R. H. S becomes constant at a given temperature.

Extent of solubility of a gas in a liquid is expressed by absorption co-efficient. It is defined as the volume of the gas (reduced to N.T. P.) dissolved in 1 c.c. of a liquid when the partial pressure of this gas above the liquid is 1atm. It is dependent on temperature.

10.6 NERNST'S DISTRIBUTION LAW

If a given solid solute is soluble in two different liquids, which are immiscible, then its distribution between these two liquids remaining side by side follow some law known as Nernst's distribution law. According to this law —

If a given solute is distributed between two immiscible solvents remaining side by side then ratio of its concentration in one solvent to that in another solvent is constant at a given temperature." Thus if C_1 and C_2 are molar concentrations of solute in solvents '1' and '2' respectively, then according to this law —

$$\frac{C_1}{C_2} = \text{constant (at a given temperature).} \qquad ....(10.9)$$

Equation (10.7) and equation (10.9) are same. Thus Henry's law is another form of Nernst's distribution law.

Nernst's distribution law can be derived from thermodynamics, using the concept of chemical potential. Derivation is beyond the scope of this book. However it should be mentioned that equation (10.6) is valid when the solute is neither associated nor dissociated (i.e. exists as monomer) in both solvents. If, however solute exists as 'n'– mer in solvent 1 but monomer in solvent 2, then (C_1/C_2) will not be constant, but $\left(\sqrt[n]{C_1}/C_2\right)$ will be constant.. Incidently this constant is called *partition co-efficient* or *distribution co-efficient*.

10.7 SOLVENT EXTRACTION METHOD

Nernst's distribution law is applied in case of solvent extraction. Let some solute is dissolved in a solvent from which it can not separated by distillation because it decomposes on heating. In such cases we are to choose another solvent which is immiscible with this solvent and where solubility of solute is more. If this solvent is added to the solution then more solute will come here. This new solution can be separated from the old by a separating funnel and if this solvent is highly volatile then solid solute will be separated simply by spreading this solution on a flat dish. It can be shown that if this second solvent is added in small quantities for several times then extraction of solute is more profitable than using large solvent at a time. Let us solve a problem.

Let we have 100 c. c. aq. solution of a solute X. It is also soluble in an organic solvent and let partition co-efficient $P = \dfrac{C_{\text{org.}}}{C_{\text{aq.}}} = 4$. We have been given 100 c.c. organic solvent for extraction. Let original 100 c.c aq. solution contained 'a' gm solute and after adding 100 c.c. organic solvent amount unextracted = 'b' g.; so amount extracted = $(a - b)$ gm.

Hence
$$\frac{C_{\text{org.}}}{C_{\text{aq.}}} = \frac{(a-b)/100}{b/100} = 4$$

$$\text{or,} \quad \frac{a-b}{b} = 4 \quad \text{or,} \quad \frac{b}{a} = \frac{1}{5}.$$

Hence 20% is unextracted, i.e. 80% extracted.

Now instead of using 100 c.c. organic solvent at a time if we use it in two equal lots of 50 c.c. each, then let after first extraction b' gm is unextracted. So

$$\frac{(a-b')/50}{b'/100} = 4$$

$$\text{or,} \quad \frac{2a-2b'}{b'} = 4 \quad \text{or,} \quad \frac{b'}{a} = \frac{1}{3}.$$

So after 1st extraction 33.3% is unextracted i.e. 66.6% is extracted. When we use second lot of 50 c.c. then 1/3 of 33.3% i.e. 11.1% will remain unextracted; so extracted amount is 22.2%. Hence total extraction is (66.6 + 22.2)% = 88.8%. This is greater than first case.

Now let us derive some formula. Let we have V litre of an aqueous solution containing 'x' gm solute. We are using an organic solvent to extract it. We use L litre of it in each lot. Let partition co-efficient = K and after first extraction x_1 gm is unextracted. Then

$$K = \frac{\text{Corg.}}{C_{\text{aq.}}} = \frac{(x - x_1)/L}{x_1/V} = \frac{(x - x_1)\,V}{x_1 L}$$

$$\text{or,} \quad x_1 = x\left(\frac{V}{KL + V}\right).$$

Proceeding in a similar manner, we can show that amount unextracted (x_2) after second extraction is

$$x_2 = x_1\left(\frac{V}{KL + V}\right) = x\left(\frac{V}{KL + V}\right)^2.$$

So after n^{th} time extraction, amount unextracted (x_n) will be –

$$x_n = x\left(\frac{V}{KL + V}\right)^n. \qquad \text{....(10.10)}$$

10.8 SOLUTION OF LIQUID IN LIQUID

Solubility of a liquid in another liquid depends on polar characters of them. The general principle is *Like likes a like*. Thus a polar liquid is soluble in polar liquid and a non polar prefers another non-polar as solvent. Depending on relative polarities three different cases may arise. These are —

(a) The liquids may be partially miscible, where one has a distinct solubility in the other. e.g. phenol-water; *n*-hexane-water etc.

(b) The liquids may be completely miscible, having no solubility limit eg. water-HCl; water-ethanol etc.

(c) The liquids may be completely immiscible e.g. water-nitrobenzene; water-aniline etc. We shall discuss them one by one —

(a) Partially-miscible liquid pair : Partially miscible liquid pairs are of three types. In first case solubility of a liquid in the other increase with temperature (e.g. phenol-water); in the second case solubility decreases with rise in temperature (e.g. triethylamine-water) and in the third case solubility increases with temperature above a certain temperature value and decreases with temperature below that temperature value. (e.g. nicotine-water).

Case-1. Phenol-water system

If we take water in a beaker at a certain temperature and add phenol to it drop by drop

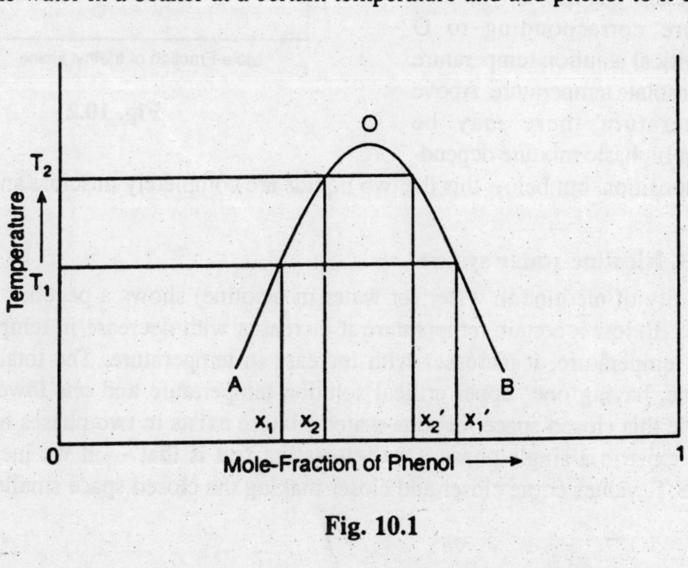

Fig. 10.1

then at the beginning phenol will be completely dissolved. But very soon the solubility will touch the limiting value. If we still add phenol we shall get a second liquid layer. While the first layer is a solution of phenol in water this new layer is a solution of water in phenol. They will remain side by side. Let x_1 be the maximum mole-fraction of phenol in water at T_1. Since dissolution of phenol in water is endothermic, so its solubility will rise with tenperature (x_2 at T_2). Thus OA is the solubility curve of phenol in water. Similarly if we gradually add water in phenol at various temperatures and plot the solubility of water in phenol vs temperature we shall get curve OB. At O (t = 68.4°C) the two curves meet. This means that phenol and water are completely miscible. This temperature is called *critical solution temperature* (or c.s.t) or *upper consolate temperature.* Below this temperature phenol and water will be completely miscible and remain in one phase if conc. of phenol is outside this curved area. But they will remain separated in two layers if conc. of phenol is within this curved area. One of these two layers will be solution of phenol in water while the other will be solution of water in phenol and there conc. will be fixed at a given temperature. Thus at temperature T_1 if some one add such amount of phenol in water that its mole-fraction is in between x_1 and x_1' then we shall get two layers, in one of which mole-fraction of phenol will be x_1 and in the other it is x_1'. Similarly at temperature T_2 if we get two layers then mole-fraction of phenol in these layers will be x_2 and x_2'. C.S.T. values are highly sensitive towards impurities. Trace amount of impurity present in a liquid can be detected by measuring its C.S.T. value with some other known liquid.

Case-2. Triethylamine-water system

Another example of partially miscible liquid pair is triethylamine-water system. However, unlike phenol-water system, here solubility decreases with increase in temperature, because the solution process is exothermic. So here temperature vs composition diagram is the inverted form of previous one. (Fig.10.2)

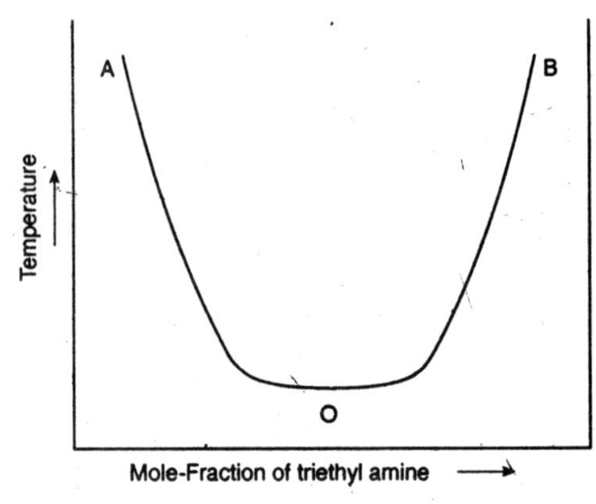

Fig. 10.2

Here OA is solubility curve of triethylamine in water and OB is solubility curve of water in triethylamine. Temperature corresponding to O (18°C) is critical solution temperature or lower consolate temperature. Above this temperature there may be uniphasic or biphasic mixture depending on composition, but below this the two liquids are completely miscible and there is only one phase.

Case-3. Nicotine-water system

Solubility of nicotine in water (or water in nicotine) shows a peculiar variation with temperature. Below a certain temperature it increases with decrease in temperature, while above this temperature, it increases with increase in temperature. The total curve is thus a closed one, having one upper critical solution temperature and one lower C.S.T. (Fig. 10.3). Inside this closed space nicotine-water mixture exists in two phases but outside this space they exist in a single phase. One interesting fact is that — if we increase pressure the two C.S.T. values come closer and closer making the closed space smaller and smaller.

At a certain pressure, the space vanishes and the two liquids becomes miscible in all proportions.

An interesting point to note is that theoretically all partially miscible liquid pairs have a upper and a lower critical solution temperature. But for some-pairs upper value is so high that liquid will be vaporised before reaching that temperature,— and thus it becomes unattainable. For some-other pairs the lower value is so low that liquid will be freezed before reaching this value. Following table gives c.s.t. values of some other pairs.

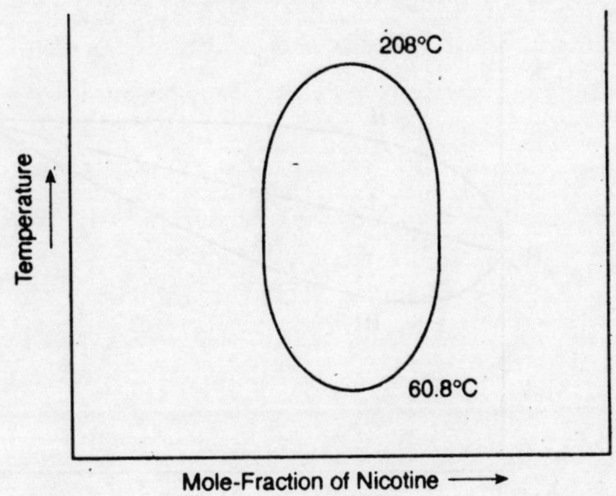

Fig. 10.3

Table 10.2

System	c.s.t. (°C)
Aniline-Hexane	59.6
Aniline-Water	167
Cyclohexane-methanol	46
Glycerol-*m*-toluidine	7 and 120

(b) Completly miscible liquid pair : If two liquids are miscible in all proportions i.e. one has no solubility limit in the other then the liquid pair is said to be completely miscible. Examples being — methanol-water; benzene-toluene; ethanol-water; HNO_3-water etc. Such solutions may be ideal or non-ideal.

Ideal solutions are those which obey Raoult's law. According to this law partial vapour pressure of one component of the liquid pair is the product of its vapour pressure in the pure state and its mole-fraction in solution. Thus if two liquids A and B from an ideal solution then their partial vapour pressure P_A and P_B should be

$$P_A = X_A P_A^0 \qquad ....(10.11)$$

$$P_B = X_B P_B^0 . \qquad ....(10.12)$$

So total pressure $P = P_A + P_B$

or, $P = X_A P_A^0 + X_B P_B^0$.

As $X_A + X_B = 1$.

$$\text{or, } P = P_B^0 + \left(P_A^0 - P_B^0\right) X_A \qquad ....(10.13)$$

$$\text{or, } P = P_A^0 + \left(P_B^0 - P_A^0\right) X_B \qquad ....(10.14)$$

Equations (10.13) and (10.14) indicate that in such case the total vapour pressure (P) of the solution is always intermediate between P_A^0 and P_B^0 and it varies linearly with mole-fraction of either of them. (Fig.10.3)

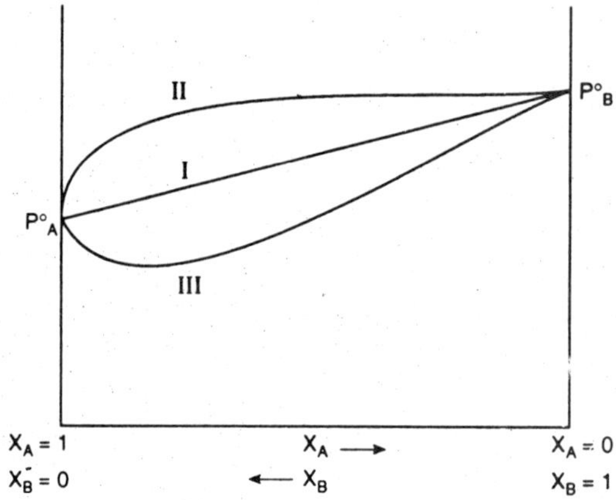

'B' is supposed to be more volatile.
Type -I : Ideal soln.
Type-II : Non-ideal showing V.P. maxima.
Type-III : Non-Ideal showing V.P. minima.

Fig. 10.4

For non ideal solution P_A is either greater than or less than $X_A P_A^0$. From Duhem-Margules equation (beyond the scope of this book) it can be proved that if $P_A > X_A P_A^0$ then P_B is also $> X_B P_A^0$; so that vapour pressure of solution is always more than either of the pure component and reaches a maximum at some composition (Fig. 10.4). Similarly if $P_A < X_A P_A^0$ then P_B is also $< X_B P_A^0$ and so total vapour pressure is always less than either of the pure liquid and reaches a minimum-value at some composition (Fig. 10.4). Lines of figure 10.4 are called *P–X diagrams*.

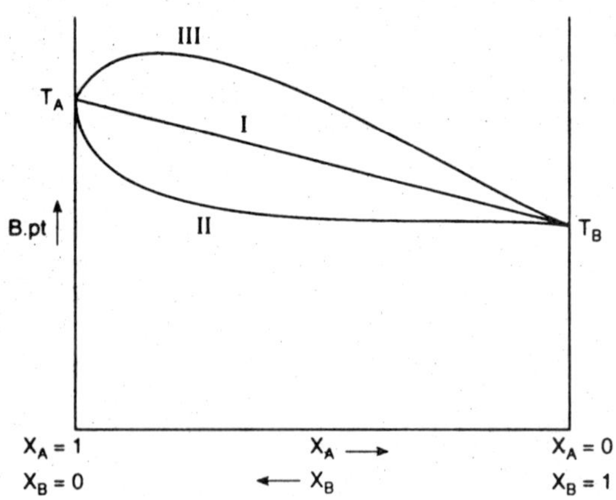

Fig. 10.5

Above P–X diagrams have been drawn at a particular temperature. Now boiling point of a liquid is the temperature at which its vapour pressure becomes equal to atmospheric pressure. Since total vapour pressure over a completely miscible liquid pair is dependent on composition, so different compositions will have different boiling points. Composition having maximum vapour pressure will have minimum boiling point and composition having minimum vapour-pressure will have maximum boiling point. Thus like vapour pressure vs composition (P–X) diagrams, boiling point Vs composition (T–X) diagrams can also be drawn (Fig. 10.5). This will be reverse of the previous diagram.

10.9 FRACTIONAL DISTILLATION

Fractional distillation is a process by which a liquid can be seperated from its solution in another liquid. Complete separation by fractional distillation is possible only for liquid

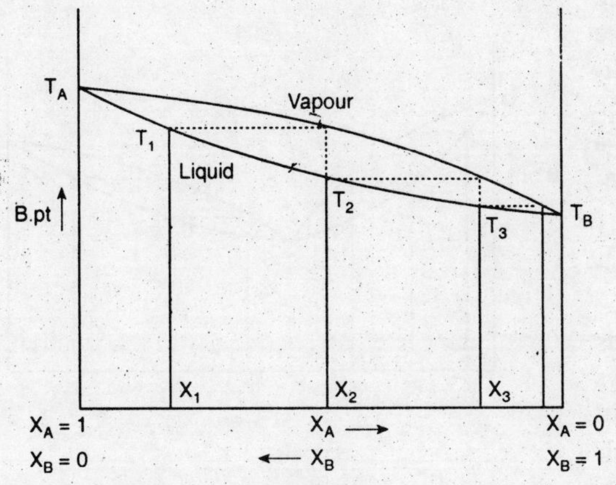

Fig. 10.6

pairs showing ideal behaviour. In such cases since the two liquids are completely miscible their composition in the liquid phase can be made whatever we like; but the vapour phase above the mixture will always be enriched by more volatile part. Hence compositions of vapour phase and liquid phase remaining in eqm. at boiling point are different and vapour phase is richer in more valatile part. A lense-shaped T–X diagram (fig. 10.6) can be drawn, where upper line indicates composition of vapour phase and lower line indicates that of liquid phase. [Upper-line is richer in 'B' i.e. more volatile part.]

Now let we have a liquid mixture where mole-fraction of A is X_1. It will boil at a temperature T_1 and the mole-fraction of the vapour remaining in eqm. with it will be X_2. If this vapour is condensed we shall get a new liquid mixture where mole-fraction of A will be X_2. This will boil at a lower temperature T_2 and the vapour above it will be richer in B (mole-fraction of A is further decreased to X_3). If this is condensed and the new liquid is again distilled we shall get vapour which is further enriched in B. In this way if the process is repeated we shall get more volatile B as distillate and less volatile A as residue. This is the principle of fractional distillation.

10.10 EXPERIMENTAL METHOD OF FRACTIONAL DISTILLATION

We have seen that during fractional distillation alternate distillation and condensation of the vapour is necessary. The vapour distilled off by the first distillation is richer in more volatile part and hence when it is condensed the new liquid boils at a lower temperature. Its boiling gives a vapour which is further richer in B and the above process is repeated. In laboratory these alternate boiling and condensation at different temperatures are done in a tall tube, known as fractionating column (Fig. 10.7). It is fitted with distilling flask where the liquid mixture of concentration X_1 (mole fraction of A) is boiled. It vaporises at T_1. Since the tube is very tall temperature of it falls as we go upwards. Thus $T_1 > T_2 > T_3 >$ ....etc. This vapour at T_1 when goes upward is condensed and trickles down the column. This will boil at a lower temperature say T_2. Thus it will again be vaporised. This new vapour will be condensed at still some upper part of the column and while

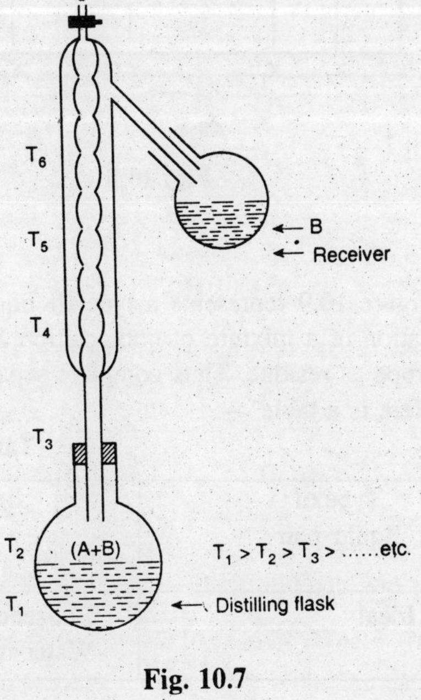

Fig. 10.7

trickling down it will be again vaporised say at T_3. Thus alternate vaporisation and condensations are done. More volatile part B will ultimately come out from the top of the column and it will be condensed in a receiver. Less volatile part will remain in distilling flask.

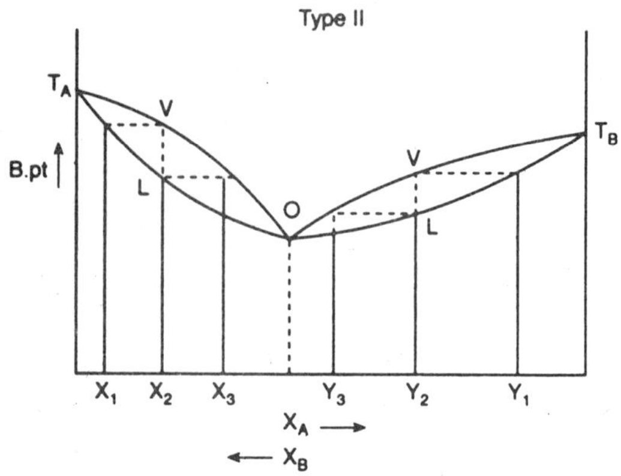

Fig. 10.8

10.11 NON-IDEAL SOLUTION-AZEOTROPIC MIXTURE

For ideal solution of two liquids composition of the vapour is always different from that of liquid mixture;vapour is richer in more volatile part. However, for non-ideal solutions (Types-II & III) showing boiling point minimum and maximum respectively, composition of vapour phase and liquid phase becomes same at the point of maximum departure. So at this point the liquid will boil with unchanged composition and hence fractional distillation will stop. Liquid mixture corresponding to this composition is called azeotropic (latin word-meaning constant boiling) mixture. So in such cases complete separation by fractional distillation is not possible. Let us examine following figures.

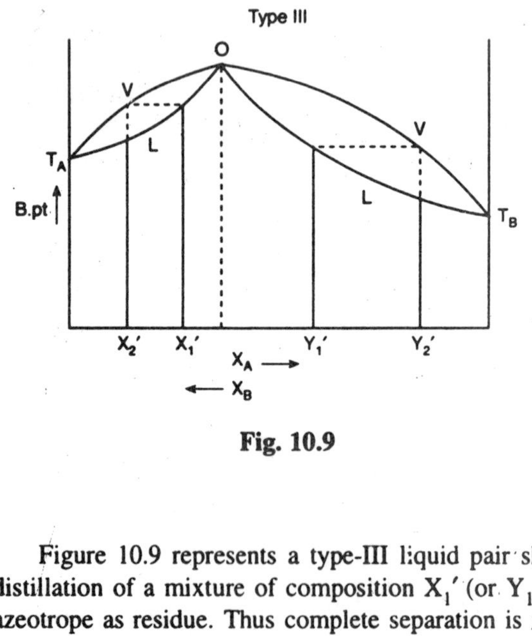

Fig. 10.9

Figure 10.8 represents boiling point vs composition diagram for type-II liquid mixtures. At 'O' composition of liquid and vapour phase is same. This is composition of azeotrope. If a liquid mixture of composition X_1 (or Y_1) is subjected to fractional distillation we shall get azeotrope as distillate and pure A (or pure B) as residue.

Figure 10.9 represents a type-III liquid pair showing boiling point maximum. Here distillation of a mixture of composition X_1' (or Y_1') will give A (or B) as distillate and azeotrope as residue. Thus complete separation is not possible.

Here is a table —

Table 10.2

Type of liquid pair	Example	Result of distillation D–Distillate R–Residue
Ideal	Benzene-toluene Water-methanol	D–More volatile part R–Less volatile part

Non-ideal boiling point minimum	Water-ethyl alcohol Acetic acid-toluene	D–Azeotrope. R–Either of the two depending on starting condition.
Non-ideal boiling point maximum	Water-HCl Water-HNO_3	D–Either of the two depending on starting condition R–Azeotrope

Water and ethyl alcohol form an azeotrope which is 95.6% with respect to alcohol. This is rectified spirit. 100% alcohol can not be obtained by fractional distillation because distillation stops after reaching azeotropic composition. Any further separation require special technique e.g. use of lime, sodium etc. That is why cost of absolute alcohol (100%) is much more than that of rectified spirit.

10.12 COMPLETELY IMMISCIBLE LIQUID PAIR-STEAM DISTILLATION

If two liquids are completely miscible then the total vapour pressure over them is equal to summation of vapour pressure of pure liquids. Since it is more than V.P. of either of the pure liquids so the mixture will boil at a temperature which is less than the boiling point of either of the two. Thus if one liquid is water, the mixture will boil below 100°C. This will prevent many organic liquids from decomposing during distillation. The principle is called steam-distillation. Generally the impure organic liquid, (immiscible with water) is mixed with water (or steam is passed over it) and the mixture is distilled. As it boils below 100°C, the organic liquid will not be decomposed. Pure organic liquid and water come as distillate. Since they are immiscible they exist as two separate layers and can be easily separated by a separating funnel. Weight ratio of water and organic liquid can be calculated as follows. If $P^0_{H_2O}$ and P^0_{org} are vapour pressure of water and organic liquid then total vapour pressure $P = P^0_{H_2O} + P^0_{org}$. The temperature at which P becomes 760m.m. will be the boiling point and in the vapour mixture mole-fractions will be —

$$X_{H_2O} = \frac{n_{H_2O}}{\Sigma n} = \frac{P^0_{H_2O}}{P}$$

$$\text{and} \quad X_{org} = \frac{n_{org}}{\Sigma n} = \frac{P^0_{org}}{P}. \qquad [n = \text{no. of moles}]$$

So taking the ratio

$$\frac{n_{org}}{n_{H_2O}} = \frac{P^0_{org}}{P^0_{H_2O}}$$

If 'g' is weight and M is molecular weight then

$$\frac{g_{org}/M_{org}}{g_{H_2O}/M_{H_2O}} = \frac{P^0_{org}}{P^0_{H_2O}}$$

$$\text{or,} \quad \frac{g_{org}}{g_{H_2O}} = \frac{M_{org}\, P^0_{org}}{M_{H_2O}\, P^0_{H_2O}} \qquad \text{....(10.15)}$$

Thus weight ratio is evaluated.

POINTS TO REMEMBER

Homogeneous mixture of two or more substances called *solution.* Relative amounts of the constituents of solution are their concentration. They can be expressed in different units. These are (a) *percentage (W/W) or (W/V)* (b) *Normality* (c) *Molarity* (d) *Molality* and (e) *Mole-fraction. Normal, molar* and *w/v percentage* units are *temperature dependent,* the rest are independent. Solution of gas in liquid follows *Henry's law* : according to which at constant T and given volume of solvent $\frac{W}{P}$ = *Constant.*

Nernst's distribution law states that $\frac{C_1}{C_2}$ = *constant,* where C_1 and C_2 are concentrations of monomeric forms of a substance distributed between two immiscible forms of a liquid remaining side by side. When it is n-merised in a solvent (say solvent '1') then $\frac{\sqrt[n]{C_1}}{C_2}$ = *Constant.* The constant is called *partition* or *distribution co-efficient.* The principle is used for solvent extraction.

Mixture of liquids may be (a) *partially miscible* (b) *completely miscible* (c) *completely miscible.* Partially miscible pairs have a temperature below / above which the liquids are completely miscible. This temperature is called *critical solution temperature (c.s.t.).*

Completely miscible liquid pairs may be *ideal* or *non-ideal.* Ideal pairs can be completely separated by fractional distillation. Non-ideal pairs, during fractional distillation produce a mixture of constant composition. This is called *azeotropic mixture.*

QUESTIONS

SHORT TYPE

1. Define the term 'concentration'. When this term becomes relevant ?
2. What is the importance of 'molal'concentration over molar concentration ?
3. Which unit of concentration is dimensionless ?
4. Prove that volume of a gas dissolved in a given volume of liquid at a given temperature is independent of its pressure.
5. What is partition co-efficient ?.
6. A solute exists as monomer in solvent I and as n-mer in solvent II; the two solvents are immiscible — what is the expression for its partition co-efficient ?
7. What is critical solution temperature ? What is its utility ?
8. State a system which shows both upper and lower consolate temperature.
9. What is an azeotrope ? How can you prove that azeotropes are mixture and net compounds ?
10. What is the utility of steam-distillation ?

ESSAY TYPE

1. State Dalton's law of partial pressure. Establish a relation between partial pressure and mole-fraction.
2. State Henry's law. Express it in the form of Nernst's law for partition co-efficient. What is absorption co-efficient ?
3. Taking any arbitrary example explain how Nernst's distribution law has been applied in the case of solvent extraction.
4. Explain the temperature vs composition diagram of phenol-water system.
5. Draw P–X and hence T–X diagrams for different types of completely miscible liquid pairs.
6. Describe with help of a T–X diagram the principle of fractional distillation. In which case complete separation of liquids is possible ?
7. Discuss the principle of steam distillation. Show that during steam distillation, the distillate generally contains large amount of organic liquid in spite of its low valatility.

PROBLEMS

1. A sample of air containing N_2 (75%); O_2 (24.8%) and CO_2 (0.2%) is in equilibrium with water at 1 atm. If absorption co-efficients are 0.022 for N_2 and CO_2 and 0.049 for O_2; find out their percentages in the dissolved gas. Given vapour pressure of water at the given state = 24 m.m.

   [*Hints.* Find partial pressure in atm. of each; total pressure = (760 – 24) = 736 m.m. = 0.968 atm.

   Volume dissolved / C.C. = Partial pressure × absorption co-efficient.]

2. Show that the following data for distribution of benzoic acid between benzene and water at 25°C shows that it is dimerised in benzene layer.

C_{H_2O}	$C_{C_6H_6}$
0.0150	0.242
0.0195	0.412
0.0289	0.970

[*Hints*. Find both $\frac{C_{org}}{C_{aq}}$ and $\frac{\sqrt{C_{org}}}{C_{aq}}$ and show that latter is almost constant.]

3. An organic compound is extracted from aqueous solution with successive quantities of 25 c.c. chloroform. The original volume of solution is 500 c.c and the distribution co-efficient of the compound is 20 between chloroform/water. Calculate the number of extractions needed for at least 95% recovery of the compount.

[*Hints*.

$X_n = X\left(\frac{V}{KL+V}\right)^n$; here $K = 20$; $L = 25$; $V = 500$; put $X_0 = 100$ and $X_n = 5$ to get $n = 4.3$, so answer is '5'.]

4. An organic compound (Molecular weight = 204) immiscible with water, is steam distilled under 1atm pressure. Vapour pressure of water at the temperature of distillation is 710 m.m. What weight per cent of organic liquid is present in the distillate.

[*Hints*.

In the distillate $\frac{g_{org}}{g_{H_2O}} = \frac{M_{org}\,P^0_{org}}{M_{H_2O}\,P^0_{H_2O}}$; $P^0_{H_2O} + P^0_{org} = 760$.]

OBJECTIVE TYPE

1. **Write T (true) or F (false) inside the boxes besides the following statements.**

(a) Normality unit is temperature dependent. ☐

(b) Molal unit is temperature independent. ☐

(c) For benzoic acid distributed between H_2O & C_6H_6, $\frac{C_1}{C_2}$ is constant. ☐

(d) Distribution law was proposed by Gibbs. ☐

(e) Phenol-water system shows lower c.s.t. ☐

(f) Ideal solution can be completely separated by fractional distillation ☐

(g) Water-HCl forms an azeotropic mixture ☐.

(h) In steam-distillation mixture always boil above 100°C. ☐

2. **Fill in the blanks with the appropriate word given in the brackett.**

(a) For (1 – 1) type electrolyte 'normal' and 'molar' values ———

[**Ans.** (i) are same (ii) are different (iii) cannot be compared]

(b) According to Henry's law if pressure is doubled, volume of dissolved gas ———

[**Ans.** (i) will remain same (ii) will be doubled (iii) will be halved.]

(c) In solvent extraction technique better result is obtained if the solvent is used ———.

[**Ans.** (i) all at a time (ii) in two parts (iii) in a number of small parts].

(d) Nicotine water system has ———.

[**Ans.** (i) upper consolate temperature (ii) lower consolate temperature (iii) both upper and lower consolate temperature].

(e) An example of ideal solution is ———.

[**Ans.** (i) Water-ethanol (ii) Water-methanol (iii) Water-HCl].

3. Match terms of column 'A' with those of column 'B'.

A	B
(a) Nernst.	(a) Water-HNO_3
(b) Henry.	(b) Water-Nitrobenzene.
(c) Azeotrope.	(c) Distribution law.
(d) Molality.	(d) W/P = constant.
(e) Steam-distillation.	(e) Temperature independent.

Answers :

1. (a) T ; (b) T ; (c) F ; (d) F ; (e) F ; (f) T ; (g) T ; (h) F.

2. (a) i ; (b) i ; (c) iii ; (d) iii ; (e) ii.

3. a – c ; b – d ; c – a ; d – e ; e – b.

Chapter 11

DILUTE SOLUTION — COLLIGATIVE PROPERTY

11.1 INTRODUCTION-DEFINITION OF COLLIGATIVE PROPERTY

Addition of a solid solute to a solvent generally lowers its vapour pressure; raises its boiling point; depresses its freezing point and develops osmotic pressure. It is found that for dilute solutions of non-volatile, non-electrolyte solid solutes, lowering of vapour pressure, elevation of boiling point, depression of freezing point and osmotic pressure depend only on concentration of solute and not on its nature. Since these properties depend only on concentration i.e. number of particles in a given volume, they are called colligative (Greek : **colligatus** – collected together) properties. We shall discuss all these four properties in detail.

11.2 LOWERING OF VAPOUR PRESSURE

(a) Reason for lowering : In a liquid there are few molecules having high translational K. E. they can overcome their neighbour's attraction and escape from the surface. Thus evaporation occurs. If the liquid is kept in a closed vessel, then very soon an equilibrium is established between liquid and vapour. Pressure due to the vapour in equilibrium with the liquid is called vapour pressure. It is a function of temperature and increases with increase in temperature (vide-Clausius-Clapeyson equation). If now some non-volatile solid solute is dissolved in the solvent, its particles will attract the solvent molecules and so escaping tendency of solvent will decrease and hence less vapour will be accumulated. Consequently vapour pressure will be lowered.

(b) Lowering - Relative Lowering - Raoult's law : Let P^0 be the vapour pressure of pure solvent and P be the vapour pressure of solution then $(P^0 - P)$ is lowering of vapour pressure and $\frac{P^0 - P}{P^0}$ is relative lowering of vapour pressure. Raoult gave the following laws regarding lowering of vapour pressure.

"For dilute solution of non-volatile, non-electrolyte solid solute lowering of vapour pressure is directly proportional to mole-fraction of solute (x_2) and relative lowering of vapour pressure is equal to mole-fraction of solute." This law is temperature independent for dilute solution.

$$\text{Thus} \quad (P^0 - P) \propto x_2 \quad \text{and} \quad \frac{P^0 - P}{P^0} = x_2 \qquad \text{....(11.1)}$$

Raoult also told — "equimolecular quantities of different non-volatile, non-electrolyte solid solutes dissolved in the same mole of same solvent (to prepare a dil. solution) will lower the vapour pressure by the same extent." This statement in fact indicates that lowering of vapour pressure is a colligative property.

(c) Ideal solution-derivation of Raoult's law : Raoult's law for vapour pressure lowering can be derived from Raoult's definition of ideal solution. [Thermodynamic derivation is beyond the scope of this book]. According to Raoult an ideal solution is a solution where partial vapour pressure of solvent or solute is the product of its vapour pressure in the pure-state and its mole-fraction in solution. If the solute is non-volatile then its vapour pressure is zero. In such case vapour pressure of solution (P) is equal to partial vapour pressure of solvent alone.

So $\quad P = x_1 P^0 \qquad [x_1 = \text{mole-fraction of solvent}]$.

But $\quad x_1 + x_2 = 1$

So $\quad P = (1 - x_2) P^0$

or, $\quad \dfrac{P^0 - P}{P^0} = x_2$.

This is Raoult's law.

(d) Application of Raoult's law : Let n_2 moles of solute be dissolved in n_1 mole solvent to make a dilute solution. Then

$$x_2 = \frac{n_2}{n_1 + n_2}.$$

So from Raoult's law —

$$\frac{P^0 - P}{P^0} = \frac{n_2}{n_1 + n_2}.$$

For dilute solution $n_2 << n_1$ then —

$$\frac{P^0 - P}{P^0} \simeq \frac{n_2}{n_1}.$$

Let 'a' gm solute (molar mass = m) is dissolved in 'b' gm solvent (molar mass = M), then

$n_2 = a/m$ and $n_1 = b/M$. So —

$$\frac{P^0 - P}{P^0} = \frac{aM}{bm}. \qquad \text{....(11.2)}$$

Thus knowing the concentration of solution (i.e., 'a' and 'b') and determining relative lowering of vapour pressure, we can determine molar mass (m) of solid solute. This is most important application of Raoult's law. Actually expression for any colligative property can be used to determine molar mass of SOLID substances.

(e) Experimental Method for Measuring $\dfrac{P^0 - P}{P^0}$: Raoult measured individual vapour pressures of pure solvent and solution by introducing solvent and solution separately in the Toricellian vacuum of a barometer and noting the depression of Hg-level. The method is not accurate as lowering of vapour pressure is very small. A very good method of

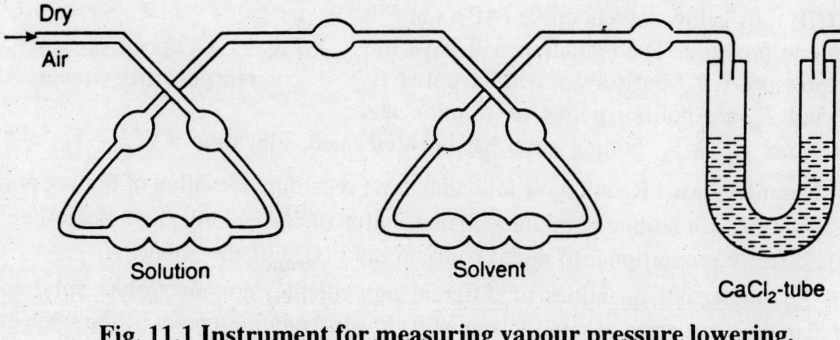

Fig. 11.1 Instrument for measuring vapour pressure lowering.

measuring relative lowering of vapour pressure is Ostwald and Walker's dynamic method. Their apparatus consists of three parts; the first consisting of few interconnected bulbs (Fig. 11.1) containing solution of known concentration. The second part which is same as first part has pure solvent inside. The third part is a U-tube containing fused $CaCl_2$. All these parts are separately weighed and are interconnected. Now a stream of perfectly dry air is passed through solution for sometime. Dry air will collect vapour from solution, till it will be saturated with its vapour pressure (P). Solution will loose some weight and weight loss is proportional to P. This air next enters solvent bulbs whose $V.P$ is more (P^0). It will collect vapour from here till it becomes saturated with respect to P^0. So its weight loss will be proportional to ($P^0 - P$). Air saturated with vapour pressure of solvent now enters $CaCl_2$ tube which absorbs all the vapour and dry air will come out of it. Its weight gain is proportional to vapour pressure of solvent (P^0). Thus we see —

Loss in weight of solution $\propto P$.

Loss in weight of solvent $\propto (P^0 - P)$ and

Gain in weight of Calcium chloride tube $\propto P^0$.

So $$\frac{\text{Loss in weight of solvent}}{\text{Gain in weight of } CaCl_2\text{-tube}} = \frac{P^0 - P}{P^0}.$$

Hence relative lowering of vapour press can be known.

(f) Critical comments : Lowering of vapour pressure will be observed only for non-volatile solutes. For volatile solutes, solute vapour will also reside over solution and hence vapour pressure may not decrease. Here

$$P = x_1 P^0_1 + x_2 P^0_2$$

x_1 and x_2 are mole-fraction of solvent and solute; P^0_1 and P^0_2 are vapour pressures of pure solvent and solute.

11.3 ELEVATION OF BOILING POINT

(a) Reason for Elevation : We know that boiling point of a liquid is the temperature at which its vapour pressure becomes equal to atmospheric pressure. Now at any temperature vapour pressure of solvent is more than that of solution. So during heating vapour pressure of solvent will touch one atmosphere mark before solution; hence boiling point of solution will be more. This is clearly understood from the adjacent figure. (Fig.11.2)

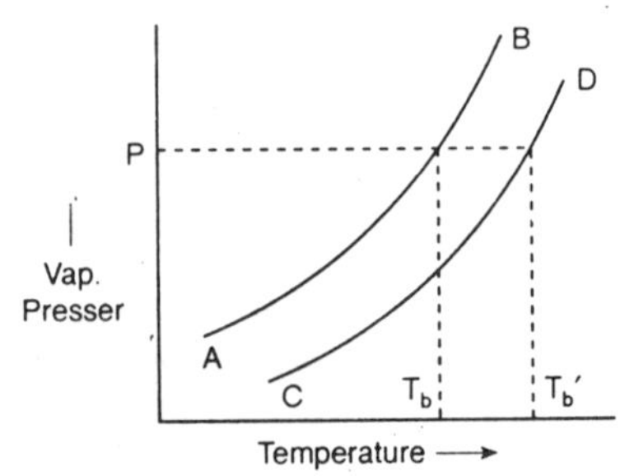

Fig. 11.2 Vapour pressure vs temperature curves.

Here we have plotted vapour pressure vs temperature curves for both solvent and solution. Solution curve (CD) runs below solvent curve (AB). Let P is atmospheric pressure. Then solution will have this vapour pressure at T_b' but solvent will have it at T_b. So T_b' and T_b are boiling points of solution and solvent. Since $T_b' > T_b$, boiling point has been elevated. Elevation $= (T_b' - T_b) = \Delta T_b$.

(b) Raoult's Law : Raoult gave following laws regarding elevation of boiling point —

(i) Elevation of boiling point for a dilute solution of non-volatile, non-electrolyte solid solute is directly proportional to molal concentration (C_m) of the solute.

(ii) Equimolecular quantities of different non-volatile, non-electrolyte solid solutes dissolved in the same gram of same solvent will elevate the boiling point by the same extent.

Thus if ΔT_b is elevation of boiling point and C_m is molal concentration, then

$$\Delta T_b \propto C_m$$

$$\text{or,} \quad \Delta T_b = K_b \cdot C_m \qquad \text{....(11.3)}$$

Here K_b is a constant, known as ebullioscopic constant or molal elevation of boiling point. Thermodynamics shows that $K_b = \frac{RT_b^2}{1000\, lv}$, where T_b = Boiling point of solvent and lv is latent heat of vaporisation /gm. Thus we see that K_b is a solvent property. Its unit is deg. kg. mole^{-1}.

(c) Application of Raoult's law : Let 'a' gm. solute (molar mass = m) is dissolved in 'b' gm. solvent (molar mass = M). Then molal concentration of solute $C_m = \frac{a \times 1000}{b \times m}$.

Now putting values of K_b and C_m in equation (11.3) we get —

$$\Delta T_b = \frac{RT_b^2}{1000\, lv} \cdot \frac{a \times 1000}{b \times m}. \qquad \text{....(11.4)}$$

Thus for a solution of known concentration in a known solvent if we can measure ΔT_b we can find out 'm' i.e., molar mass of the solute. Thus this colligative property is also used to determine molar mass of solute.

(d) Experimental methods for measuring ΔT_b : Several methods are available for measuring elevation of boiling point (ΔT_b). Two such methods will be described here. These are (i) Landsberger-Walker method and (ii) Cotrell's method.

(i) Landsberger-Walker method :

The apparatus used in this method is shown in fig.11.3. An inner tube (graduated) is taken. Inside it solvent or solution is taken. In a separate flask (not shown) solvent is boiled and its vapour is sent to this inner tube. The vapour enters in to solvent through a rose head and condenses here rejecting latent heat of condensation. This boils the solvent. A thermometer (accuracy 0.01°K) is dipped here to note the boiling temperature. Excess vapour comes out of this inner tube through a hole and goes to a condenser. After recording the boiling point of solvent supply of solvent vapour is temporarily cut off and small amount of solute (weighed) is added in the form of a pellet to make a dilute solution. Then the same process is repeated and boiling point of solution is measured. From these two values, elevation of boiling point can be known. This method does not give very good result. Actually elevation of boiling point is very small and ordinary thermometer cannot measure this small difference accurately. For this purpose Cotrell used Backmann's thermometer.

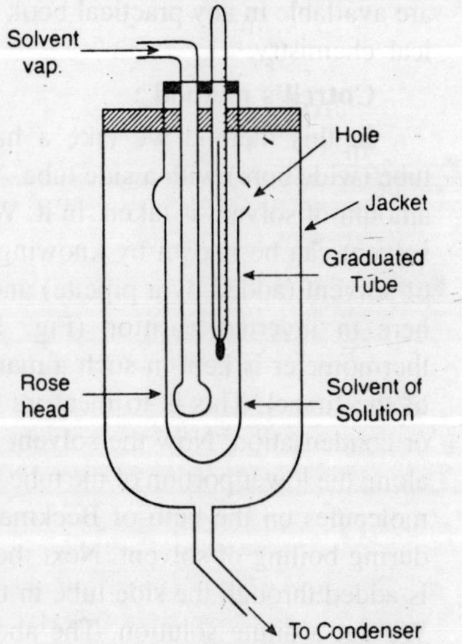

Fig. 11.3 Landsberger-Walker's instrument.

(ii) Cotrell's method :

Description of Beckmann's thermometer :

Beckmann's thermometer does not measure temperature of a substance. It measures temperature difference between two states. So its correct name is Beckmann's differential thermometer. It is a long tube (almost one metre) having two mercury reserviors. The lower reservior is like ordinary thermometer (Fig. 11.4) but the upper reservior is a U-tube. They are joined by a capillary. The length of the capillary is divided in to six equal large divisions and each large division is divided in to one hundred small divisions, so that a reading of 0.01°K can be taken very accurately. Before using Beckmann's thermometer it must be 'set'. The term setting means adjusting the amount of mercury in the lower reservior to such an amount that during experimental temperature mercury stands almost at the middle of the scale and it gets a scope of rising or falling through 3° during experiment like elevation of boiling point or depression of freezing point. Adjustment of Hg in the reservior is done by exchange between lower and upper reservior. Details are available in any practical book of physical chemistry.

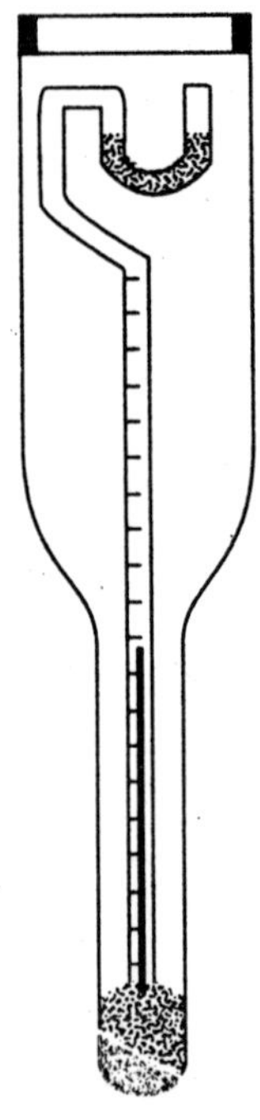

Fig. 11.4 Beckmann's thermometer.

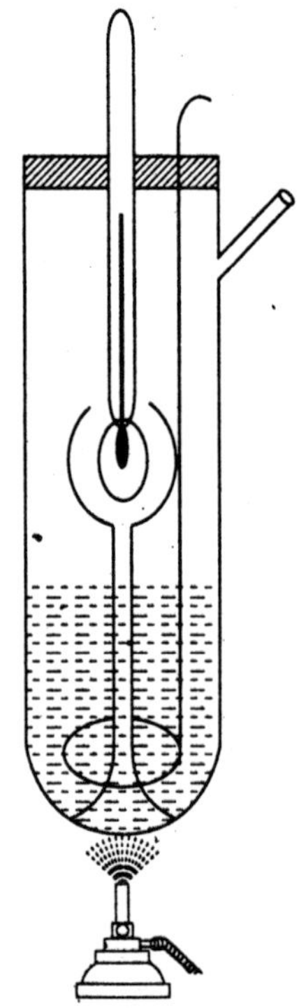

Fig. 11.5 Cottrell's instrument.

Cotrell's method :

In this method we take a hard-glass tube (wide bore) with a side tube. Weighed amount of solvent is taken in it. Weight of solvent can be known by knowing volume of solvent (added by a pipette) and its density. A funnel whose stem is bifurcated is kept here in inverted position (Fig. 11.5). A stirrer is also kept. Properly set Beckmann thermometer is kept in such a manner that its bulb is just in between the bifurcated ends of the funnel. This is to measure exact boiling temperature, avoiding either superheating or condensation. Now the solvent is boiled by a micro-burner which is constantly moving along the lower portion of the tube for uniform heating when the liquid boils it sprays vapour molecules on the bulb of Beckmann thermometer and we note the position of Hg-thread during boiling of solvent. Next the burner is removed and weighed amount of solid solute is added through the side tube in the form of pellet. We get a dilute solution. The above experiment is then repeated. New position of the Hg-thread in the thermometer is noted. Difference between the two positions give elevation of boiling point.

(e) Critical Comments : So far we have been saying that boiling point of a liquid is elevated when some solute is dissolved in it. However, if the solute is volatile then both liquid and vapour will contain solvent and solute. It can be shown by calculation that if mole-fraction of solute in the vapour phase is greater than mole-fraction of solvent in same phase, then there will be a depression of boiling point instead of elevation !

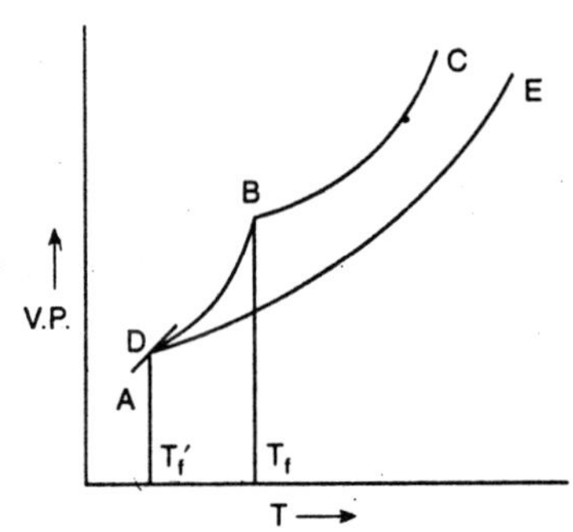

Fig. 11.6 V.P. vs T curves.

11.4 DEPRESSION OF FREEZING POINT

(a) Reason : Freezing point of a solvent is the temperature at which its vapour pressure is equal to that of solid. In the adjacent figure (Fig. 11.6) we have plotted vapour pressure vs temperature curves for solid solvent (AB); liquid solvent (BC) and dilute solution (DE). At B (temp : T_f), vap. pressure of solid = vap.pressure of solvent. So T_f is freezing point of solvent. At D (temperature : T_f'), vap. pressure of solid = vap. pressure of solution. So T_f' is freezing point of solution. Since $T_f' < T_f$ freezing point is depressed. Depression of freezing point = $(T_f - T_f') = \Delta T_f$.

(b) Raoult's Law : Raoult put forward two laws regarding depression of freezing point. These are similar to his laws regarding elevation of boiling point. Thus the laws are —

(i) Depression of freezing point (ΔT_f) of a dilute solution of non-volatile, non-electrolyte solid solute is directly proportional to molal concentration (C_m) of solute.

(ii) Equimolecular quantities of different non-volatile, non-electrolyte solid solutes dissolved in the same gram of same solvent will depress the freezing point by same amount.

Thus from first statement —

$$\Delta T_f \propto C_m$$

$$\text{or} \quad \Delta T_f = K_f \, . \, C_m \quad \text{....(11.5)}$$

Here K_f is a constant, known as cryoscopic constant or molal depression of freezing point constant. It is equal to depression of freezing point for a one molal solution. Its unit is deg. kg. mole^{-1}. From thermodynamics it can be proved that –

$$K_f = \frac{RT_f^{\,2}}{1000\, l_f} \text{ ,where } T_f = \text{Freezing point of solvent and } l_f = \text{latent heat of fusion/gm.}$$

Thus K_f is a solvent property.

(c) Application of Raoult's Law : Let 'a' gram solute (molar mass. = m) be dissolved in 'b' gram solvent (molar mass = M). Then molal conc. of solute $C_m = \dfrac{a \times 1000}{b \times m}$. So equation (11.5) becomes

$$\Delta T_f = \frac{RT_f^{\,2}}{1000\, l_f} \cdot \frac{a \times 1000}{b \times m}.$$

Hence if 'a' 'b' 'T_f' and l_f are known then measuring depression of freezing point we can determine molar mass of solute. This is the most important application of Raoult's law.

(d) Experimental Methods for determining depression of freezing point : Depression of freezing point is generally so small that ordinary thermometer fails to measure accurately such a small change in temperature. And so the cryoscopic method uses Beckmann's differential thermometer. However, for certain selected solutes camphor can be used as solvent and as K_f value of camphor is very high (nearly 40). ΔT_f value is also high and hence ordinary thermometer can be used in such cases. Rast did this and the method is known as Rast's method. We are discussing both methods here.

(i) Cryoscposic method (Beckmann's method) :

The instrument consists of a hard-glass tube (with wide bore) having a side tube. Inside it weighed amount of solvent is taken. A properly set Beckmann thermometer and a stirrer are kept inside. This tube is then kept inside another tube of wider bore and then it is dipped in a freezing mixture whose temperature is at least 6°C below the freezing point of solvent. Air gap between two tubes serves as a mantle and helps in uniform cooling. Stirrer is continuously stirred and when the solvent is solidified stirring can not be done. Position of the Hg-thread in Beckmann thermometer is noted. Now the inner tube is brought outside and rubbed with palm so that the solvent is liquefied again. Weighed amount of solute in the form of pellet is introduced through the side tube and a solution is made. Above experiment is then repeated and position of Hg-thread in Beckmann thermometer is noted. Difference between two readings gives ΔT_f.

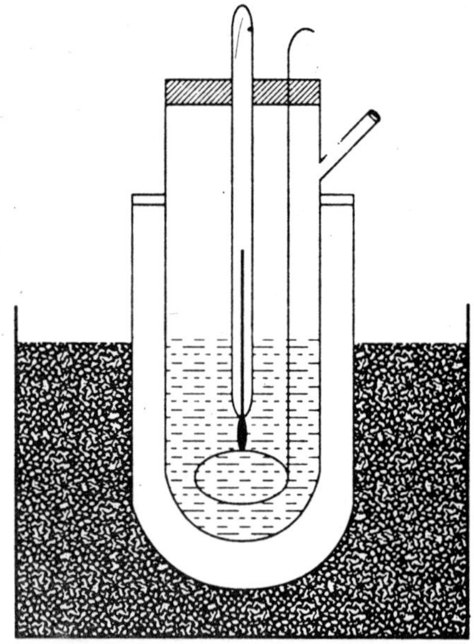

Fig. 11.7 Cryoscopic method.

(ii) Rast's method :

Commonly used solvents have low K_f - value. Thus ΔT_f for dilute solution is very low e.g. $K_f = 1.86$ for water and hence for 0.1 molal aq. sol^n. $\Delta T_f = 0.186$. But camphor has $K_f = 40$; so for 0.1 molal solution in camphor $\Delta T_f = 4°$; which can be measured conveniently by an ordinary thermometer. Rast did this for certain solutes which are soluble in camphor. Since camphor is usually solid, so instead of finding liquid camphor's freezing point he determined melting point of solid camphor, which is also freezing point of liquid camphor. Then he mixed known amount of solute with known amount of camphor and determined melting point of this solid mixture. This is freezing point of solution. The difference between two readings will give ΔT_f. Besides camphor some other solvents with high K_f – values are — camphor quinone ($K_f = 45.7$); Hexachloroethane ($K_f = 47.1$); tetrabromomethane ($K_f = 86.7$) etc.

(e) Critical comments : During freezing of a **dilute** solution generally solid solvent comes out and there is a depression of freezing point. However, if there is a solid-solid solubility and a solid solution crystallises then unexpected thing may occur. Thermodynamics shows that if mole-fraction of solute in solid solution is greater than mole-fraction of solvent in solid-solution then instead of depression, there will be an elevation of freezing point.

11.5 OSMOTIC PRESSURE

(a) Semi-permeable membrane : If in a beaker, containing some solution, we pour some water slowly, then solute particles from lower part (solution) of this mixture diffuse to upper part (water) against gravity to make a homogeneous mixture. This is the general rule of diffusion. Now if a solution is kept separated from solvent by fish's bladder; pig's bladder etc. then these membranes prevent flow of solute. To bring homogeniety here solvent enters to solution side through the membrane. Since these membranes allow solvent to pass through but prevents solute particles, they are called semi-permeable membrane. They may be (i) natural and (ii) artificial. Natural membranes are animal bladder; some fruit-shells etc. They can not withstand high pressure. That is why artificial membranes have been prepared. This can be made by depositing copper-ferrocyanide in the pores of a porous wall.

An earthen pot is filled with $K_4Fe(CN)_6$ solution. It is then placed inside a large beaker. The annular space between them is then filled with $CuSO_4$ solution. $CuSO_4$ tries to enter in to the earthen pot through its pores and $K_4Fe(CN)_6$ tries to come out through the same path. They react inside the pores and $Cu_2Fe(CN)_6$ is deposited. After sufficient time all the pores will be filled up. The earthen pot is then cleaned and dried. It will serve now as an excellent semi-permeable membrane.

(b) Osmosis and Osmotic pressure : Abbe Nollet took a thistle funnel; tied its mouth with a piece of pig's bladder (semi-permeable membrane) and then took sugar solution inside it. He next kept it in hanging condition with its mouth, dipped in water. He found that level of sugar solution is rising, which means that solvent water is spontaneously entering in to solution through semi-permeable membrane. Such a spontaneous flow of solvent towards solution, when they are separated by a semi-permeable membrane is called osmosis. Osmosis also occurs, when a dilute and a concentrated solution are separated by a semi-permeable membrane. In such case solvent from dilute side will enter to the concentrated side.

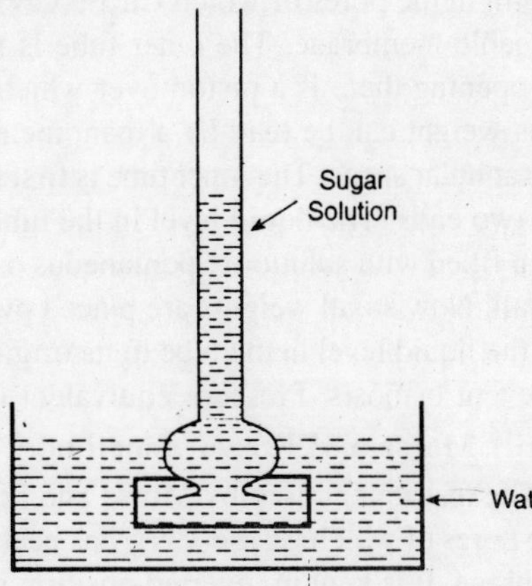

Fig. 11.8 Abbe Nollet's experiment.

When a solvent and solution are kept separated by a semi-permeable membrane, then minimum amount of pressure that is to be applied on solution side to check osmosis is called osmotic pressure.

(c) Reason for Osmosis : Reason for osmosis can be given both from kinetic and thermodynamic consideration. According to former solvent molecules on both sides of the membrane are exerting lateral pressure on it; but on solution side this pressure is less because solvent molecules are attracted by solute particles. So osmosis occurs and osmotic pressure is the difference in lateral pressures of solvent on solvent side and solution side.

According to thermodynamics chemical potential of a solution is always less than that of solvent. So, when solution and solvent are kept separated by a semi-permeable membrane, solvent enters in to solution to equalise chemical potential.

(d) Measurement of Osmotic pressure : There are many methods for measuring osmotic pressure of a solution. We shall discuss here only two among them. These are (i) Berkeley and Heartley's method and (ii) Morse and Frazer's method.

(i) Berkeley and Heartley's method :

This method is very simple and a quick measurement of osmotic pressure can be done by this method. The instrument consists of two concentric tubes; the inner tube is made of

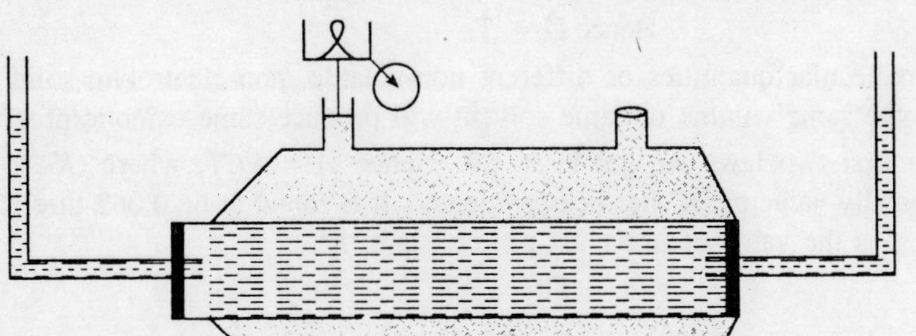

Fig. 11.9 Berkeley and Heartley's method.

porcelin in the pores of which $Cu_2Fe(CN)_6$ has been deposited. So its wall will act as semi-permeable membrane. The outer tube is made of gun–metal having two openings. In one such opening there is a piston over which weights can be placed and pressure, equivalent to this weight can be read by a manometer. By the other opening liquid can be introduced in the annular space. The inner tube is first filled with solvent by one of the two tubes attached at its two ends. The liquid level in the tubes is noted. The annular space between two tubes is then filled with solution. Spontaneous osmosis will occur and liquid level in the side tubes will fall. Now small weights are placed over the piston. We place such weight which brings back the liquid level in the tube to its original position. This is the minimum weight required to prevent osmosis. Pressure equivalent to this weight is read. This is osmotic pressure.

(ii) Morse and Frazer's method :

The instrument used by Morse and Frazer is shown in fig. 11.10. 'A' is a porcelin tube in the pores of which $Cu_2Fe(CN)_6$ has been deposited i.e., wall of 'A' acts as semi-permeable membrane. It is kept in inverted position inside another tube 'B' which has two open mouths. To its lower mouth, mouth of 'A' is attached through a cork, and another tube (U-shaped) is attached with 'A' by another cork. 'B' is now filled with solution, whose osmotic pressure is to be measured. To its upper mouth a Hg-manometer is attached. Tube 'A' is now filled with solvent through the side tube, spontaneous osmosis will take place. This will continue till the difference in pressure on the two limbs of manometer balances the osmotic pressure. From the manometer reading osmotic pressure can be read.

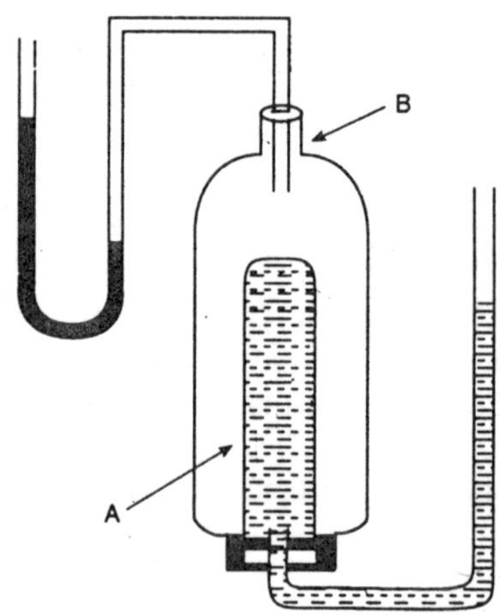

Fig. 11.10 Morse and Frazer's method.

(e) Laws of Osmotic pressure : Laws of osmotic pressure had been given by vant' Hoff. The laws can be studied under three parts. These are —

At constant temperature, Osmotic pressure (Π) of a dilute solution of non-volatile, non-electrolyte solid solute is proportional to molar concentration (C) of solute.

$$\text{So} \quad \Pi \propto C.$$

For a dilute solution of non-volatile, non-electrolyte solid solute at a given concentration, osmotic pressure (Π) is directly proportional to absolute temperature (T).

$$\text{Hence } \Pi \propto T.$$

Equimolecular quantities of different non-volatile, non-electrolyte solid solutes dissolved in the same volume of same solvent will produce same osmotic pressure.

From first two laws we get $\Pi \propto CT$. Hence $\Pi = KCT$, where 'K' is a constant. Experimentally value of 'K' has been evaluated. It is found to be 0.082 litre-atmos mole^{-1} deg^{-1}. This is the value of universal gas constant 'R'.

$$\text{So} \quad \Pi = CRT. \qquad ....(11.6)$$

(f) Application of the law — analogy with ideal gas equation : So we have seen that $\Pi = CRT$, where C = molar concentration. If 'n' moles of solute is dissolved in V litre of solvent then $C = \frac{n}{V}$. In such case

$$\Pi = \frac{n}{V} . RT \quad \text{or} \quad \Pi V = nRT. \qquad ....(11.7)$$

If weight of solute = 'g' gm. and its molar mass = m, then $n = g/m$,

$$\text{so} \quad \Pi V = \frac{g}{m} RT. \qquad ....(11.8)$$

Equation (11.8) can be employed to determine molar mass of solid solute. So, once again we see that expression for colligative property is being used to determine molar mass of solid solute.

A look at equation (11.7) [$\Pi V = nRT$] tells us about the striking similarity of this equation with ideal gas equation. So osmotic pressure of a solution is equal to pressure of an ideal gas, whose volume and temperature are equal to volume and temperature of solution and whose number of moles is equal to number of moles of dissolved solute. This similarity is due to the fact that a dissolved solute will behave as a gas. Actually when a solute is dissolved in a solvent the solid solute breaks down in to molecular dimension and particles become free to move anywhere within the volume of the solution. For dilute solution particle density is small. So they will behave as an ideal gas with its volume equal to that of solution.

(g) Plasmolysis — Isotonic solution : Walls of animal or vegetable cells are semi-permeable membrane. The fluids inside cells are solution of sugar etc. So such a cell, when placed inside a solution (whose osmotic pressure may be greater or less than it) then solvent may enter in the cell or come out from the cell depending on respective osmotic pressures. Thus there may be a shrinkage or swallowing of cells. From this an idea about its osmotic pressure can be obtained. Such processes of comparison of osmotic pressures are called plasmolysis.

During plasmolysis if two solutions cause same amount of shrinkage of a particular cell wall then they are said to be iso-tonic. Thus iso-tonic appears to be same as iso-osmotic. In many cases two iso-tonic solutions have same osmotic pressure but there are exceptions also. Shrinkage of a cell wall depends not only on concentration of solution inside and outside but also nature of cell membrane, which are not perfectly semi-permeable. Thus same shrinkage does not mean that their conc, and hence osmotic pressure are exactly same. Blood plasma i.e., cellular fluid of RBC has an osmotic pressure of about 7.4 atmosphere. A 0.84% NaCl solution is iso-tonic with it and that is why it is used as normal saline.

(h) A fun : An interesting demonstration of plasmolysis can be done by making an experiment with an egg. The hard outer cover of an egg ($CaCO_3$) is removed by dipping it in HCl. The thin white membrane comes out with the fluid inside. It is now placed in water. Osmosis will occur, water will begin to enter inside and ultimately the egg will burst.

(i) Abnormal colligative properties : We have said at the beginning that expressions for colligative properties are valid for dilute solutions of non-volatile, non-electrolyte solid solute. Let us try to know the reason behind it.

Dilution :

All the expressions for colligative properties are derived assuming ideal-behaviour of solution. And ideal solution means dilute solution where molar properties are considered to be equal with partial molar properties.

Volatility :

Solutes are further supposed to be non-volatile. Because if solutes are volatile, then vapour of solutes will also reside over solvent. Consequently there may be a rise of vapour pressure, instead of lowering; also there may be depression of boiling point instead of elevation.

Electrolyte :

vant' Hoff found that electrolytes do not obey equations for colligative properties. In fact, for electrolytes, experimental values of colligative properties are greater than theoretical value. Thus for electrolytes $\Pi = iCRT$; $\Delta T_b = i\, K_b . C_m$;

$\Delta T_f = i\, K_f\, C_m$ and $\dfrac{P^0 - P}{P^0} = ix_2$. The term '$i$', known as vant' Hoff's factor is actually —

$$i = \frac{\text{Experimental value of colligative property}}{\text{Theoretical value of that property}}.$$

The reason for such an increase in colligative property values is given on the basis of theory of electrolytic dissociation. Due to dissociation, number of particles increases in the solution and since colligative properties are particle-dependent, so their values also increase. An expression for degree of dissociation can be obtained. Let one molecule of a particular electrolyte dissociates in to 'n' particles; then if its degree of dissociation is 'α' and we start with 1 mole electrolyte then we get $(1 - \alpha)$ mole undissociated electrolyte and '$n\,\alpha$' mole dissociated particles. So total moles obtained = $(1 - \alpha) + n\alpha$. As there was initially '1' mole, so this is the factor, by which number of moles and hence colligative property has increased. Hence it is vant' Hoff's 'i' factor. So

$$(1 - \alpha) + n\,\alpha = i$$

$$\text{or,}\quad \alpha = \frac{i - 1}{n - 1}. \qquad ....(11.9)$$

Thus from a measurement of colligative property, 'α' can be determined.

Association : Instead of dissociation, if there is association e.g. dimerisation, trimerisation etc. then number of particles decreases and colligative property values will also decrease.

POINTS TO REMEMBER

Solutions of *solid* solutes in *liquid* solvents have some properties, which depend only on number of particles present in a given amount of solvent. These properties are *colligative properties.* They are — lowering of vapour pressure; elevation of boiling point; depresson of freezing point and osmotic pressure. For the first three the laws relating the properties with concentration were given by *Raoult* and for the last one the law was given by *vant Hoff.* The expressons are —

(i) $\dfrac{P^0 - P}{P^0} = x_2$ (ii) $\Delta T_b = K_b . C_m$ (iii) $\Delta T_f = K_f . C_m$ and (iv) $\Pi = CRT$. The laws will be valid for *dilute* solution of *non-volatile, non-electrolyte* solid solutes. All these laws are helpful for measuring molar mass of solid solute. For *electrolytes* R.H.S. of above equations are to be multiplied by 'i', known as *vant Hoff's factor.*

QUESTIONS

SHORT TYPE

1. What are colligative properties ? Why are they so called ?
2. Why vapour pressure of a liquid is lowered on dissolving a non-volatile solute ?
3. What is an ideal solution ?
4. Why and when elevation of boiling point occurs ?
5. How molecular weight of a solute is known from elevation of boiling point measurement.
6. What is ebullioscopic constant ? What is its unit ?
7. What is cryoscopic constant ? Why it is called a solvent property ?
8. What is a Beckmann thermometer ? What is its peculiarity ?
9. Can there be any depression of boiling point ?
10. Can there be any elevation of freezing point ?
11. What are semi-permeable membranes ?
12. What is osmosis ? Why it occurs ?
13. Show the similarity between osmotic pressure equation and ideal gas equation.
14. What is an isotonic solution ?
15. What is vant' Hoff's factor ?
16. Derive a relation between 'i' and 'α' for an electrolytic solution.

ESSAY TYPE

1. State and explain Raoult's law for vapour pressure lowering. How the law can be utilised to measure molar mass of a non-volatile solute ?
2. Describe any method of your choice for measuring elevation of boiling point. How this data is used to know molar mass ?
3. Derive a relation between elevation of boiling point and depression of freezing points.
4. Describe a method for determining freezing point depression by ordinary thermometer. Discuss Raoult's law for depression of freezing point.
5. Describe an method for determining osmotic pressure. Explain vant' Hoff's law for osmotic pressure.
6. Why volatile and electrolyte solutes disobey the expressions for colligative properties.

PROBLEMS

1. The freezing point of a solution of 0.684 gm. of cane-sugar in 100 gms. of water is – 0.037°C. Calculate the freezing point constant of water. Given molecular weight of cane-sugar is 342.

[*Hints.* $\Delta T_f = K_f \cdot \dfrac{a \times 1000}{b \times m}$.

Here $\Delta T_f = 0.037$; $a = 0.684$

$b = 100$ $m = 342$ $K_f = ?$].

2. A solution of 10 gms. of an organic compound in a litre solution gave an osmotic pressure of 1.18 atm. at 0°C. Calculate the molecular weight of the compound.

[*Hints.* $\Pi V = \frac{g}{M} RT$; $\Pi = 1.18$; $V = 1$ litre; $g = 10$; $T = 273°A$;

$R = 0.082$; $M = ?$]

3. Under a pressure of 760 m.m. ether boils at 35°C. By dissolving 1 gm. of napthalene ($C_{10}H_8$) in 8 gms of ether the boiling point of the ether is raised by 0.205°C. Calculate the latent heat of evaporation of ether per gram.

[*Hints.* $$\Delta T_b = \frac{RT_b^2}{1000 l_v} \cdot \frac{a \times 1000}{b \times m}$$

$\Delta T_b = 0.205$; $T_b = 308°A$; $a = 1$gm.; $b = 8$gm.; $m = 128$; $l_v = ?$]

4. At 25°C the vapour pressure of a solution containing 9 gms. of a substance in 90 gms. of water is 22.87 mm. of mercury. Calculate the molar mass of the substance taking the molar mass of water as 18 and the vapour pressure of pure water at 25°C = 23.55 mm. of mercury.

[*Hints.* $$\frac{p^0 - p}{p^0} = x_2 = \frac{a/m}{a/m + b/M};$$ $P^0 = 23.56$; $a = 9$; $b = 90$;

$P = 22.87$; $M = 18$; $m = ?$]

5. The vapour pressure at 25°C of a solution containing 4.5 gms. of a substance dissolved in 45 gms. of water is 22.88 m.m. of mercury. The vapour pressure of water at 25°C is 23.56 mm. of mercury. Calculate the molar mass of the substance.

[*Hints.* See problem '4']

6. A solution of cane-sugar (molar mass = 342) Containing 5 gms. of sugar per litre had the same osmotic pressure as a solution of urea having 0.887 gm. of urea per litre at the same temperature. Find the molar mass of urea.

[*Hints.* $$\frac{g_1}{M_1} \cdot RT = \frac{g_2}{M_2} RT$$

or $\frac{g_1}{M_1} = \frac{g_2}{M_2}$; $g_1 = 5$; $g_2 = 0.887$; $M_1 = 342$; $M_2 = ?$]

7. Calculate the freezing point depression of glucose solution containing 1.8 gm. of glucose per 100 gms. of water. Molar mass of glucose = 180; $K_f = 1.86$.

[*Hints.* $$\Delta T_f = K_f \times \frac{a \times 1000}{b \times m}$$
$$= 1.8 \times \frac{1.8 \times 1000}{100 \times 180}]$$

8. The vapour pressure of a solution containing 18 gms. of a solid dissolved in 180 gms. of water is 22.87 m.m. of mercury at 25°C. Calculate the molar mass of the solid taking the molar mass of water as 18. and the vapour pressure of water at 25°C as 23.55 m.m. of mercury.

[*Hints.* See problem '4']

OBJECTIVE TYPE

1. Fill in the blank with appropriate term given in the brackett.

(a) Lowering of vapour pressure is proportional to ———.

[Ans. (i) molar conc. (ii) mole-fraction (iii) molality]

(b) Boiling point may not be elevated if the solute is ———.

[Ans. (i) electrolyte (ii) non-electrolyte (iii) volatile]

(c) At freezing point vapour pressure of solvent becomes equal to that of ———.

[Ans. (i) solution (ii) solid solute (iii) atmosphere]

(d) Osmotic pressure equation is analogous to ———.

[Ans. (i) ideal gas equation (ii) Nernst's equation (iii) Raoult's law]

(e) Copper ferricyanide acts as a ———.

[Ans. (i) Boiling point elevator (ii) semi-permeable membrane (iii) Osmotic pressure creator]

2. Write T (true) and F (false) against the following statements according to their nature.

(a) Osmotic pressure for 0.1 M NaCl is more than that of 0.1 M glucose solution at same temperature.

(b) If a solute is dimerised in a solvent then experimental value of colligative property will be half the theoretical value.

(c) K_b and K_f are constants independent of solvent.

(d) Molar concentration is temperature independent.

(e) Blood is isotonic with 0.85% NaCl solution.

3. Match terms of column A with those of column B.

A	B
(a) Osmotic pressure.	(a) Colligative.
(b) Freezing point depression.	(b) $PV = nRT$.
(c) Paticle dependent	(c) Electrolyte.
(d) $\Pi = CRT$	(d) Rast
(e) $\Pi = iCRT$	(e) vant Hoff

Answers :

1. (a) ii; (b) iii; (c) ii; (d) i; (e) ii.

2. (a) T; (b) T; (c) F; (d) F; (e) T.

3. a – e; b – d; c – a; d – b; e – c.

Chapter 12

HETEROGENEOUS EQUILIBRIUM — PHASE RULE

12.1 INTRODUCTION

If all participating members of an equilibrium exist in same state of matter i.e. (all are gases; or all exist in homogeneous solution) then the equilibrium is said to be homogeneous equilibrium. Such an equilibrium is best studied by applying law of mass action. However, when the participating members of an equilibrium exist in different phases [e.g. $CaCO_3$ (s) $\rightleftharpoons$ CaO (s) + CO_2 (g) or Ice (s) $\rightleftharpoons$ Water (l)] then the equilibrium is said to be heterogeneous equilibrium. Such an equilibrium can be best studied by applying Gibb's phase rule. Before understanding phase rule we must first know some definitions.

12.2 PHASE

Let in a closed vessel containing water and its vapour only we put some pieces of ice. Then there exist solid, liquid and gaseous state. Each has well-defined boundary and each one is easily mechanically separable. Each will be termed as a phase. So definition of phase is — *Any part of a heterogeneous system, which is itself homogeneous, which has well-defined boundary and which is mechanically separable from remaining medium is called a phase.* Mixture of gases always form one phase, as there exists no boundary between them. Mixture of liquids may form one phase or more depending on whether they are miscible or not and mixture of solids always produce more than one phase — excepting the cases of alloy.

12.3 COMPONENT

Let us consider the equilibrium $CaCO_3$ (s) $\rightleftharpoons$ CaO (s) + CO_2 (g). There are three constituents; but if composition of any two are told that of the third is known by doing some arithmetic because —

$$\begin{aligned} CaCO_3 &= CaO + CO_2 \\ CaO &= CaCO_3 - CO_2 \\ CO_2 &= CaCO_3 - CaO \end{aligned}$$

So knowing composition of any two, that of the third is known. We shall call it a two component system. Again we consider the equilibrium NH_4Cl (s) $\rightleftharpoons$ NH_3 (g) + HCl (g) though here also we have three constituents we shall call it one-component system, because here concentration of NH_3 and HCl are same. And equimolecular gaseous homogeneous mixture of NH_3 and HCl can well be regarded as NH_4Cl (g) and so there is only component which is NH_4Cl. However, if some extra NH_3 (g) or HCl (g) is introduced so that $C_{NH_3} \neq C_{HCl}$ then this system will also be two component. Thus component can be defined as — *The minimum number of chemical individuals in terms of which composition of all the phases can be known.* Number of components of a heterogeneous equilibrium is equal to *number of constituents present minus the number of equations connecting them.* Let us look at the table in the next page.

Table 12.1

Eqm.	No. of constituents	Equations	No. of components
1. $CaCO_3$ (s) $\rightleftharpoons$ CaO (s) + CO_2 (g)	3	$K = p_{CO_2}$	3 – 1 = 2
2. NH_4Cl (s) $\rightleftharpoons$ NH_3 (g) + HCl (g)	3	$K = p_{NH_3} \cdot p_{HCl}$ and $p_{NH_3} = p_{HCl}$	3 – 2 = 1
3. Eqm (2) with extra NH_3 gas	3	$K = p_{NH_3} \cdot p_{HCl}$	3 – 1 = 2

12.4 DEGREE OF FREEDOM

Let we have one mole of an ideal gas enclosed in a cylinder. Its equation of state is $PV = RT$. Though there are three variables viz. P, V and T yet number of independent variables is 2, because if any two are made independent, the rest becomes dependent. We shall call the system as bivariant i.e. its degree of freedom = 2. Again consider liquid water in a closed vessel containing its vapour (**no air**) above it. There are two variables vapour pressure and temperature and at equilibrium one is related with the other by Clausius-Clapeyron equation. [$\ln P = -\frac{L_v}{RT}$ + constant]. Thus if any one of them is independent; the other is dependent. This is a univariant system or its degree of freedom is one. Now let us know some interesting fact. If we want to have an equilibrium among ice, water and water vapour (**no air**) then it is possible only at a particular temperature and pressure ($t = 0.0098$°C, $P = 4.58$ m.m.). If either of them is slightly changed, one of the phases will disappear. So the system is non-variant i.e. degree of freedom is zero. Hence degree of freedom **is the minimum number of independent variables whose values can be changed without affecting the equilibrium i.e. without altering the number of phases.**

12.5 PHASE RULE

Phase rule was put forward by W. Gibbs. The rule did not find much recognition during life-time of its discoverer. But much later it was found to be a very important tool in analysing different metallurgical problems. The law can be best stated in terms of mathematical language. It states — *summation of number of phases (P) and degree of freedom (F) of a heterogeneous system exceeds the number of components (C) by two.* So

$$P + F = C + 2 \quad ....(12.1)$$

or

$$F = C - P + 2 \quad ....(12.2)$$

The law can be derived from thermodynamics using the concept of chemical potential; but derivation is beyond the scope of this book.

12.6 PHASE DIAGRAM

Phase diagram is a diagram from which we can know how many and what types of phases of an equilibrated system will be found to remain at equilibrium under a given set of experimental conditions i.e. pressure temperature etc. For one component system the diagram is two-dimensional, variables being pressure and temperature and for two component systems, the diagram should be three dimensional, where variables are pressure, temperature and concentration. However, such a system is generally studied either at constant pressure or at constant temperature to make the diagram two-dimensional.

12.7 ONE COMPONENT SYSTEM

(a) Water-System : Best example of one-component system is water system. Water may exist in three different phases, ice, water and vapour. When three phases remain side by side ($P = 3$) then degree of freedom $F = 1-3+2 = 0$ i.e. system becomes non-variant. For a two-phasic equilibrium ($P = 2$) $F = 1$, system is univariant. And when single phase exists ($P = 1$), system is bivariant. These are explained by the following figure.

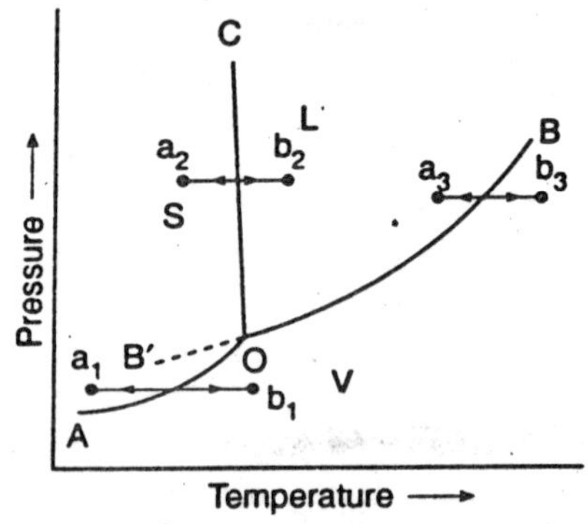

Fig. 12.1 Phase diagram of water.

In the above diagram (Fig. 12.1) BO is the vapour-pressure vs temperature diagram of water. Since a liquid boils when its vapour pressure becomes equal to atmospheric pressure, BO curve can also be regarded as indicating variation of boiling points of liquid water with atmospheric pressure. Such a curve will vanish at critical temperature (here 374°C) after which vapour can never be liquefied by applying pressure. Just like BO curve, OC curve represents variation of melting point of ice with atmospheric pressure. As m.pt. decreases with increase in pressure. This curve is slightly inclined towards pressure axis. Similarly AO, curve represents variation of sublimation temperature at various pressures.

Let us now examine the diagram by analysing the lines, point and area. The diagram has three lines OB, OC and OA. Line OB represents temperature and pressures at which liquid and vapour remain at equilibrium (different boiling point at different pressures). Line OC represents temperature and pressures at which liquid and solid will remain at equilibrium (different melting points at different pressures). Finally line OA represents solid vapour equilibrium. Three lines meet at point 'O'. ($t = 0.0098$°C; $P = 4.58$ m.m). So here three phases solid liquid and gas ($P = 3$) remain at equilibrium. This point is called *tripple point*. Areas market as '*S*', '*L*' and '*V*' represent solid, liquid and vapour zones. If the given *P* and *T* value fall in '*S*' zone, we shall see only solid phase and similarly at *P* & *T* belonging to *L* or *V* zone we shall find liquid and vapour phases. So we see that in phase diagram of one component system area represents one phase, line indicates two phasic equilibrium and point indicates a three phasic equilibrium. [Dotted line OB′ is the vapour-pressure vs temperature curve for supercooled water]. Another thing to be noted is that breaking of a phase diagram indicates appearance of a new phase.

Let a particular pressure, temperature value is represented by point 'a_1'. From the diagram we see that we shall find only solid phase at this state. If now we heat the system keeping '*P*' constant we can reach point 'b_1'. Since it is in the vapour zone, heating will vaporise the solid i.e. sublimation will occur. But if initial state is 'a_2' (solid-state) and if we reach 'b_2' by heating at constant pressure, the solid will melt. Thus heating of a solid will cause melting if it is done above point 'O' i.e. tripple point pressure, but heating below tripple point pressure will cause sublimation. If our prescribed state is a_3 then the state is liquid and here heating may reach point b_3 i.e. vapour-state. Snow, which fall in hilly area is by condensation of vapour below tripple point, (atmospheric pressure is much low at the peak of mountains). Since here vapour directly enters into solid phase, the snow that falls is dry i.e. not contaminated with liquid water like ice.

(b) CO_2-system : The phase diagram of CO_2 is similar to that of water system. A rough sketch is shown below :

OA, OB and OC are equilibrium pressure vs temperature diagrams for solid-vapour; liquid-vapour and solid-liquid equilibrium. 'O' is tripple point where three phases, solid, liquid and vapour exist in equilibrium. Tripple point pressure for CO_2 system is 5.11 atm. Since it is much above atmospheric pressure, condensation of vapour at ordinary pressure will directly yield solid, known as dry ice. Similarly heating of solid CO_2 at 1 atm. (i.e. below tripple point pressure) will yield vapour instead of liquid. 'OB' curve will vanish at critical temperature for CO_2 (31.2°C) above which there can not be any liquid-vapour equilibrium.

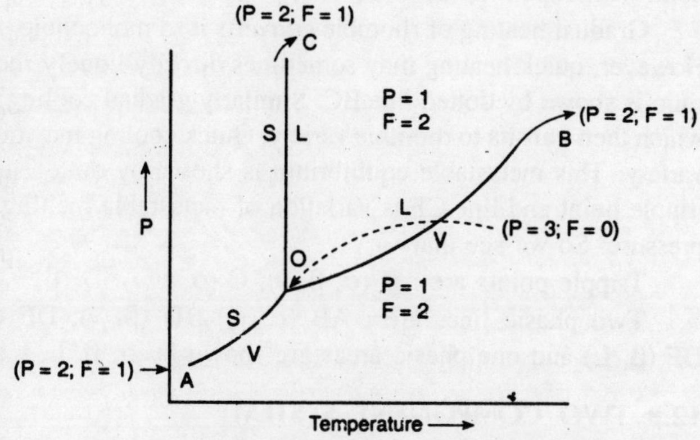

Fig. 12.2 Phase diagram of CO_2.

(c) Sulphur-system : Sulphur is available in a number of forms. However, following forms are stable during interphasic transitions. These are rhombic (α); monoclinic (β); liquid and vapour. Before drawing its phase diagram let us do some home-work.

(i) If four phases co-exist then $P = 4$; then $F = 1 - 4 + 2 = -1$; which is absurd. So four phases can never exist in equilibrium.

(ii) If three phases exist in equilibrium., then $P = 3$ and hence $F = 1 - 3 + 2 = 0$; which is tripple point condition. Now there are four such possibilities, which are (α, β, l); (α, β, v); (α, l, v) and (β, l, v).

(iii) If two phases exist in equilibrium then $P = 2$ and $F = 1 - 2 + 2 = 1$. Six such possibilities are there. These are (α, β); (α, l); (α, v); (β, l); (β, v) and (l, v).

(iv) When only one phase exists alone, then $P = 1$ and $F = 1 - 1 + 2 = 2$. Four such possibilities are there. These are α; β; l and v.

During our study of water system, we have seen that one phase is represented by a bound area; two phases by lines and three phasic equilibrium by point. So from the above home work we see that phase diagram of sulphur will consist of four areas, six lines and four points. These are shown in figure 12.3.

In the above diagram (fig. 12.3), line 'AB' indicates equilibrium between rhombic (α) and vapour sulphur. Gradual heating causes transition of rhombic to monoclinic variety at 95°C (P = 0.01mm). So phase diagram breaks and BD line indicates monoclinic-vapour equilibrium. At D (t = 120 C, P = 0.025 mm) monoclinic sulphur melts and once again the line breaks indicating appearance of a new phase, which is

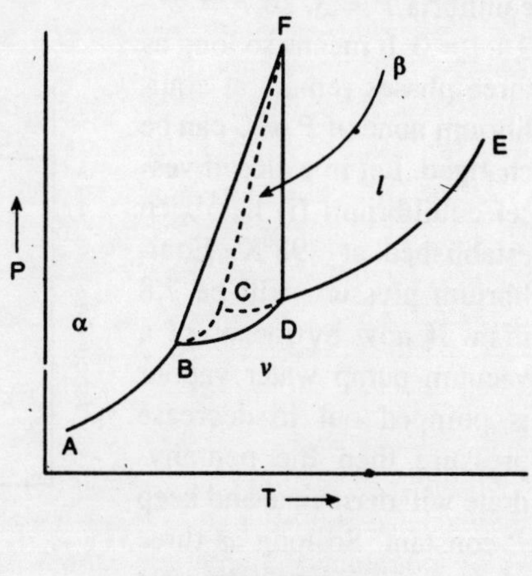

Fig. 12.3

liquid sulphur. Line DE represents liquid-vapour equilibrium. Line BF indicates variation of transition temperature ($\alpha \rightarrow \beta$) with pressure, while line DF indicates variation of melting point of monoclinic ($\beta \rightarrow L$) with pressure.

Gradual heating of rhombic converts it to monoclinic, which on further heating melts. However, quick heating may sometimes directly liquefy rhombic. This metastable equilibrium is shown by dotted line BC. Similarly gradual cooling of liquid gives monoclinic first which then transits to rhombic variety. Quick cooling may directly convert liquid to rhombic variety. This metastable equilibrium is shown by dotted line DC. Thus 'C' is metastable tripple point and line CF is variation of metastable melting point of rhombic sulphur with pressure. So we see that —

Tripple points are : B (α, β, v); C (α, v, l), D (β, v, l) and F (α, β, L)

Two phasic lines are : AB (α, v); BD (β, v); DE (L, v); BF (α, β); CF (α, L); DF (β, L) and one phasic areas are shown by α, β, L, v symbols in the diagram.

12.8 TWO COMPONENT SYSTEM

In a two component system, there are three variables viz. pressure, temperature and concentration. So the phase diagram becomes three-dimensional and its study will be very difficult. To case the situation we always study it keeping one variable constant and thus it becomes (a) P vs T diagram (C constant) (b) P vs C diagram (T constant) or (c) T vs C diagram (P constant). A two component heterogeneous system may consist of (i) binary liquid -liquid mixture (ii) Solid-gas mixture or (iii) Solid-liquid mixture. Of these case (i) has been discussed in chapter 10 (solution) we shall now discuss cases (ii) and (iii).

12.9 SOLID-GAS EQUILIBRIUM

Equilibrium between a salt hydrate and water vapour is an example of solid-gas equilibrium. As an illustration let us consider the case of $CuSO_4$. Following three equilibrium are possible.

(i) $CuSO_4, 5H_2O\ (s) \rightleftharpoons CuSO_4, 3H_2O\ (s) + 2H_2O\ (g)$

(ii) $CuSO_4, 3\ H_2O\ (s) \rightleftharpoons CuSO_4, H_2O\ (s) + 2H_2O\ (g)$

and (iii) $CuSO_4\ , H_2O\ (s) \rightleftharpoons CuSO_4\ (s) + H_2O\ (g)$

Such a system is generally studied at constant temperature and effect of pressure is observed. So here phase rule is $F = C - P + 1$. (as T is constant). In each of the above three equilibria $P = 3$, so $F = 2 - 3 + 1 = 0$. It means so long as three phases remain at equilibrium none of P or C can be changed. Let in a closed vessel equilibrium (i) has been established at 298°K. Equilibrium pressure will be 7.8 m.m. If now by means of a vacuum pump water vapour is pumped out to decrease pressure then the pentahydrate will dissociate and keep P constant. So long as three phases remain, pentahydrate will keep pressure same. But

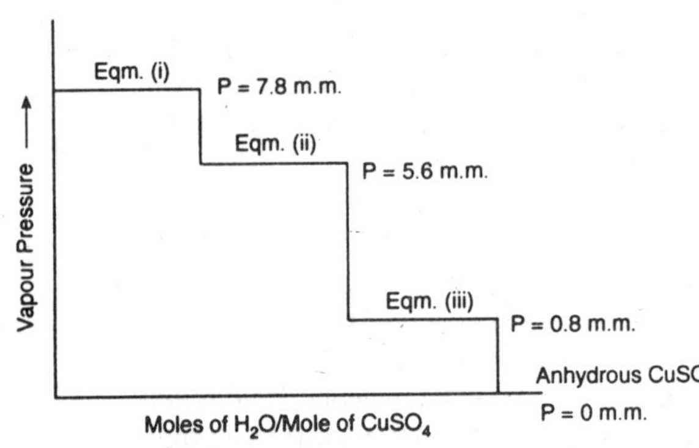

Fig. 12.4

ultimately pentahydrate will be finished and there will be two phases. So *F* becomes 1 and now pressure can be changed. But when pressure falls to 5.6 m.m. equilibrium (ii) is established and again due to three phasic equilibrium *F* becomes zero. So pressure will not change till there will be trihydrate. Continuous pumping out of vapour will force trihydrate to dissociate to keep pressure constant. Ultimately trihydrate will be finished and then pressure falls. When pressure becomes 0.8 m.m. equilibrium (i) is established and the process is repeated again. These are illustrated in the following figure.

12.10 EFFLORESCENCE AND DELIQUESCENCE

If the equilibrium vapour pressure of a salt hydrate is greater than the partial pressure of water vapour in the atmosphere, then salt will release water molecules when exposed to water. This phenomena is called *efflorescence*. During such phenomena the salt may become anhydrous ultimately. Examples of salts which effloresce are Na_2CO_3, $10H_2O$; Na_2SO_4, $10H_2O$ etc. However extent of efflorescence depends on atmospheric condition. If the atmosphere is very dry (so that $p_{H_2O} < 7.8$ m.m) then a salt like $CuSO_4$ may even be efflorescent.

If the partial pressure of water vapour in the atmosphere is greater than the vapour pressure of a saturated solution of salt then when the solid is exposed to atmosphere it will collect water vapour and ultimately form a solution. The phenomena is called *deliquescence*. Example being the case of $CaCl_2$. Vapour pressure of its saturated solution is 4.5 m.m. It is much below the partial pressure of water vapour in atmosphere in the rainy season. So $CaCl_2$ absorbs water vapour from atmosphere. Common salt contains some $CaCl_2$ in it. That is why it becomes moistened in rainy season. Both efflorescence and deliquescence depend on place and season. In a very dry atmosphere $CaCl_2$ will not deliquescence.

12.11 TWO COMPONENT SOLID-LIQUID SYSTEM

If two substances A and B are completely miscible in the liquid state then progressive cooling of their liquid mixture may lead to any one of the following phenomean depending on nature of A and B.

(i) Solid A (or solid B) may be crystallised first and ultimately a solid-solid mixture (eutectic) will come. Whether solid A or solid B will come first that depends on starting condition. Examples are Bi-Cd; KI – H_2O system.

(ii) Either solid A (or solid B) or a compound A*x* B*y* may appear first. Ultimately an eutectic of A (or B) and $A_x B_y$ will appear. This compound is stable during melting (conguent melting point). Examples are Sn-Mg ; Fe_2Cl_6-H_2O system.

(iii) Case (ii) like incident, but the compound is not stable and breaks before melting (incongruent melting point). Example is Na_2SO_4-H_2O ; NaCl-H_2O.

(iv) Neither pure A, nor pure B separates but a solid solution of A and B separate. A and B are partially miscible in the solid-state. Example : Ag-Cu system.

(v) Case (iv) but A and B are completely miscible in the solid state. Example : Pt-Au system.

Phase diagrams of such systems can be drawn and studied by two methods which are (a) solubility method and (b) thermal analysis method. The former method is based on depression of freezing point principle and the latter is based on time-temperature cooling curve. Since detailed discussions are beyond the scope of this book we shall discuss the solubility method to explain phase-diagrams of some selected systems.

12.12 SALT-WATER SYSTEM

If a very dilute solution of KI in water is colled then during freezing solid ice separate out. Separation of ice from solution will make the solution richer in KI and as depression of freezing point is directly proportional to concentration of solute so the freezing point of the remaining solution will be less than the original one. Progressive cooling will lower the freezing point of the solution along path A B (fig 12.5).

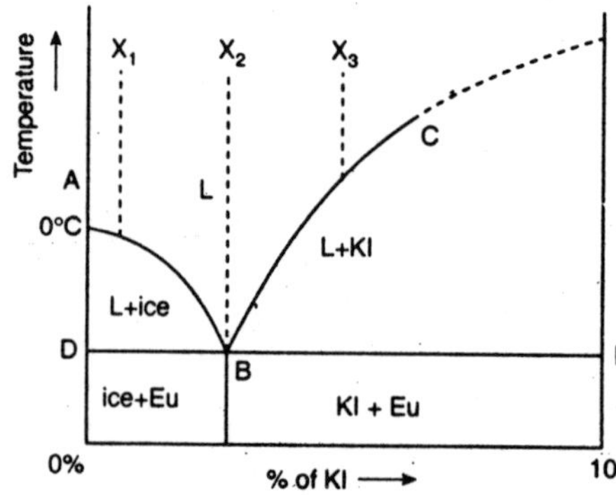

Fig.12.5

Solution will be richer and richer in KI. When concentration reaches point B the entire liquid containing both water and KI solidifies en masse. This new solid mixture is called eutectic mixture. When one component of eutectic mixture is water as in this case, then it is also known as *cryohydrate*.

On the contrary if some one cools a KI solution, whose concentration is more than point B, then cooling will first crystallise out KI. Solution becomes dilute with respect to KI. Progressive cooling will separate more and more KI and concentration will follow the line C B. At 'B' again the entire liquid will be solidified *en masse*. So cooling of liquid with composition X_1 will first give ice and ultimately eutectic; cooling of liquid with composition X_2 will directly give eutectic and cooling of liquid of composition X_3 will first give KI crystal and ultimately eutectic.

12.13 FREEZING MIXTURE

We know that when salt is added to ice, it melts and temperature falls considerably. Such mixtures are called *freezing mixtures*. Actually when solid salt is added over ice, contaminated with droplets of water we have three phases (P = 3). It is a two component system (C = 2). So degree of freedom $F = C - P + 1 = 2 - 3 + 1 = 0$. [As pressure is constant, we use reduced phase rule $F = C - P + 1$]. So no variable should change. If any one is changed forcibly some phase must disappear. But addition of salt changes concentration of solution as salt is dissolved in contaminated water. To cause disappearance of solid salt (and thus reduce number of phases) more and more ice will melt and for this purpose it will collect latent heat from surrounding and so temperature will fall.

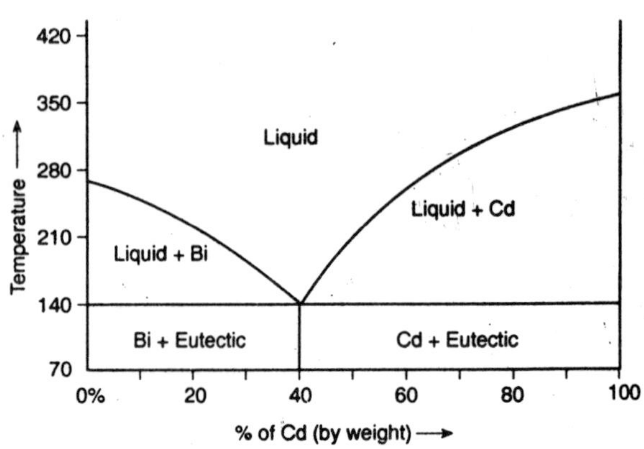

Fig. 12.6

12.14 ALLOYS AS SIMPLE EUTECTIC

The general phase diagram of simple eutectic system is as shown in figure 12.5. Many alloys e.g. Bi-Cd; Pb-

Sb etc. show such diagrams. The following (fig. 12.6) is the phase diagram of Bi-Cd system. Melting point of bismuth is 271°C while that of Cd is 321°C. Eutectic has 40% Cd by weight and eutectic temperature is 140°C.

12.15 OTHER SYSTEMS

Besides simple eutectic many other systems are known (as has been discussed in section 12.11.) Details are beyond the scope of this book. However, it may be mentioned that systems where the components combine to form stable compounds (congruent melting point) show phase diagrams, which are combination of diagrams of simple eutectic systems. Thus Mg and Sn combine to form a stable compound Mg_2Sn and so its phase diagram is combination of two simple eutectics. If two compounds are produced the total phase diagram will consist of diagrams of three eutectic systems. Fig. 12.7 is phase diagram of two component solid-liquid system, where one compound is produced.

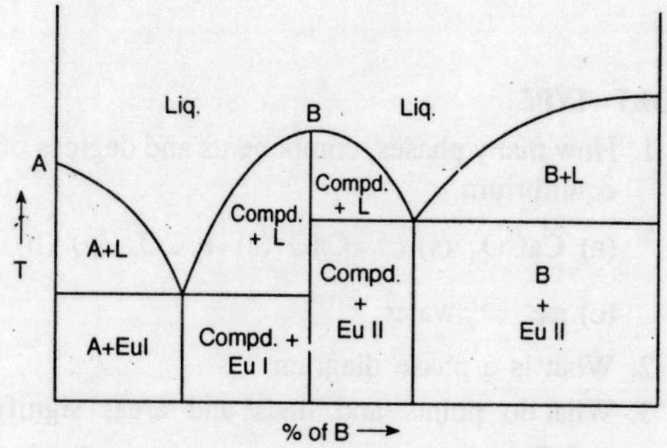

Fig.12.7

A → Melting point of A.

B → Melting point of compound ($A_x B_y$)

C → Melting point of B.

Zones are labelled and other things of the diagram are self-explanatory. There will be two eutectics here Eu I consists of A and compound and Eu II consists of B and compound.

POINTS TO REMEMBER

Heterogeneous equilibrium is best studied by phase rule : which is $P + F = C + 2$; P (i.e., phase) means number of distinctly different substances separated by well defined boundaries; C (component) is the minimum number of constituents sufficient to give composition of the mixture and F (degree of freedom) is the number of independent variables viz. P, T, concentrations etc. For one component systems variables are P & T and show a two dimensional phase diagram can be drawn by plotting eqm. P & T values. In the diagram area means one phase lines represents two phasic eqm. and points (tripple points) are three phasic eqm. where $F = 0$. Heating of a solid below tripple point will cause sublimation. For two component system variables are P, T & C. Diagrams are made two dimensional either by studying at constant P or at constant T.

QUESTIONS

SHORT TYPE

1. How many phases, components and degrees of freedom are present in the following equilibrium.

   (a) $CaCO_3$ (s) $\rightleftharpoons$ CaO (s) + CO_2 (g) (b) NH_4Cl (s) $\rightleftharpoons$ NH_3(g) + HCl (g)

   (c) ice $\rightleftharpoons$ water
2. What is a phase diagram ?
3. What do 'points' and 'lines' and 'areas' signify in phase diagram of one component system.
4. Why ice melts but iodine sublimes when heated at atmospheric pressure ?
5. Can four phases of a one component system co-exist at equilibrium.
6. Sodium carbonate effloresece but calcium chloride deliquesce why ?
7. What are eutectic ? Are they mixture or compound ?
8. A mixture of salt and ice is used as freezing mixture — why ?

ESSAY TYPE

1. Explain the terms 'Phase', component and 'degree of freedom'. What is Phase rule?
2. Draw the phase diagram of water and explain it with special reference to lines; point and area of the diagram.
3. Draw and interprete the phase diagram of sulphur.
4. Explain the phase diagram of KI-Water system. What is a cryohydrate ?
5. What is a tripple point ? What is the difference between tripple point and melting point ?

OBJECTIVE TYPE

1. Write 'T' against true and 'F' against false statements.

(a) Phase rule was given by vant Hoff.

(b) Tripple point is non-variant.

(c) Snow and ice are different according to phase diagram.

(d) Eutectics have very high melting point.

(e) For one component systems four phase can never exist at equilibrium.

2. Fill up the gaps.

(a) On heating iodine sublimes because its tripple point pressure is ——— than atmospheric pressure, (more/less)

(b) Phase diagram of sulphur consists of ——— tripple points. (three/four)

(c) Breaking of phase diagram indicates ——— of new phase. (appearance/disappearance)

(d) Sn–Mg system is an example of ——— melting point. (congruent/incongruent)

(e) Salt can ——— melting point of ice. (increase/decrease)

3. Match terms of column A with those of column B.

A	B
(a) Tripple Point.	(a) Easy melting.
(b) Congruent melting.	(b) Phase rule.
(c) Eutectic.	(c) Two component system.
(d) Gibbs.	(d) Simple Eutectic.
(e) *Pb–Sb* system.	(e) One component system.

Answers :

1. (a) F ; (b) T ; (c) T ; (d) F ; (e) T .

2. (a) more; (b) four; (c) appearance ; (d) congruent.

3. a – e; b – c; c – a; d – b; e – d.

Chapter 13

CHEMICAL KINETICS

13.1 INTRODUCTION

Regarding any chemical reaction, two basic questions can be asked. These are — (i) how far the reaction proceeds and (ii) how fast does it proceed ? Answer to question (i) has been given in the chapter of chemical equilibrium; answer to question (ii) will be given in the present chapter of chemical kinetics.

13.2 TYPES OF REACTIONS

Depending on speed, chemical reactions can be classified into following three classes.

(a) **Very fast reactions :** These are reactions proceeding with very high velocity e.g. explosion reactions; precipitation reactions etc. Generally they are completed within fraction of a second.

(b) **Very slow reaction :** These are reactions which proceed at such a slow rate that we simply do not measure their rate as our life is not so long ! Thus it takes more than one hundred fifty years to produce one drop of water if H_2 and O_2 are mixed at ordinary temperature and pressure in absence of catalyst.

(c) **Slow reactions :** These are reactions, which proceed with moderate velocity. All organic reactions and many inorganic reactions e.g. substitution in inorganic complex compounds fall in this category.

We generally study kinetics of reactions of type (iii).

13.3 UTILITY OF STUDY

Main application of study of chemical kinetics is to know the mechanism of a chemical reactions. Out of a number of proposed mechanisms, the correct one can be selected and the incorrect can be discarded by the help of kinetics study.

13.4 RATE OF A REACTION

The term 'rate' of a chemical reaction means amount of reactant disappearing or amount of product appearing per unit time. Let us consider a reaction $R = P$.

Let at time 't_1' and 't_2' concentration of reactant were 'C_1' and 'C_2' and of product were 'X_1' and 'X_2'. Then *it may appear* that

$$\text{rate with respect to reactant} = -\frac{C_2 - C_1}{t_2 - t_1} = -\frac{\Delta C}{\Delta t}$$

and $$\text{rate with respect to product} = \frac{X_2 - X_1}{t_2 - t_1} = -\frac{\Delta X}{\Delta t}.$$

But such results are wrong. As rate is a function of time so rate at 't_1' ≠ rate at 't_2'. Hence above calculation is erroneous. Amount of error becomes smaller and smaller as time difference becomes smaller and smaller.

So correct expression for rate = $\underset{\Delta t \to 0}{Lt}\left(-\frac{\Delta C}{\Delta t}\right) = -\frac{dc}{dt}$ or $\underset{\Delta t \to 0}{Lt}\left(\frac{\Delta X}{\Delta t}\right) = \frac{dx}{dt}$.

Thus rate is always expressed in the derivative form ($-dc/dt$) or (dx/dt). Its value at any moment can be determined from the slope of conc. vs. time curve at that moment.

13.5 RATE EQUATION AND RATE CONSTANT

The general form of rate equation for any reaction is $-\frac{dc}{dt} = kC^n$.

This is according to law of mass action. Here C means concentration of reactant at the time of measuring rate. 'k' and 'n' are constant. 'n' is called *order* and 'k' is rate constant or specific reaction rate. Unit of 'k' is $(\text{conc.})^{1-n}\,\text{Time}^{-1}$.

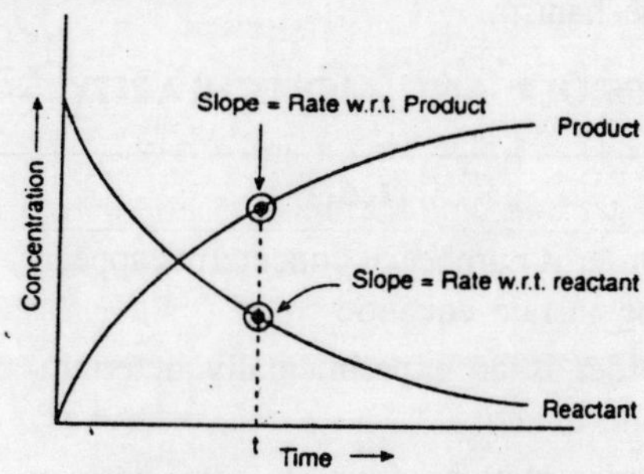

Fig. 13.1 Conc. vs time plot.

13.6 ORDER OF A REACTION

In the following lines we write some arbitrary reactions along with their rate equations as expected from law of mass action.

	Reaction	Rate equation
1.	$A \rightarrow$ Product	Rate $\propto C_A$
2.	$2A \rightarrow$ Product	Rate $\propto C_A{}^2$
3.	$A + B \rightarrow$ Product	Rate $\propto C_A \cdot C_B$
4.	$3A \rightarrow$ Product	Rate $\propto C_A{}^3$
5.	$A + 2B \rightarrow$ Product	Rate $\propto C_A \cdot C_B{}^2$
6.	$A + B + C \rightarrow$ Product	Rate $\propto C_A \cdot C_B \cdot C_C$

Just a casual look on rate equations may influence one to tell that reactions (1) is of first order; reactions (2) and (3) are of second order and remaining reactions are of third order. This is because 'order' is defined as the number of concentration terms appearing in the rate equations. However, hair-splitting analysis may reveal different picture. Let us consider the following reaction —

$$C_{12}H_{22}O_{11} + H_2O = \text{Glucose} + \text{Fructose}.$$

Plain application of law of mass action will compell us to write, rate = $k.C_{H_2O} \cdot C_{sugar}$ and we shall say that it is a second order reaction. However, experiment shows that rate does not depend on concentration of water and hence it is a first order reaction.

The decomposition reaction $2HI \rightarrow H_2 + I_2$ may appear as second order; but if the reaction is carried out on Au-surface at high pressure range, rate is found to be constant for any concentration of HI. So here rate does not depend on any concentration term and hence it is zero order. Thus we see that order is purely an experimentally determined quantity and it may be zero; whole number fraction or even negative with respect to some reactant.

13.7 MOLECULARITY OF A REACTION

The term molecularity of a reaction is the number of molecules participating in each elementary step of a reaction. All reactions take place in a number of steps and so each step should be assigned seperate molecularity. Sometimes molecularity of the slowest step (popularly known as rate determining step i.e. r. d. s.) is taken as overall molecularity of the reaction. Thus knowledge of molecularity requires a knowledge of all intermediate steps i.e., mechanism.

13.8 ORDER AND MOLECULARITY — A COMPARISON

Order	Molecularity
1. Order is number of conc. terms appearing in rate equation.	1. Molecularity is number of molecules taking part in each step of reaction.
2. Order is an experimentally determinable quantity.	2. Molecularity is a theoretical quantity.
3. Order may be fraction, –ve; zero or whole number.	3. Molecularity is always a whole number.
4. Knowledge of order does not require knowledge of mechanism.	4. Knowledge of molecularity requires a knowledge of mechanism.
5. Order depends on external factors e.g. P, T. , governing the reaction.	5. Molecularity does not depend on such factors.

13.9 FIRST ORDER KINETICS

A first order reaction is a reaction, whose rate determining step consists of only one reactant molecule e.g.

$$A \rightarrow \text{Product.}$$

If 'C_0' is the starting concentration of 'A' and 'C' is its concentration after time 't' then rate ($-dc/dt$) at this time will be —

$$-\frac{dc}{dt} = k_1 dt \qquad [k_1 = \text{rate constant}]$$

or $$-\frac{dc}{c} = k_1 dt .$$

Integrating between limits we get —

$$-\int_{c_0}^{c} \frac{dc}{c} = k_1 \int_0^t dt$$

or $$-\ln \frac{c}{c_0} = k_1 t \qquad ....(13.1)$$

or $$c = c_0 . e^{-k_1 t} . \qquad ....(13.2)$$

Above equation indicates that plot of concentration of reactant with time will be an asymptotic plot and hence concentration will be zero only after infinite time. So the reaction is incomplete.

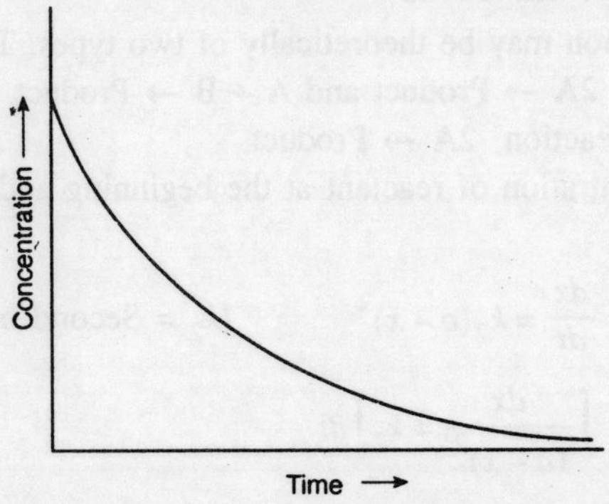

Fig. 13.2 Conc. vs time plot for 1st order reaction.

Rate equation for first order reaction can be derived in another way. Let '*a*' be the starting concentration of '*A*' and '*x*' amount of it decomposes after time '*t*'. Hence remaining concentration of 'A' is $(a - x)$; so rate with respect to product is —

$$\frac{dx}{dt} = k_1(a - x)$$

or $$\int \frac{dx}{(a - x)} = k_1 \int dt$$

or $$-\ln(a - x) = k_1 t + \text{constant}$$

when $t = 0$, then $x = 0$; so constant $= -\ln a$.

Hence $$\ln \frac{a}{a - x} = k_1 t \qquad \text{....(13.3)}$$

or $$k_1 = \frac{2.303}{t} \log \frac{a}{a - x}. \qquad \text{....(13.4)}$$

[N. B. Equations 13.1 and 13.3 are same as $c = a - x$ and $c_0 = a$.]

13.10 HALF-LIFE PERIOD

Half-life period is the time required for decomposition of half the initial amount. Thus when $t = t_{1/2}$, $x = a/2$. So from equation 13.4 we get —.

$$k_1 = \frac{2.303}{t_{1/2}} \log \frac{a}{a - a/2}$$

or $$t_{1/2} = \frac{2.303}{k_1} \log 2 = \frac{0.693}{k_1}. \qquad \text{....(13.5)}$$

Thus we see that half-life is a constant quantity.

13.11 EXAMPLES OF FIRST ORDER REACTION

There are many examples of first order reactions, both in gaseous and solution phase. In the gaseous phase, decomposition of N_2O_5 is a well known examples. [$N_2O_5 \rightarrow 2NO_2 + \frac{1}{2}O_2$]. In the solution phase decomposition of H_2O_2, catalysed by I^- ; Fe^{3+} ions etc; hydrolysis of cane sugar; ester catalysed by H^+ ions are some examples.

13.12 SECOND ORDER KINETICS

A second order reaction may be theoretically of two types. These are

$$2A \rightarrow \text{Product and } A + B \rightarrow \text{Product.}$$

Let us consider the reaction $2A \rightarrow$ Product.

Let 'a' be the concentration of reactant at the beginning and 'x' amount of it reacts after time 't'. Then —

$$\frac{dx}{dt} = k_2(a-x)^2 \qquad [k_2 = \text{Second order rate constant}]$$

or $$\int \frac{dx}{(a-x)^2} = k_2 \int dt$$

or $$\frac{1}{(a-x)} = k_2 t + \text{constant.}$$

When $t = 0$, $x = 0$, so constant $= \frac{1}{a}$.

Hence $$\frac{1}{(a-x)} - \frac{1}{a} = k_2 t$$

or $$k_2 t = \frac{x}{a(a-x)}$$

or $$k_2 = \frac{1}{t} \cdot \frac{x}{a(a-x)}. \qquad ....(13.6)$$

13.13 CRITICAL COMMENTS

Like first order reaction, second order reaction is also incomplete. Because at the time of completion 'x' becomes equal to 'a' and so 't' becomes infinity. Hence second order reaction is also an incomplete reaction. To find half-life we see that after time for half-decomposition ($t_{1/2}$) 'x' becomes '$a/2$'. So —

$$k_2 = \frac{1}{t_{1/2}} \cdot \frac{a/2}{a(a-a/2)}$$

or $$\frac{1}{t_{1/2}} = \frac{1}{k_2} \cdot \frac{1}{a}. \qquad ....(13.7)$$

Thus here half-life is inversely proportional to initial concentration.

13.14 SECOND ORDER REACTION WITH TWO DIFFERENT REACTANTS

Let the second order reaction be of the type $A + B \rightarrow$ Product where A and B are two different reactants. Here if starting concentration is same for each (let 'a') then if 'x' amount of each reacts, in time 't' then remaining concentration of A and B are '$(a - x)$' for each. So the rate equation becomes $\frac{dx}{dt} = k_2(a-x)^2$ and on solving we get the same equation i.e. 13.6.

However, if starting concentration are different say 'a' for A and 'b' for B then if 'x' amount of each reacts in time 't', remaining concentration are $(a - x)$ for A and $(b - x)$ for B. Rate equation becomes —

$$\frac{dx}{dt} = k_2(a-x)(b-x)$$

or $$\frac{dx}{(a-x)(b-x)} = k_2 dt$$

or $$\left[-\frac{1}{(a-b)}\int\frac{dx}{a-x} + \frac{1}{a-b}\int\frac{dx}{b-x}\right] = k_2\int dt$$

or $$\frac{1}{a-b}\left[\ln(a-x) - \ln(b-x)\right] = k_2 t + \text{constant}$$

or $$\frac{1}{a-b}\ln\frac{a-x}{b-x} = k_2 t + \text{constant}.$$

But when $t = 0$, $x = 0$ so constant $= \dfrac{1}{a-b}\ln\dfrac{a}{b}$.

Hence $$\frac{1}{a-b}\ln\frac{(a-x)}{(b-x)} - \frac{1}{a-b}\ln\frac{a}{b} = k_2 t$$

or $$k_2 = \frac{1}{t(a-b)}\ln\frac{b(a-x)}{a(b-x)}. \qquad ....(13.8)$$

13.15 THIRD ORDER REACTION

A third order reaction is theoretically of three types. These are —

3A → Product; A + 2B → Product; and A + B + C → Product

We shall here consider only the first case. we write

3A → Products.

Let initial concentration of reactant be 'a' and after time 't', 'x' amount of it reacts. Then rate at this time is —

$$\frac{dx}{dt} = k_3(a-x)^3$$

or $$\int\frac{dx}{(a-x)^2} = k_3\int dt$$

or $$\frac{1}{2(a-x)^2} = k_3 t + \text{constant}.$$

When $t = 0$, then $x = 0$, so constant $= \dfrac{1}{2a^2}$.

Hence $$\frac{1}{2}\left[\frac{1}{(a-x)^2} - \frac{1}{a^2}\right] = k_3 t$$

or $$k_3 = \frac{1}{2t}\left[\frac{1}{(a-x)^2} - \frac{1}{a^2}\right] \qquad ....(13.9)$$

13.16 CRITICAL COMMENTS

Like first and second order reactions third order reaction is also incomplete. Because if at completion time, $x = a$ and 't' becomes α.

To find half-life period for third order reaction we put $x = \frac{a}{2}$ and get $t_{1/2} = \frac{3}{2k_3} \cdot \frac{1}{a^2}$.

Thus here half-life $\propto a^{-2}$.

13.17 ZERO ORDER REACTION

A zero order reaction is a reaction where rate does not depend on time and is thus independent of concentration of reactant. Rate equation will there be —

$$-\frac{dc}{dt} = k_0$$

$$\text{or} \quad -dc = k_0 \, . \, dt.$$

If c_0 is starting concentration of reactant and c is concentration at time 't', then

$$-\int_{c_0}^{c} dc = k_0 \int_{0}^{t} dt$$

$$\text{or} \quad k_0 = \frac{c_0 - c}{t}. \qquad \text{....(13.10)}$$

Such a reaction will go to completion. Because at the time of completion $c = 0$ and $t = \frac{c_0}{k_0}$, which is a finite quantity.

At time for half-decomposition ($t_{1/2}$), $c = \frac{c_0}{2}$ and so $\frac{c_0}{2k_0}$. Thus here $t_{1/2} \propto c_0$.

13.18 HALF-LIFE AND INITIAL CONCENTRATION

From the expressions for half-life of reactions of different orders we find the following relation.

Order	'$t_{1/2}$' is proportional to
0	Initial concentration i.e. I. C.
1	(I. C.)^0
2	(I. C.)$^{-1}$
3	(I. C.)$^{-2}$

So as a general relation we can write that for reaction of nth order

$$t_{1/2} \propto (\text{I. C.})^{1-n}. \qquad \text{....(13.11)}$$

13.19 PSEUDO-UNIMOLECULAR REACTION

Let us consider a second order reaction A + B = Products. Its rate equation is —

$$\frac{dx}{dt} = k_2 \, C_A \, . \, C_B$$

where 'k_2' is second order rate constant. Now if this reaction is carried out by taking 'A' in large excess e.g. $C_A : C_B \geq 10$ then during the period of reaction concentration of 'A' remains more or less same, but concentration of 'B' changes significantly. As here C_A is approximately constant, rate equation becomes $\frac{dx}{dt} = k' C_B$. Here k' is a new constant having value $k_2 C_A$. Now the rate equation is of first order reaction. Thus a normally second order reaction is behaving like a first order reaction. Such reaction will be called pseudo first order or pseudo-unimolecular reaction. The value of pseudo first order rate constant k' depends on amount of the substance taken in large excess. In a similar manner a third order reaction A + 2B → product can be made pseudo first order by taking 'B' in large excess and it can be made pseudo-second order by taking 'A' in large excess.

13.20 EXAMPLES

Following are some examples of reactions of different orders.

Order	Examples
Zero	Decomposition of HI on Au-surface at high pressure; Iodination of acetone is zero order w. r. t. I_2.
First	Decomposition of N_2O_5 ; Catalytic decomposition of H_2O_2.
Second	Saponification of $CH_3COOC_2H_5$; $H_2 + I_2 = 2HI$
Third	$2NO + Cl_2 = 2NOCl$; $2FeCl_3 + SnCl_2 = SnCl_4 + 2FeCl_2$.

13.21 EXPERIMENTAL METHODS FOR DETERMINING 'ORDER'

We have already stated that order is an experimentally determinable quantity. Following methods are available for the determination of order of a reaction.

(a) Trial and error method (b) Graphical method (c) Half-life method and (d) Vant-Hoff's differential method.

All these methods are applicable for reactions involving single reactant. When however number of reactants is more then one, then above methods can be applied after some modification. This technique of application is known as (e) Ostwald's isolation technique. We are now discussing the methods one by one.

(a) Trial and error method : Experiment in chemical kinetics means collection of a set of x vs t values; where 'x' is the amount of product formed (or reactant reacted) at some arbitrary time 't'. Now, we have different equations for rate constants of different orders e.g. eqn. (13.4) for 1st order; (13.6) for 2nd order; (13.9) for 3rd order etc. If our experimentally obtained x vs t values fit well in equation (13.4) we shall say that the reaction is of first order. If it does not fit in equation (13.4) but fits well in equation (13.6) then the reaction is of second order. The method is rather approximate as there is no scope of knowing fractional order.

(b) Graphical method : This method is also matching test method. However here we take the help of graph. Equation (13.4) indicates that for 1st order reaction plot of log $\frac{a}{a-x}$ vs t is a straight line through origin; eqn. (13.6) states that for 2nd order reaction plot of $\frac{x}{a(a-x)}$ vs t is a straight line through origin. So with the help of experimental data we are to try and see what type plot is given by our data and then we can know the order. Thus basically this is also a trial and error method.

(c) Half-life method : We have seen that for a reaction of nth order time for half decomposition ($t_{1/2}$) follows the relation

$$t_{1/2} \propto C^{1-n}. \qquad [C = \text{Initial concentration}]$$

So if a particular reaction is studied at two different initial concentrations C_1 and C_2 and if T_1 and T_2 are half-lives at these concentrations then

$$\frac{T_1}{T_2} = \left(\frac{C_1}{C_2}\right)^{1-n}. \qquad ....(13.12)$$

Thus by determining T_1 and T_2 at concentration C_1 and C_2 we can know order 'n'.

(d) Vant Hoff's differential method : The general expression for rate of a reaction of nth order is —

$$-\frac{dc}{dt} = kc^n.$$

Taking logarithm of both sides we get

$$\log(-dc/dt) = \log k + n \log c.$$

Now if $-dc_1/dt$ and $-dc_2/dt$ are rates of the reaction when concentration of reactant is c_1 and c_2 then

$$\log(-dc_1/dt) = \log k + n \log c_1$$

and $$\log(-dc_2/dt) = \log k + n \log c_2.$$

On solving we get

$$n = \frac{\log\{(-dc_1/dt)/(-dc_2/dt)\}}{\log c_1/c_2}. \qquad ....(13.13)$$

If we determine concentration of reactant at various time and plot c vs t then slopes of the curve at c_1 and c_2 will be the values of $(-dc_1/dt)$ and $(-dc_2/dt)$. Hence order 'n' can be known.

This method and the previous one are accurate methods as we are using the general form of equation and so exact order can be known.

(e) Ostwald's isolation method : The difficulty with the above methods is that they are applicable for reaction involving one reactant as we plot concentration of reactant against time. However, when number of reactant is more than one we face the difficult question — 'whose concentration is to be plotted ?' In such cases Ostwald adopted a modification technique, known as Ostwald's isolation method. Let our reaction is A + B → Product and the rate equation is Rate = $k. C_A^{\alpha} . C_B^{\beta}$. So order is ($\alpha + \beta$). We first take A in large excess so that C_A remains almost constant and reaction becomes pseudo β^{th} order. We measure C_B at different times and by adopting either method (c) or (d) determine β taking 'B' alone as reactant. Next we perform the same reaction again but this time we take 'B' in large excess, so that the reaction becomes pseudo α^{th} order. 'α' is determined in the same manner. Thus over all order ($\alpha + \beta$) is known.

13.22 TEMPERATURE AND RATE CONSTANT : ARRHENIUS EQUATION

Rate of every reaction is found to be increased with increase in temperature. Now rate = $k.c^n$. So increase in rate with increase in temperature means rate constant 'k' increases with temperature. The ratio $\left(k_{T+10}/k_T\right)$ is called temperature co-efficient of reaction rate. The ratio $k_{35°C} : k_{25°C}$ is generally 2 ~ 3. This means that for 10°C rise in temperature rate constant is doubled or trebled. It was Hood, who first gave an empirical relation between rate constant (k) and absolute temperature (T). He found that plot of ln k vs $\frac{1}{T}$ is a straight line with negative slope. So the proposed equation is

$$\ln k = A - \frac{B}{T}. \quad \text{[A and B are constant]} \qquad ....(13.14)$$

The first theoretical equation connecting rate constant and temperature was derived by Arrhenius. Derivation of Arrhenius equation is beyond the scope of this book. His equation is

$$k = A.e^{-E/RT} \qquad ....(13.15)$$

Here A is called frequency factor or pre-exponential factor and 'E' is activation energy or energy of activation. If we take logarithm of Arrhenius equation we shall get Hood's empirical equation.

13.23 ACTIVATION ENERGY AND FREQUENCY FACTOR

The two important terms appearing in Arrhenius equation are activation energy (E) and frequency factor (A). Idea about these two terms can be obtained from following consideration.

We know that a chemical reaction means breaking of some old chemical bonds and formation of new bonds. Now the question is how these bonds are being broken. It was thought that collision between molecules is responsible for bond breaking. But number of molecules colliding per c. c. of gas per second can be calculated. So number of molecules reacting per c. c. per second i.e. rate of reaction can also be calculated. It was found that calculated rate is always much higher than experimentally observed rate. Also increase in temperature by 10°C increases collision frequency by 2 — 3%, whereas such increase in temperature increases rate by 100 — 200%. Hence collision theory was about to be rejected. At this time Arrhenius introduced the idea of activation energy. According to him each of the molecules present in an assembly does not possess same energy. Molecules possessing energy ≥ E will be called *active molecules*. This 'E' is called *activation energy* and its value depends on type of reaction. Molecules having energy < E are passive molecules. While collision between two active molecule leads to chemical reaction, that between two passive molecules is fruitless. There is an equilibrium Passive $\rightleftarrows$ Active. This equilibrium shifts towards right as temperature increases and so rate increases with temperature. If n is total number of molecules and n_E is number of active molecules then $n_E = n.e^{-E/RT}$ [Boltzmann's equation]. So if Z is total collision frequency and Z_E is collision frequency among active molecules then $Z_E = Z.e^{-E/RT}$. Since reaction takes place due to collision among active molecules, so rate of a reaction is proportional to Z_E i.e. Rate $\propto Z.e^{-E/RT}$

This is comparable with Arrhenius equation $k = A.e^{-E/RT}$. Comparison of the two justifies the name of A as frequency factor.

POINTS TO REMEMBER

This chapter on chemical kinetics discusses about rate of a reaction; its measurement under different conditions and their applications. The utility of such study is knowing the mechanism of chemical reactions. Rate of a reaction is measured at some specific instant as it is a function of time (excepting zero-order reaction). Order, is the number of concentration terms found in empirical rate equation; for reaction A + B $\rightarrow$ product, it rate is found to be proportional to $C_A{}^{\alpha}.C_B{}^{\beta}$ then $(\alpha + \beta)$ is the order. Molecularity is the number of molecules taking part in each elementary step of the reactions, excepting zero-order are incomplete and time for half-decomposition i.e. $T_{1/2}$ is related with initial concentrate (I.C.) by the reaction $T_{1/2} \propto (\text{I.C.})^{1-n}$.

Rate and rate constant of a reaction increases with increase in temperature. According to Arrhenius $k = A.e^{-E/RT}$; k = rate constant; A = Frequency factor and E = activation energy. 'E' is the minimum energy that must be possessed by a molecule for undergoing chemical reaction.

QUESTIONS

SHORT TYPE

1. Can rate of a reaction be expressed as $-\frac{\Delta C}{\Delta t}$?
2. What is specific reaction rate ? Why it is so called ?
3. What is the dimension of rate constant ?
4. Rate of reaction A + B → Product; is found to be Rate = $k.C_A{}^{\alpha}.C_B{}^{\beta}$ — what is the order ?
5. For a first order reaction, one hour is needed for decaying from c to c/2; how much time will be needed for decaying from c/2 to c/4 ?
6. If above question is for a second order reaction — what will be your answer ?
7. What is the special characteristic of zero-order reaction ?
8. What is a pseudo-unimolecular reaction ?
9. What is the general relation between half-life and initial concentration ?
10. What is temperature co-efficient of reaction ?
11. What is activation energy ?
12. What do you understand by frequency factor of a reaction ?

ESSAY TYPE

1. Distinguish between order and molecularity ?
2. Derive an expression for rate constant of a first order reaction and show that it is an incomplete reaction.
3. Derive an expression for rate constant of a second order reaction with two reactants having same initial concentration. Show that here half-life is inversely proportional to initial concentration.
4. Derive an expression for rate constant of zero order reaction and discuss about its (i) half-life (ii) completion time.
5. Describe some methods for experimental determination of order of a reaction.
6. Describe an accurate method for determining order of a reaction with three reactants.
7. How does rate of a reaction vary with temperature. Discuss Arrhenius equation.
8. Briefly develop the concept of activation energy and temperature.

PROBLEMS

1. In the decomposition of hydrogen peroxide equal volumes of the decomposing mixture are titrated against standard $KMnO_4$ solution at regular intervals and following data are obtained.

Time (in minutes)	0	10	20	30
Volume of $KMnO_4$ used in c.c.	25	16	10.5	7.08

Show that it is unimolecular reaction.

[*Hints.*

For unimolecular reaction $k_1 = \frac{2.303}{t} \log \frac{a}{a-x}$.

Here $a \propto V_0$; $(a-x) \propto V_t$.

So $k = \frac{2.303}{t} \log \frac{V_0}{V_t}$.

Put $t = 10; 20; 30$ etc. and $V_t = 16; 10.5; 7.08$; show k_1 is same.]

2. During the hydrolysis of ethyl acetate by NaOH using equivalent concentrations. 25 c.c. of the reaction mixture was titrated at regular intervals against standard acid. The volume of acid used by the unchanged NaOH are given below :

Time (in min.)	0	5	15	25	35
Volume of acid in c.c.	16	10.24	6.13	4.32	3.41

Show that the reaction is of the second order.

[*Hints.*

$$k_2 = \frac{1}{t} \cdot \frac{a}{a(a-x)}$$

a ≡ 16 when $t = 5$ $(a-x) \equiv 10.24$.

So $x \equiv 5.76$ find k_2

Similarly for $t = 15$; 25 etc find k_2. Show they are same.]

3. The half-life periods for the decomposition of ammonia (catalytically) are as follows :

Pressure in mm	50	100	200
Half- life period in hrs.	3.52	1.92	1

Find the order of the reaction.

[*Hints.*

$$\frac{T_1}{T_2} = \left[\frac{C_1}{C_2}\right]^{1-n} \qquad \text{So, } \frac{1.92}{3.52} = \left[\frac{100}{50}\right]^{1-n}$$ Find 'n'.]

4. If one percent decomposes in the first minute in a unimolecular reaction, calculate how much would remain undecomposed after the first hour.

[*Hints.*

For 1% decomposition $\frac{a}{a-x} = \frac{100}{99}$.

So $k_1 = \frac{2.303}{1} \log \frac{100}{99} / \text{min.}$

After 1 hr $k_1 = \frac{2.303}{60} \log \frac{100}{(a-x)}$. Find $(a-x)$.

This is % of undecomposed amount.]

5. If $K_1 = K_2 = K_3$ for three reactions being respectively uni, bi and ter-molecular, when concentration is expressed in moles/litre, what will be the above relation if the conc. unit is mole/c.c. ?

[*Hints.*

When C is M/L; Unit of k is $mole^{1-n}\ litre^{n-1}\ time^{-1}$.

And $mole^{1-n}\ litre^{n-1}\ time^{-1} \equiv mole^{1-n}\ 10^{3(n-1)}\ c.c.^{n-1}\ time^{-1}$.

For first order in conc. unit is changed rate constant is same. For 2nd order if conc. unit is changed to mole/cc then it is $10^{3(2-1)}$ times increased for 3rd order it is 10^6 times increased.]

OBJECTIVE TYPE

1. Write T against true and F against false statements.

(a) Reactions of higher order ((≥ 4)) are absent.

(b) Molecularity may be fraction.

(c) Order of a reaction may be negative.

(d) For exothermic reaction rate decreases with increase in temperature.

(e) Frequency factor is related with collision frequency.

2. Fill in the gaps.

(a) Slope of eoncentration vs. time curve at any time is ———.

(b) Only zero order reactions go to ———.

(c) For first order reactions half life is a ——— quantity.

(d) For reaction involving more than one reactant ——— is used for determining order.

(e) Molecules having energy equal to or greater than are capable of reacting.

3. Match terms of column A with those of column B.

A	B
(a) Molecularity.	(a) Experimental quantity.
(b) Half-life.	(b) Qualification for reaction.
(c) Activation energy.	(c) $(Conc.)^{1-n}$ Time^{-1}.
(d) Rate constant.	(d) Proportional to $(Conc.)^{1-n}$.
(e) Order.	(d) Mechanism.

Answers :

1. (a) T; (b) F; (c) T; (d) F; (e) T.

2. (a) rate. (b) completion (c) constant (d) Ostwald's isolation method (e) activation energy.

3. a – e; b – d; c – b; d – c; e – a.

Chapter 14

CATALYSIS — ADSORPTION

14.1 INTRODUCTION — DEFINITION OF CATALYST

We know that chemical reactions are nothing but breaking of some old bond and formation of new bonds. Berzelius found that presence of some foreign materials (i.e. not reactant and product) accelerates the rate of chemical reaction. He thought that they are loosening the old bonds and are thereby increasing the rate. Thus he picked up the term catalyst for these foreign materials (Greek ; Kata = wholly; lein = loosen). Thus he defined catalysts as those substances which accelerate rate of chemical reactions without undergoing any change in mass and composition.

14.2 NAMING OF CATALYSTS

At present a catalyst is defined as a substance which alters the rate of a reaction without undergoing any chemical change. Depending on mode of action, catalyst are named as follows —

(i) Positive catalyst (ii) Negative catalyst (iii) Auto catalyst and (iv) Induced catalyst.

Positive catalysts are those which accelerate the reaction rate by lowering activation energy. There are numerous examples of such catalyst. MnO_2,during decomposition of $KClO_3$, V_2O_5 during combination of SO_2 and O_2 and Ni during hydrogenation of unsaturated oils act as positive catalyst.

Negative catalyst or inhibitors are those, which retard reaction rate by creating obstacles in the path of the reaction e.g. by blocking chain propagation covering adsorption sites of solid catalyst etc. Thus H_2SO_4 acts as a negative catalyst during decomposition of H_2O_2; antiknock compounds e.g. lead-tetraethyl are inhibitors in the oxidation processes of fuels.

Sometimes product of a reaction begins to catalyse that reaction. Thus at the initial stage when product was absent rate remain slow. But as products begin to be accumulated the reaction rate becomes faster due to catalytic activity of the product produced. Example being the reaction between $KMnO_4$ and oxalic acid. The reaction is catalyst by Mn^{2+} ion, which is a product of the reaction.

Sometimes it is found that a reaction, which normally does not occur by itself, occurs in an atmosphere where similar such reaction is taking place. Such phenomena have been termed as induced catalysis. Thus Na_3AsO_3 solution is not oxidised by aerial oxygen, but Na_2SO_3 solution is oxidised arerially. If now a mixture of the above two solutions are kept in air both solutions are oxidised to Na_3AsO_4 and Na_2SO_4 respectively.

14.3 CRITERIA OF CATALYSIS

Following are the criteria of a reaction catalysed by some catalyst.

(a) A catalyst remain unchanged in mass and composition at the end of the reaction. Only thing which can change is its physical state. Thus a catalyst introduced as powder may be regenerated as lump. Bright metal rods introduced as catalyst get some blisters over their surface at the end.

(b) A very small amount of catalyst is sufficient to catalyse a reaction. Thus trace amount zymase catalyse the fermentation of sugar. Very small amount of Fe-powder can produce large amount ammonia in the Haber process.

(c) A catalyst does not alter the equilibrium of a reaction. Let us consider a reaction R → P done in absence and presence of catalyst 'C'. In its absence standard free energy change $\Delta G^0 = (G^0{}_P - G^0{}_R)$ and in its presence $\Delta G^0 = (G^0{}_P + G^0{}_C) - (G^0{}_R + G^0{}_C) = (G^0{}_P - G^0{}_R)$. So ΔG^0 is same. But $\Delta G^0 = - RT \ln K$. So eqm. constant K is not altered. Thus catalyst does not alter equilibrium constant and hence the yield. It only helps to reach equilibrium quickly. We should bear in mind that difference between a catalytic and non catalytic reaction is like the difference between a local and express train. Both take us to the same destination but of course there is a difference in time. Now eqm. constant $K = \frac{k_1}{k_2}$, where k_1 and k_2 are forward and backward rate constants. A catalyst does not change 'K' but increase rate, so catalyst increases both k_1 and k_2 by the same extent. Hence a catalyst which accelerates forward reaction also accelerates backward reaction. That is why nickel is used both for hydrogenation and dehydrogenation purpose; Al_2O_3 is used both for hydration and dehydration etc. Again rate constant $k = A.\, e^{-E/RT}$ (Arrhenius equation). If a catalyst tries to increase k it can do it by lowering activation energy 'E'. But in order to do this reaction path i.e. mechanism should be changed. So the conclusion is that *catalyst changes the mechanism of a reaction in such a manner that activation energy is lowered.*

(d) Whether presence of catalyst is essential or not for the occurrence of a reaction is a controversial point. According to some authors feasibility of a reaction is determined from thermodynamics, which says that if ΔG for a reaction is negative, it must occur. Rate of a reaction is determined by the value of activation energy. If this is very high rate will be abnormally slow and it may appear that it is not taking place. In such cases catalysts increase their rates by lowering activation energy. It does not mean that presence of catalyst is a must for the initiation of a reaction. However, there are some other authors who think that there are few reactions where presence of catalyst is a must for their initiations. In fact no mechanism for such reactions can be given in absence of catalyst. As an example they cite the case of mutarotation of glucose, whose mechanism demands presence of both acid and base as catalyst. The mutarotation of tetramethyl glucose does not occur either in pyridine (base) or in meta-cresol (acid), but takes place in a mixture of the two.

14.4 TYPES OF CATALYST

Catalysts are classified mainly into two types, which are (i) homogeneous and (ii) heterogeneous. There is yet another types of catalyst, which are of biological origin e.g. enzymes. However, as enzymes exist in colloidal solution and colloids are heterogeneous suspension of micro particles, enzyme catalysis is also heterogeneous or more appropriately micro-heterogeneous catalysis.

In homogeneous catalysis, reactant; product and catalyst all exist in the same phase. Examples are 'NO' for the combination between SO_2 and O_2; H^+ ion for the hydrolysis of cane sugar etc.

In heterogeneous catalysis reactants, products are in gaseous or liquid forms but the catalyst is solid. Example being Pt for combination between SO_2 and O_2; Ni, in the hydrogenation of oil etc.

14.5 THEORY OF CATALYSIS

In general there are two theories to explain catalytic reactions. These are (a) intermediate compound formation theory and (b) adsorption theory. Though (a) generally explains homogeneous catalysis and (b) explains heterogeneous catalysis, hair-splitting analysis will show that the two theories are basically same.

(a) Intermediate compound formation theory : The theory believes that the first step of a catalytic reaction is the formation of an intermediate by the reaction of catalyst with some reactant. Subsequently this intermediate gives product. As examples let us consider the following reactions —

(i) Reaction between SO_2 and O_2 catalysed by NO :

The proposed mechanism is

$$\frac{1}{2}O_2 + NO = NO_2 \text{ (Intermediate)}$$

$$SO_2 + NO_2 = SO_3 + NO$$

As proof we can say that though SO_2 ; O_2 (reactants); SO_3 (product) and NO (catalyst) are colourless the reaction mixture assumes a brown colour during the reaction, due to formation of intermediate NO_2.

(ii) Decomposition of $KClO_3$ catalysed by MnO_2:

The catalytic decomposition of $KClO_3$ is believed to occur via the following paths

$$2KClO_3 + 2MnO_2 = 2KMnO_4 + Cl_2 + O_2$$

$$2KMnO_4 = K_2MnO_4 + MnO_2 + O_2$$

$$K_2MnO_4 + Cl_2 = 2KCl + MnO_2 + O_2$$

To support the above mechanism following facts can be mentioned. Oxygen evolved during this reaction have a faint smell of bleaching during this reaction (due to contamination by the intermediate Cl_2). Also if during the course of the reaction everything in the reaction is vessel of added to water taken in a large beaker and the mixture is stirred we shall see pink colour, which proves the presence of $KMnO_4$ as an intermediate.

(b) Absorption theory : Later in this chapter we shall discuss adsorption, which is nothing but sticking of foreign molecules on the surface of solid or liquid. These processes are exothermic. Reactions between gases catalysed by solids can be well explained by adsorption theory. It believes that before reaction, the reactant molecules are adsorbed on some suitable sites over the solid catalyst. This brings them in close proximity, which is a necessary condition for the occurrence of a reaction. Also the heat liberated during adsorption is taken up by the reactant molecules so that effective activation energy becomes less. Thus if 'E' is activation energy in absence of catalyst and 'E_a' is heat of adsorption then effective activation energy in presence of catalyst is $(E - E_a)$. Hence rate will be faster. Support towards such a theory can be given from the following facts.

Adsorption has certain selectivity; solid Ni has preference for H_2 while solid Al_2O_3 for H_2O molecule. That is why if we pass ethanol vapour over Ni catalyst and Al_2O_3 catalyst we get different products.

$$\underset{\text{(Vapour)}}{C_2H_5OH} \xrightarrow[-H_2]{\text{Ni Catalyst}} CH_3CHO$$

$$\underset{\text{(Vapour)}}{C_2H_5OH} \xrightarrow[-H_2O]{Al_2O_3 \text{ Catalyst}} C_2H_4$$

14.6 PROMOTERS AND POISONS

There are some substances, which increase activity of a catalyst e.g. molybdenum increases the activity of Fe catalyst in the Haber process for the manufacture of ammonia. These substances are called catalytic promoters. On the contrary there are some substances, which decrease or even destroy the activity of catalyst. They are called catalytic poisons. Interestingly things, which are poisonous to human body e.g. H_2S, As_2O_3 cyanogen compounds etc. are also poisonous to catalysts. Existence of promoters and poisons are indirect evidences for adsorption theory. Promoters actually modify the surface, make it more rough and increase number of active spots. Poisons occupy the active spots and reactant molecules do not find necessary sites for adsorption and hence rate in diminished.

14.7 ACID-BASE CATALYSIS

One of the oldest known catalytic reactions is acid -base catalysed reaction. According to Arrhenius acid-catalysed reactions are those, which are catalysed by H^+ ions e.g. inversion of cane sugar; hydrolysis of esters etc. Later Bronsted- Lowry theory was proposed and we came to know that many anions e.g. $H_2PO_4^-$; cations of weak base e.g. NH_4^+ act also as acid. So now we use the term general acid catalysis. General acid catalysis means any reaction which is catalysed by H^+ ; undissociated acid (e.g. CH_3COOH) or cations of weak base.

Similarly reactions which are catalysed by OH^- (e.g. conversion of acetone to diacetone alcohol decomposition of nitramide etc.) are base-catalysed reaction. Here also after the advent of Bronsted-Lowry theory we have extended the scope of base catalysed reaction. Now the term general base catalysis refer to any reaction catalysed by OH^- ; undissociated base (e.g. NH_4OH) or anion of weak acid (e.g. CH_3COO^- ion). Decomposition of nitramide $[NH_2NO_2 \rightarrow N_2O + H_2O]$ is catalysed by OH^- and CH_3COO^- ion.

Finally there are reaction which require presence of both acid and base. They are known as acid-base catalysed reaction. Mutarotation of glucose is an example, which is

$$HG \xrightarrow[\text{Proton donor (acid)}]{+H^+} \overset{+}{HGH} \xrightarrow[\text{Proton Acceptor (base)}]{-H^+} GH$$

14.8 ENZYME CATALYSIS

Enzymes are complex protein molecules produced by bacteria. They are natural catalyst and the fastest known catalysts. They catalyse many bio-chemical processes. e.g. inversion of cane sugar is done by invertase; zymase causes conversion of glucose to ethanol and urease can initiate the process of hydrolysis of urea. Since enzymes exist as colloid and colloid is a heterogeneous suspension of micro particles, enzyme catalysis is also known as micro-heterogeneous catalysis. Characteristics of enzyme-catalysed reactions are —

(i) These are fastest reaction. No artificial catalyst can enhance the rate like enzyme catalysis.

(ii) Enzymes are highly specific. One enzyme can catalyse only one reaction. Thus urease, which catalyses *brilliantly the hydrolysis of urea fails to catalyst hydrolysis of methyl urea.*

(iii) Unlike ordinary catalysts, enzymes are consumed during reaction; enzyme producing bacteria regenerate them.

(iv) Since bacteria are living organisms they must be served food for survival. Generally NH_4NO_3 is added for this.

(v) Enzymes require presence of another known as co-enzyme. Though it can be compared with catalytic promoters, but it should be remembered that promoters are not essential, but co-enzymes are essential.

(vi) Since enzymes are colloids and colloids co-agulate at high temperature; enzyme-catalysed reactions should be carried out never above ~ 35°C.

(vii) They act well at moderate pH.

Mechanism of both acid-base and enzyme catalytic reaction is intermediate compound formation theory.

14.9 ADSORPTION-DEFINITION-DIFFERENCE WITH ABSORPTION

Surface of any solid or liquid is always in a state of strain due to unsaturated force. A molecule in the bulk is equally attracted to all sides by its neighbours and so net attractive force on it is zero. But a molecule on the surface feels an unbalanced backward force, because there are none in front of it to counterbalance this. As a result of this unbalanced backward attraction, the surface molecules have a tendency to attract foreign particles in order to relieve strain. This tendency leads to a phenomena, known as *adsorption*, which is sticking or adherence on the surface. This is different from absorption, which means penetration in the bulk. Thus sponze absorbs water, but charcoal absorbs ammonia. However, adsorption is always accompanied by certain extent of absorption and vice-versa. That is why we sometimes use non-committal term *sorption* meaning both.

14.10 SOME TERMS

The substance on whose surface adsorption occurs is called *adsorbent* and the substance, which is adsorbed is called *adsorbate*. Extent of adsorption is always expressed as (x/m) where 'x' is amount of adsorbate adsorbed by 'm' gm. of adsorbent. 'x' can be expressed in any unit e.g. if the adsorbate is a gas 'x' may be volume unit; if it is a dissolved solute it may be moles/litre unit etc.

14.11 TYPES OF ADSORPTION

Study of adsorption of gases on metal surface reveals that easily liquifiable gases are strongly adsorbed particularly at high pressure. Now these are gases with high critical temperature $T_c\left(=\dfrac{8a}{27Rb}\right)$, high Boyle temperature $\left(=\dfrac{a}{Rb}\right)$ and high inversion temperature $\left(=\dfrac{2a}{Rb}\right)$. So they have high $a : b$ value. Since 'a' stands for Van-der-Waals attraction and 'b' for repulsion, so it was thought that Van-der-Waals force is responsible for adsorrption. Subsequently it was called van-der-waals adsorption or physical adsorption. However, it was later found that extent of adsorption not only depends on nature of adsorbate but also on nature of adsorbent. So there is some selectivity which gives rise the concept of chemical bond formation. It was later proved that after adsorbing oxygen on tungsten surface, if it is heated then not only oxygen but also tungsten oxide (WO_3) is desorped. Thus chemical force is also responsible for adsorption. Such adsorptions will be called *chemical adsorption* or simply *chemi-sorption*.

14.12 DISTINCTION BETWEEN TWO TYPES OF ADSORPTION

A comparison between physical and chemical adsorption can be done as follows.

Physical Adsorption	Chemi-sorption
1. It is caused by physical force viz. van-der-waals force.	1. It is caused by chemical force.
2. It is characterised by low heat of adsorption (~ 5 k.cal/mole).	2. It is characterised by high heat of adsorption (~15 to ~20 K.cal/mole).
3. The process is reversible; on heating desorption occurs.	3. It is irreversible process. On heating surface compound is distilled off.
4. It becomes prominent at high pressure.	4. Pressure change does not affect it.
5. Adsorbed layer is multimolecular.	5. Adsorbed layer is unimolecular in thickness.

14.13 ADSORPTION ISOTHERM

Extent of adsorption depends on

(a) Nature of both adsorbent and adsorbate.

(b) Nature of surface

(c) Pressure (or concentration) of adsorbate and

(d) Temperature.

The relation between extent of adsorption and equilibrium pressure (or concentration) at a given temperature is called adsorption isotherm.

14.14 FREUNDLICH ISOTHERM

Freundlich proposed the following adsorption isotherm, which is an empirical relation.

$$\frac{x}{m} = K.P^n \text{ (or } K.C^n) \qquad .....(14.1)$$

where 'x' is the amount of adsorbate adsorbed by 'm' gm of adsorbent at equilibrium pressure P (or concentration C). 'K' and 'n' are constants with $0 < n < 1$. If we take logarithm of both sides of equation (14.1) we get

$$\log \frac{x}{m} = \log K + n \log P. \qquad .....(14.2)$$

So plot of $\log \frac{x}{m}$ against $\log P$ (or $\log C$) is a straight line with slope 'n' and intercept '$\log k$'. The isotherm has been experimentally verified in the low concentration range.

14.15 LIMITATION OF FREUNDLICH'S ISOTHERM

Equation (14.2) states that $\frac{x}{m}$ will increase with 'P' and there is no limit for such an increase. However experimental plot is something otherwise. (Fig. 14.1).

It is found that at very low pressure range $\frac{x}{m}$ varies linearly with P, it then becomes concave towards P-axis and ultimately assumes a flat run. Freundlich's isotherm fails to explain this.

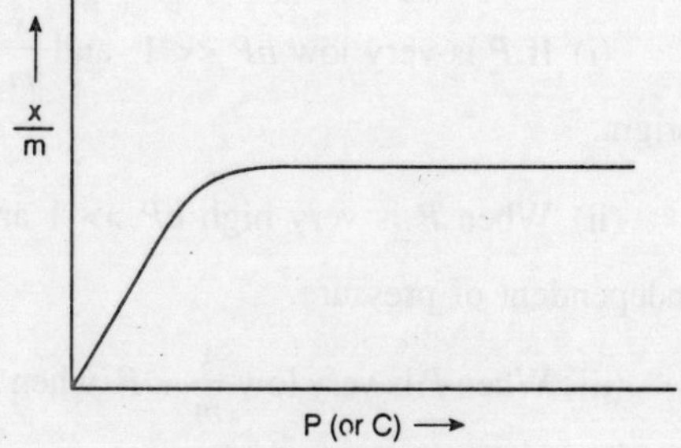

Fig. 14.1 Plot of $\frac{x}{m}$ vs. C.

14.16 LANGMUIR'S ADSORPTION ISOTHERM

The first theoretically derived adsorption isotherm is of Langmuir. To derive his isotherm Langmuir made the following assumptions.

(a) Adsorption takes place on some selected sites or spots known as active spots. These are points with very high unsaturated force. Generally corners; edges, crevices scratches etc. are such spots.

(b) Adsorbed layer is 'one-molecule-thick'. i.e. on a particular spot only one molecule will be adsorbed.

(c) Adsorption is a dynamic process i.e. adsorption and desorption are taking place simultaneously. At equilibrium the two rates become same.

Let us consider a solid surface over which adsorption equilibrium has been established. Let at equilibrium 'θ' fraction of the surface has been covered by adsorption, which means $(1 - \theta)$ fraction is bare. Since adsorbed layer is unimolecular in thickness, any new adsorption will occur on this $(1 - \theta)$ portion and desorption will take place from 'θ' fraction. If equilibrium pressure is P, then

rate of adsorption $\propto (1 - \theta)$

and $\propto P$

i.e. rate of adsorption $= k_1 (1 - \theta)P$. $[k_1 = \text{constant}]$

Also rate of desorption $\propto \theta$ only

i.e. rate of desorption $= k_2\theta$. $[k_2 = \text{constant}]$

At equilibrium $k_1 (1 - \theta) P = k_2\theta$

or $$\theta = \frac{k_1 P}{k_1 P + k_2}.$$

Extent of adsorption $\frac{x}{m}$ is evidently proportional to θ.

So $\frac{x}{m} = k_3 \theta$ $[k_3 = \text{constant}]$

or $$\frac{x}{m} = \frac{k_3 k_1 P}{k_1 P + k_2} = \frac{\frac{k_3 k_1}{k_2} . P}{1 + \frac{k_1}{k_2} . P} = \frac{aP}{1 + bP}$$

where 'a' and 'b' are new constants.

The equation $$\frac{x}{m} = \frac{aP}{1 + bP} \qquad .....(14.3)$$

is called Langmuir's adsorption isotherm. It explains the figure (14.1) for —

(i) If P is very low $bP << 1$ and $\frac{x}{m} = aP$ i.e. plot of $\frac{x}{m}$ vs P will be linear through origin.

(ii) When P is very high $bP >> 1$ and $\frac{x}{m} = \frac{aP}{bP} = \frac{a}{b}$; i.e. $\frac{x}{m}$ becomes constant and independent of pressure.

(iii) When P is very low $\frac{x}{m} \propto P$, when P is very high $\frac{x}{m} \propto P^0$, so when P is intermediate $\frac{x}{m} \propto P^n$ where $0 < n < 1$. This is Freundlich's adsorption isotherm. So **Freundlich's isotherm is a special case of Langmuir's adsorption isotherm.**

14.17 APPLICATION OF ADSORPTION

The phenomena of adsorption has found so many applications both in the laboratory and industry. Some application are listed below.

(a) They are used for the production of high vacuum. Silica gel, activated charcoal introduced in a vessel in very cold condition adsorbs all the gas molecules and produce high vacuum.

(b) All gas mask devices consist of adsorbent like active charcoal which adsorbs poisonous gases and thus purify the breathing air.

(c) Adsorbents like animal charcoal remove colouring matters from solutions. Thus during manufacture of cane-sugar brown colour of sugar is removed by treating its solution with animal charcoal.

(d) Now a days resins are used as ion-exchangers. These are also adsorbents.

(e) Chromatographic analysis is also done on the principle of adsorption.

POINTS TO REMEMBER

Substances, which influence rate of reactions without any change in its mass and composition are catalysts. Positive catalysts accelerate rate of reactions by lowering activation energy. Negative catalysts inhibits rates of reactions by decreasing or blocking number of reaction-sites. Catalysts cannot change the equilibrium yield. Intermediate compound formation theory and theory of adsorption are the two theories to explain the mechanism. The only natural catalyst is enzyme catalysis, which is microheterogeneous and highly specific.

Adsorption is the phenomena of sticking on the surface. It is different from absorption, which is penetration in the bulk. Extent of adsorption is expressed as (x/m) where 'x' is amount of adsorbate adsorbed by 'm' gm. of adsorbent. According to Freundlich $\frac{x}{m} = k.c^n$ $[0 < n < 1]$ and according to Langmuir $\frac{x}{m} = \frac{aP}{1+bP}$ [a & b are constant]. These relations are adsorption isotherm.

QUESTIONS

SHORT TYPE

1. Why catalysts are called catalysis ?
2. Give example of auto catalysis ?
3. How a positive catalyst accelerate the rate ?
4. How a negative catalyst decreases the rate ?
5. Explain induced catalysis.
6. A certain reaction R $\rightarrow$ P gives x % product at equilibrium in absence of catalyst; how will this amount change if a catalyst is introduced ? Other conditions are same.
7. Distinguish between adsorption and absorption.
8. Heat of adsorption for a certain substance is 18 k.calmole^{-1}. What type of adsorption it is ?
9. What type of adsorption becomes prominent at high pressure ?
10. How can you determine Freundlich's constants ?

ESSAY TYPE

1. Explain the criteria of catalysis.
2. Write down the theories of catalysis and show that the two are basically same.
3. What are the characteristics of enzyme-catalysis ?.
4. Write notes on acid-base catalysis.
5. Distinguish between physical and chemical adsorption.
6. Derive Langmuir's isotherm and hence arrive at Freundlich's isotherm.
7. Write some applications of adsorption.

OBJECTIVE TYPE

1. Write T against true and F against false statements.

(a) Negative catalyst increases activation energy.
(b) Positive catalyst decreases activation energy.
(c) Co-enzymes can be compared with promoters.
(d) Adsorption and absorption are same.
(e) Freundlich's isotherm is empirical.

2. Match terms of column A with these of column B.

A	B
(a) Anti knock compound.	(a) Langmuir.
(b) Mutarotation of glucose.	(b) Van-der-Waals.
(c) Adsorption isotherm.	(c) Berzelius.
(d) Physical adsorption.	(d) Unimolecular layer.
(e) Catalysis.	(e) Negative catalysts.
(f) Chemical adsorption.	(f) Acid-base catalysis.

Answers :

1. (a) F; (b) T; (c) F; (d) F; (e) T.
2. a – e; b – f; c – a; d – b; e – c; f – d.

Chapter 15

COLLOIDS

15.1 INTRODUCTION — IDEA OF CRYSTALLOID AND COLLOID

Thomas Graham studied diffusion of different solutions through membranes. He found that so called solutions belong to either of the two classes. In one class the dissolved substances can pass through both ordinary filter paper and parchment membrane and in another class though the substance pass through ordinary filter paper they are retained by the membranes. The substances of the first class are generally crystal in the solid state while those of the latter class are amorphous and glue-like in the solid state. Graham named them as crystalloid and colloid. (Greek; Kolla = glue; eoids = like).

15.2 TRUE DEFINITION

Graham's definition of crystalloid and colloid is no longer valid. Because NaCl, which is a typical crystalloid according to Graham behaves as a colloid in benzene solution. Also soap, which is glue-like in the solid-state and hence a colloid produces a true solution in ethanol. So, we see that same substance can behave both as a crystalloid and a colloid depending on condition. Hence a substance itself should not be called as a colloid or crystalloid; colloid is a 'state of matter'. Now which state of a matter will be called colloid? The correct definition is given on the basis of particle size. When a solid is added to a liquid; then the liquid molecules continuously strike on it and reduces its size by erosion. Question is what will be the size of the smallest particle ? If it is in between $1A^0 - 10A^0$, then these small particles find place in the intermolecular space of solvent molecules and are uniformly distributed through out the solution by thermal motion. What we get is a true solution. If however, the particle size is >1000 A^0, then they will sink to the bottom of the vessel by gravitational force i.e. in spite of thorough mixing sedimentation will slowly occur. When however, particle size is in between 10 A^0 to 1000 A^0 then they are too big to be accomodated in the intermolecular space but too small to be sedimented by gravity. In such a case the result is a heterogeneous suspension. This state is colloidal state. So the true definition of colloid is that — *A colloid is a twö-phasic heterogeneous mixture where one phase is dispersed in another phase, the particle size of the dispersed phase is in between* 10 A^0 *to* 1000 A^0.

15.3 TYPES OF COLLOIDS

Depending on nature of dispersed phase and disperison medium colloids can be classified as follows —

Dispersed Phase	Dispersion Medium	Name of the Colloid/Example
Solid	Liquid	Sol/paint
Liquid	Liquid	Emulsion/Milk
Gas	Liquid	Foam/Shaving cream
Solid	Solid	Solid sol/Alloy
Liquid	Solid	Gel/Butter
Gas	Solid	Solid foam/Cork
Solid	Gas	Smoke/Dust in air
Liquid	Gas	Aerosol/Fog

15.4 CLASSIFICATION OF COLLOIDS

Colloids can be classified either on the basis of preparatory methods or on the basis of property.

On the basis of preparatory method we call colloids as (a) **Dispersion colloid** (b) **Association** or condensation colloids and (c) **Molecular** colloids.

(a) **Dispersion colloid :** If a large particle is broken down to colloidal dimension by some means then the colloid will be called a dispersion colloid.

(b) **Association colloid :** In true solution particle size is in between $1A^0 - 10A^0$. If somehow some of such particles are associated together to enter into colloidal size ($10A^0 - 1000A^0$), then the colloid so obtained will be called association colloid.

(c) **Molecular colloid :** There are some high polymer, whose one molecule is of colloidal size, e.g. starch, protein molecules etc. They never form true solution such colloids are called molecular colloids.

Depending on property, sols i.e. solids dispersed in liquids can be classified in to two classes. These are (a) Lyophobic sols and (b) Lyophillic sols. Lyophobic (lyo = solvent ; phobia = hating) sols are those, where the dispersed phase dislike the medium in which it is dispersed.But lyophillic (phillia = loving) colloids are those where dispersed phase exhibits a strong affinity for the dispersion medium. A comparison between them can be drawn as follows —

Lyophobic	Lyophillic
1. These are generally sols of inorganic substances.	1. These are mostly sols of organic substances.
2. Special processes are taken up for preparation.	2. They are prepared simply by mixing.
3. As particles hate the medium they are unstable.	3. Particles are covered by solvent layers which impart extra-stability.
4. Co-agulated easily, either by heating or adding electrolyte.	4. Co-agulation is difficult-addition of solvent-removing agent is necessary.
5. Properties of solvent e.g. surface tension, viscosity etc. are not greatly altered.	5. Properties of solvent are greatly affected by sol-particles.
6. Exhibit Tyndall effect.	6. Do not exhibit Tyndall effect.

15.5 PREPARATION OF SOLS

Lyophillic sols may be prepared by simply warming the solid with the dispersion medium (liquid). But lyophobic colloids can not be prepared so easily. Methods adopted for the preparation of lyophobic sols can be divided into two classes. These are (a) **Dispersion** method and (b) **Condensation** method.

(a) **Dispersion method :** This is the method of breaking down large particles into colloidal dimension. This can be done by (i) mechanical means (ii) electrodispersion and (iii) peptisation.

(i) Mechanical dispersion :

In this method the solid along with the liquid (dispersion medium) is fed into a colloid mill ; which is nothing but two rotating discs very close to each other. They rotate in opposite directions at a very high speed. The solid particles are ground down to colloidal size and are dispersed in the dispersion medium.

(ii) Electrodispersion, Bredig's are method :

This is the method adopted for preparation of sols of noble metals e.g. Pt; Au; Ag etc. Two wires of the desired metal are dipped in water taken in a beaker. Their terminals inside

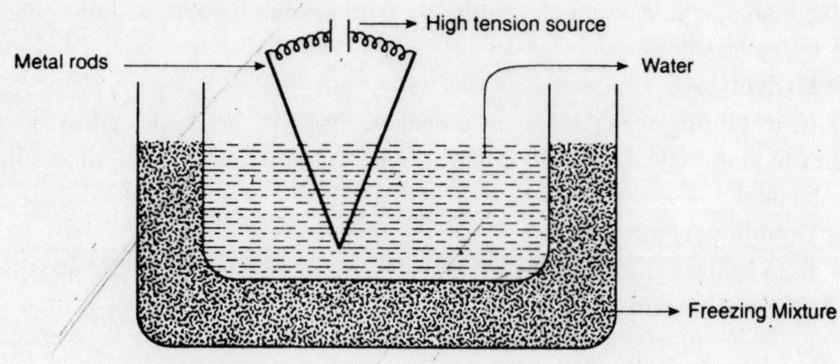

Fig. 15.1 Electrodispersion method.

water are very close. Now if they are connected to high-tension source then an electric spark will occur. Because of the very high heat some metal atoms are vaporised and the vapours are condensed in the surrounding medium to come to colloidal dimension. The entire vessel is kept over freezing mixture, so that water is not vapourised. Thus metal sols are obtained.

[*N.B. Metal sols have high medicinal value. Ag-sol is a preventive against gastroenterics. It is said that when Alexander invaded India, many of his ordinary soldiers died of gastro-related disease but none of his officers were attacked by this disease. This is because officers were used to take their food and drink in silver-vessel and so they ate trace amount of colloidal silver everyday. This gave them protective power against the deadly disease.*]

(iii) Peptization :

If a freshly prepared precipitate is struck (or shaken) by the solvent molecules along with some electrolyte having common ion, then the ppt.absorbs common ion and split into colloidal dimension. This method of sol preparation is called *peptization.* Thus if a fresh ppt. of $Fe(OH)_3$ is shaken with water and little amount of $FeCl_3$ we get a sol of $Fe(OH)_3$. The electrolyte having common ion ($FeCl_3$ here) is called *peptizing agent.* Other examples are preparation of AgCl sol by shaking fresh ppt.of AgCl with water and HCl; preparation of $Cu_2Fe(CN)_6$ sol by shaking it with water and $K_4Fe(CN)_6$, etc.

(b) Condensation Method : This method consists of associating molecules of a true solution to colloidal dimension. It can be done in many ways. Some are described below :

(i) Change of solvent :

If a solid is soluble in liquid A but insoluble in liquid B and if liquids A and B are miscible then by adding calculated amount of B in its solution in A we can get a sol. As an example — a sol of phenolphthalien can be obtained by adding some amount of water to alcoholic solution of phenolphthalien.

(ii) Oxidation reaction :

Sols can be obtained by chemical reaction. Oxidation is one such reaction. If in a solution of $Na_2S_2O_3$; trace amount of very dilute HNO_3 (oxidising agent) is added then immediate oxidation will take place and a milky *S*-sol is obtained. During qualitative analysis if the supplied sample consists of oxidising agent then when H_2S is passed in group-II, a sol of sulphur is generally produced which misleads students who take it as a ppt. of group-II.

(iii) Reduction :

A silver sol can be produced by reducing dil. $AgNO_3$ solution by mild reducing agent, like tannic acid, ethanal etc. Similarly gold sol can be obtained by reducing $HAuCl_4$ solution by hydrazine.

(iv) Hydrolysis :

If, to boiling water taken in a beaker, drops of $FeCl_3$ are added, it immediately hydrolyses to give $Fe(OH)_3$ sol. In this way sols of hydroxides of chromium and aluminium can be obtained.

(v) Double decomposition :

If, to a cold solution of As_2O_3 slow stream of H_2S is passed then a yellow coloured sol. of As_2S_3 is obtained.

15.6 DIALYSIS

We have seen that during preparation, sols become contaminated with electrolytes. Though presence of trace amount of electrolytes is essential for the stability of a colloid, excess electrolyte will co-agulate the colloid. Removal of excess electrolytes from colloidal solution is known as dialysis.

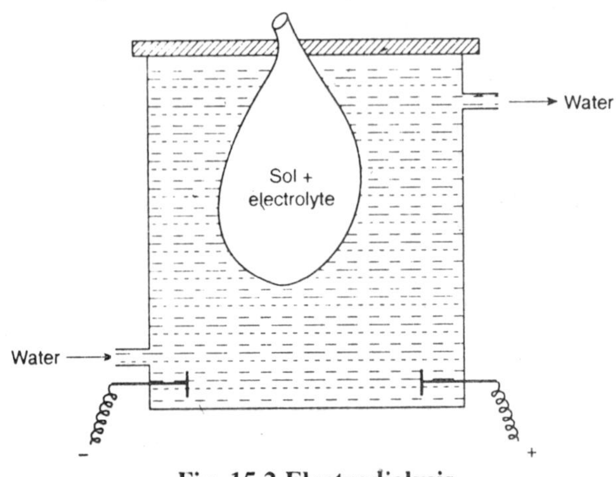

Fig. 15.2 Electrodialysis.

At the beginning of this chapter we have seen that colloids can not pass through parchment membranes but electrolytes can pass. This property is used in dialysis.

The sol, contaminated with electrolyte is taken in a parchment bag and is kept in hanging condition inside a vessel where there is water. Electrolytes will come out; the process is called *dialysis*. To enhance the rate of dialysis, the water outside is allowed to flow, so that concentration of electrolyte in water outside the dialyser remains almost zero always. This will be called *flow dialysis*. Rate of dialysis can further be increased by inserting two electrodes and connecting them to the terminals of a battery. Electrical attraction towards ions of the electrolyte will accelerate dialysis. Such a technique is called *electrodialysis*.

[*N.B. In human body kidney functions as a dialyser. Blood is a colloid, which is contaminated by electrolytes e.g. uric acid etc. during physiological processes. These are toxic chemicals; they are removed by dialysis, when blood passes through kidney. That is why for patients with damaged kidney artificial dialysis becomes necessary.*]

15.7 ULTRA-FILTRATION

Colloidal particles can pass through ordinary filter-paper because of their micro-size. However, if an ordinary filter-paper is impregnated with collodion or a regenerated cellulose such as cellophane then pores are reduced in size considerably. Now it can retain sol particles. Such a filter-paper is called *ultra-filter*. It can now be used for separation of sol-particles from electrolytes. The process of such separation is called *ultra-filtration*.

15.8 PROPERTIES OF COLLOIDS

Properties of colloids can be studied under the following headings. (a) Physical property (b) Colligative property (c) Optical property (d) Kinetic property and (e) Electrical property.

(a) Physical property : Shape, size and colour of colloids come under the heading of physical property. Regarding shape nothing concrete can be told. A sol-partical may be rod-like; disc-like; fibreous and spherical also. Their dimension (length; width; thickness; diameter) may vary from $10A^0$ to $1000A^0$. Thus size is also uncertain. Colour of a sol depends on wavelength of light scattered by it, which again depends on size of sol-particle. Since size vary within a wide length colour of a sol may also be different. Thus a Ag-sol may be orange-yellow; orange-red; purple or violet depending on particle size. A Au-sol may be violet; blue or even red.

(b) Colligative property : If a certain amount of solid forms a true solution in some solvent but forms a sol in some other solvent then since sol-particles are larger than particles of a true solution, so number of particles will be less in case of solution. We know colligative properties are dependent on number of particles present. So osmotic pressure of the solution will be less than that of true solution.

(c) Optical property : If a true solution is taken in a tube made of transperent material and if in a dark room a beam of light is allowed to pass through it then if some one places a microscope perpendicular to the direction of light, the field of microscope will appear as dark. Since particles of a true solution are of molecular size they are not obstacles in the path of the light. We say that a true solution is optically empty. But, if instead of true solution, we take a sol and repeat the same experiment, we shall see some bright moving spots. Since sol particles are larger than particles of true solution they are obstacles to the path of light of shorter wave lengths i,e. blue-violet. Thus blue-violet lights are scattered. This is known as Tyndall effect or Faraday-Tyndall phenomena. The beautiful blue colour of the sky is due to Tyndall effect. Here the dust particles are scatterers. If dust particles are totally removed from the atmosphere, sky will appear as dark as pitch. That is why astronouts see

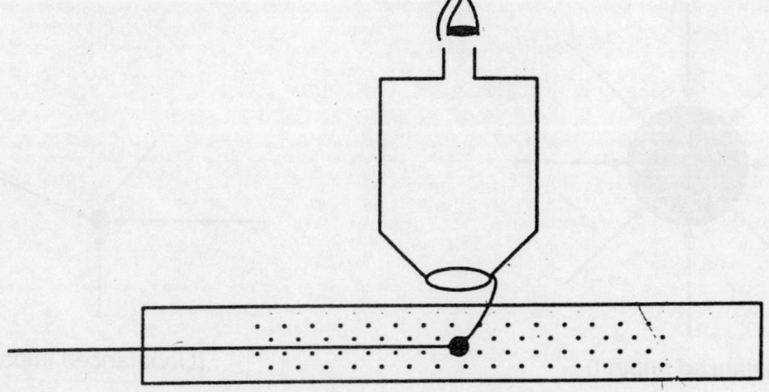

Fig. 15.3 Tyndall effect.

dark colour of the outer space. In a cinema house the path of the light coming from projector machine appears as bluish; this is also due to scattering of light i.e. Tyndall effect.

ULTRA-MICROSCOPE

Zsigmondy used this Tyndall phenomena to set up an instrument known as ultra-microscope. Solution is taken in a glass vessel. An intense beam of light is focused on it and a microscope is palced perpendicular to its path. Sol-particles appear as bright spots against a dark background. Larger halos around the sol-particles become visible. Thus it

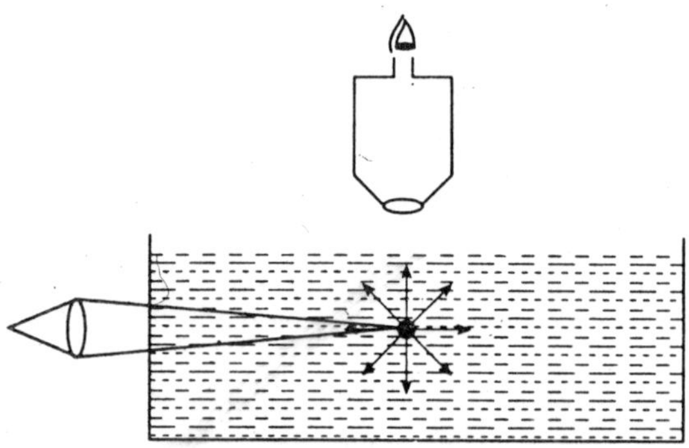
(ultra microscope)

(Microscopic view; halo against dark back-ground)

Fig. 15.4

shows the presence of sol particles, confirms the mixture as heterogeneous although does not give any idea regarding exact shape and size.

(d) Kinetic property : Robert Brown found that it pollen grains are suspended over water then the particles execute a random, zig-zag, erratic motion. He initially thought that this may be due to convection current. But the motion remains uninterrupted even inside a thermostat. This motion, known as Brownian motion is also found in case of sol particles suspended in a dispersion medium. Zsigmondy's ultra-microscope shows the particles as **dancing** bright spots. Explanation of Brownian movement was given by Einstein on the basis of kinetic molecular theory. According to this theory if any foreign particles is kept inside a liquid then liquid molecules will continuously strike it from all sides. If the particle is large then at any moment the net impact on it is zero. But as the particle becomes smaller

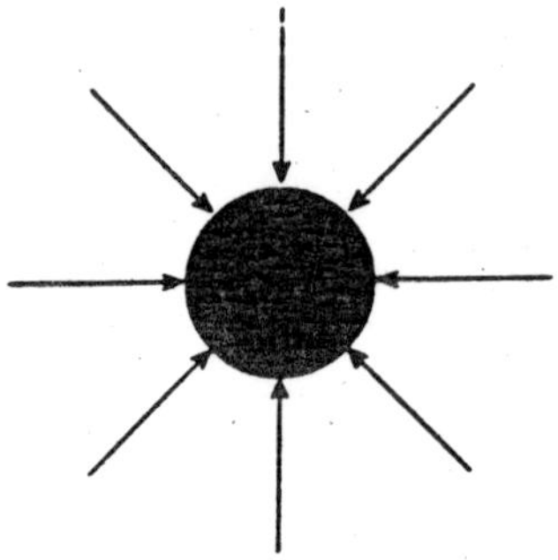
(Balanced impact)

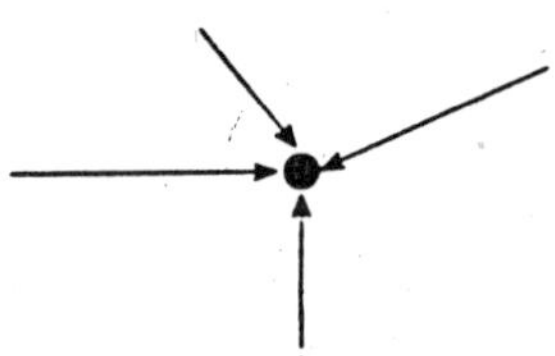
(Unbalanced impact)

Fig. 15.5

and smaller probability of unbalanced impact increases more and more. And it is this unbalanced impact which produces motion of the particle.

The fate of the particle becomes similar to the fate of a football during the play. Thus Brownian movement is a proof for the kinetic molecular theory. It is also an indirect proof for the existence of molecules. Because of the Brownian movement the sol particles do not find time to assemble and thus they are not sedimented. Thus it can be taken as one of the reasons for the stability of colloid.

(e) Electrical Properties : If a sol is taken in a U-tube and two electrodes are introduced in the two limbs, then when some potential is applied then all sol-particles move towards a particular electrode. If they move towards cathode, it will be called cataphoresis; when they move towards anode, it is anaphoresis; the general name is electrophoresis (Fig. 15.6 A). If however movement of the particles is prohibited by inserting a membrane at the middle of the tube, then the dispersion medium, moves in opposite direction. The phenomena is electro-osmosis. (Fig. 15.6B).

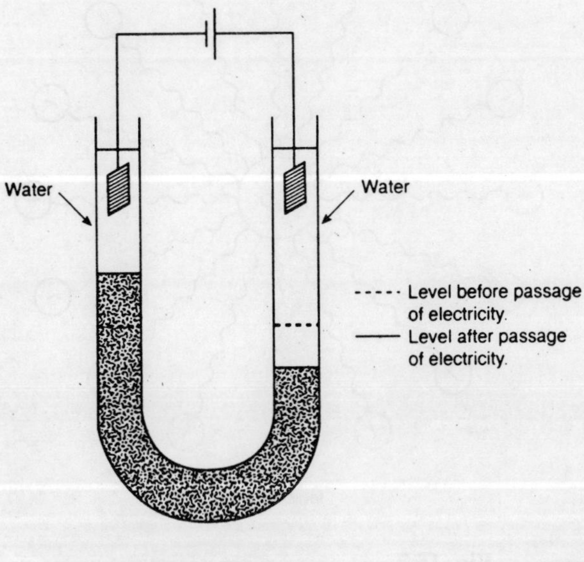

Fig. 15.6A Electrophoresis.

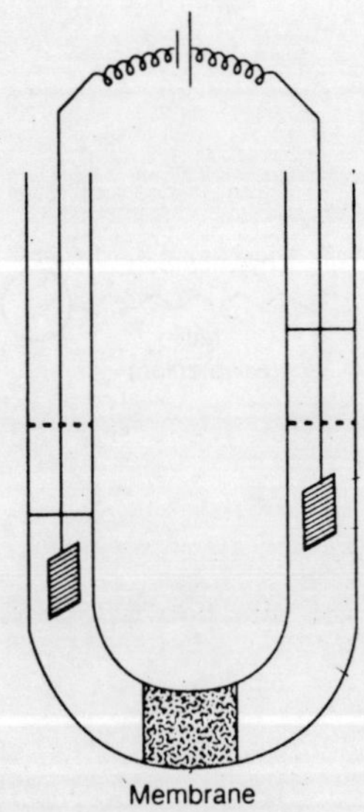

Fig. 15.6B Electro-osmosis.

Again if a sol is allowed to flow in a tube then between any two points in its path a difference in electrical potential is found. This is called *streaming potential*. Finally if a sol is taken in a tall jar; thoroughly shaken and then allowed to reach equilibrium by some settling process then between any two points of the jar, a difference in potential is found. This is called *sedimentation potential*. All these four phenomena indicate that sol particles carry charge and all carry same type of charge.

15.9 ORIGIN OF ELECTRICAL CHARGE

During study of properties of colloid we have seen that sol particles carry charge. Now the question is — what is the origin of electrical charge ? A sol particle can acquire electrical charge in three different ways, described below.

(a) If the particle is an electrolyte e.g soap, detergent etc. which are called *colloidal electrolyte* then it can have charge by self-dissociation.

(b) If the particle possesses both $-NH_2$ and $-COOH$ group e.g. amino acids then depending on pH of the medium it can have different charges and

(c) The particle may have charge by selective adsorption.

(i) Self-dissociation :

Soaps (or detergent) ionises in water to give Na^+ (or K^+) and an anion which is of a long chain fatty acid. [For soap acid group is $-CO_2H$ and for detergent it is $-SO_3H$; detergents do not form scum in hard water because of soluble Ca or Mg salt of the sulphonic acid]. If very little amount of soap is added to water then it forms a true solution; but if we go on increasing soap concentration, then at a particular concentration, known as critical micelle concentration (c.m.c.) anions aggregate to form an association, known as ionic micelle which is of colloidal size. The anions have a water-repelling hydrocarbon tail and polar COO^- (or SO_3^-) end. These polar ends surround the water molecule keeping their hydrocarbon tails away from water. A micelle is thus produced (Fig. 15.7). It will have charge of the polar head.

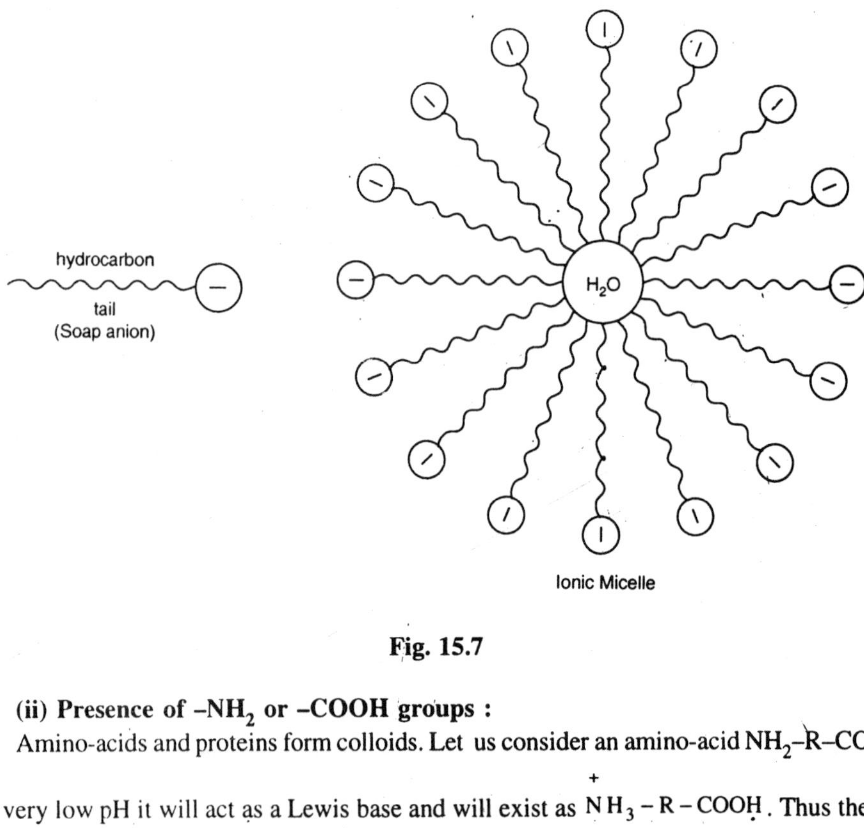

Fig. 15.7

(ii) Presence of $-NH_2$ or $-COOH$ groups :

Amino-acids and proteins form colloids. Let us consider an amino-acid NH_2–R–COOH.

At very low pH it will act as a Lewis base and will exist as $\overset{+}{N}H_3-R-COOH$. Thus the sol-particle will have + ve charge, and will move towards cathode during electrolysis. At high pH it will act as a Bronsted acid and will exist as NH_2–R–COO^- ; the particle will have –ve charge and will move towards anode. So at some intermediate pH it will not have any charge and electrophoretic mobility will be zero. This pH is called **isoelectric point.** At this pH particle exists as $\overset{+}{N}H_3-R-COO^-$. Such ions having two types of charges at two ends are called **Zwitter ions.**

(iii) By selective adsorption :

We know that surface of any solid has a tendency to adsorb foreign particles. When a sol is produced by dispersion method, then there is a tremendous increase in surface area and so adsorbing tendency increases. During adsorption the solid sol particles show some selectivity. To form chemisorption they generally prefer common ion. Thus if a $Fe(OH)_3$ solution is produced by hydrolysis of $FeCl_3$ then it absorbs Fe^{3+} ion and become +vely charged. If a As_2S_3 sol is produced by passing H_2S, particles adsorb S^{2-} ions and becomes negative. AgI sol, produced by adding little amount of KI to large amount of $AgNO_3$ is +vely charged due to adsorption of Ag^+ but AgI sol, produced by adding little amount of $AgNO_3$ to large amount of KI is negative due to adsorption of I^- ion.

15.10 ELECTRICAL DOUBLE LAYER AND ZETA POTENTIAL

We have just seen that sol particles bear charge. Because of this they attract ions of opposite charge from the dispersion medium. For electroneutrality the particles will like an equivalent amount of opposite charge standing by its side. However, thermal motion of the solution prevents if from getting required amount. An electrical double layer (fixed) is produced (Fig. 15.8) though charge on the sol particle is not fully balanced. Because of this residual force a diffuse layer follows the fixed layer. This diffuse layers carry both types of charges though charges of opposite (to that of sol particle) nature are greater in number. The potential difference between fixed layer and this diffuse layer is called *electrokinetic potential* or *zeta potential.*

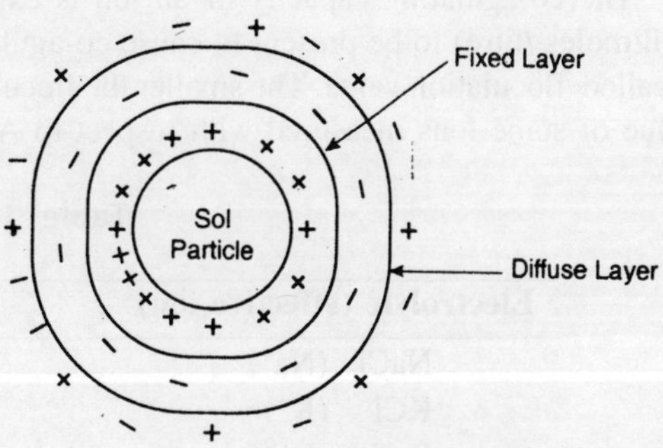

Fig. 15.8 Electrical double layer.

15.11 STABILITY OF A COLLOID

A question may be asked — why a colloid is stable inspite of the fact that particles are in suspended condition ? Why the particles are not sedimented ? The answer is that — particles are so small that weak gravitational force (= mg ; m = mass of a very very small particle) can not sodiment them. Again it may be asked-why the particles are not aggregating — what prevents them from coming together and form a big associate ? The answer lies mainly on its electrical property and to some extent kinetic property. Brownian movement continuously takes the particle from one position to another and thus probability of particles coming together is minimised. However, the main thing is electrical property. We have seen that each sol particle is surrounded by a diffuse layer, having mainly opposite charge. When two sol particles approach each other then a strong repulsive force between two diffuse layers prevent them from forming an association. Thus the sol becomes stable. The greater the width of this diffuse layer, the greater is the zeta potential and the greater is the stability.

15.12 CO-AGULATION — SCHULZE-HARDY RULE

We have seen that electrical charge on sol-particles is mainly responsible for its stability. So if to a sol we add electrolytes from outside then it will also attract ions of diffuse layer and thus width of diffuse layer will decrease and hence zeta potential will also decrease. When zeta potential falls to a certain minimum value known as critical zeta potential, the particles can overcome the barrier and come closes to form an association of large size, which is then sedimented, by gravity. This is known as *co-agulation.* Thus co-agulation is the sedimentation of a sol by adding electrolyte. There are many examples of co-agulation. Formation of delta is caused by slow co-agulation of turbid river-water by electrolytes of sea-water. Co-agulation of blood (clotting) can be done by rubbing the wound with alum. If alum powder is added to muddy water it becomes clear and clay, mud etc. are sedimented.

Since foreign ions added from outside are causing co-agulation, so co-agulating capacity must depend on valence of the added ion. Schulze and Hardy stated that **'the more is the valency of the ion, the more is its co-agulating capacity.' This is Schulze-Hardy rule.** Thus to co-agulate As_2S_3 sol (–ve) co-agulating power of $Al^{3+} > Ba^{2+} > Na^+$. For co-agulation of $Fe(OH)_3$ sol, co-agulating powers of ions are as follows — $[Fe(CN)_6]^{3-} > SO_4^{2-} > Cl^-$.

The co-agulating capacity of an ion is expressed by its minimum concentration (in millimoles /litre) to be present to cause co-agulation of a sol in just 2 hours. This amount is called floculation value. The smaller the floculation the greater is its capacity. Floculation value of some ions measured with respect to As_2S_3 (–ve) sol are given below.

Table 15.1

Electrolyte (effective ion)	Floculation value
NaCl (Na^+)	51
KCl (K^+)	50
$BaCl_2$ (Ba^{2+})	0.69
$AlCl_3$ (Al^{3+})	0.093

A series showing the co-agulating capacities of different cations and anions have been made. The series is called Lyotropic series or Hoffmeister series. It is written below –$Th^{4+} > Al^{3+} > H^+ > Mg^{2+} > Cd^{2+} > Sr^{2+} > Ba^{2+} > Na^+ > K^+ > Rb^+ > Cs^+$ and for anions citrate^{3-} > Tartarate^{2-} > Sulphate^{2-} > Acetate^{-1} etc.

15.13 LYOPHILLICS ARE PROTECTIVE COLLOID-GOLD NUMBER

We have already known that lyophillic sols are more stable than lyophobic. Because of their strong affinity towards dispersion medium, each particle is covered by a sheath of solvent layer which gives them extra stability. For their co-agulation, not only removal of charge, but also removal of solvent layer becomes necessary. These sol particles are not only stable but also they can impart stability towards unstable lyophobic sol. Thus to a lyophobic sol, if trace amount of lyophillic sol is added then the sol-particles of the former adsorb the latter which forms a coating of solvent around lyophobic sol particle and thus protect it from co-agulation. Lyophillic colloids are therefore called as *protective colloid.* Protective capacities of different lyophillic colloids are different. This is expressed in terms of **gold number.** It has been found that to 10 c.c. of a standard gold sol (0.0053 – 0.0058%), which is lyophillic if 1 c.c. 10% NaCl is added the sol is co-agulated. It can be protected from co-agulation by adding lyophillic sol. The minimum amount (in milli grams) of lyophillic sol. required for this purpose is its gold number. Thus **gold number can be defined as the minimum amount (m. gm.) of lyophillic sol. to be added to 10 c.c. of a standard gold sol (0.0053 to 0.0058%) in order to check its co-agulation (indicated by colour change from red to blue) during addition of 1 c.c. 10% NaCl.** The minimum the gold number, the maximum is its co-agulating capacity. Gold numbers of certain lyophillic sols are listed below.

Table 15.2

Lyophillic Sol	Gold Number
Gelatin	0.005
Egg albumen	0.08
Gum-arabic	0.01
Potato-starch	25

15.14 EMULSIONS

When a liquid is dispersed in another liquid the resultant colloid is called *emulsion*. Emulsions are of two types (i) Oil in water type, where water is dispersion medium e.g. milk and (ii) Water in oil type, where oil is dispersion medium e.g. grease where dispersion medium is lubricating oil. For the preparation and stability of an emulsion, addition of a third substance is necessary. This is called *emulsifier* or *emulsifying agent*. Its function is to lower the surface tension at the interface between two liquids. It also acts as a bridging meterial between the two liquids.

15.15 GELS — THIXOTROPY — SYNERESIS

When a liquid disperses in a solid then the colloid thus formed is known as *gel*. Gels may be of two types (i) elastic and (ii) non-elastic. **Elastic gels** are those, which change their shape on application of force and gets back the original shape when force is removed e.g. gelatin. **Non-elastic gels** are rigid e.g. silica gel.

Elastic gels show imbibtion when immersed in solvent and swelling occurs. Some gels when shaken thoroughly are converted to sols and these sols on standing convert to gel.

This sol $\rightleftharpoons$ gel transformation is called thixotropy.

Many gels during standing undergo shrinkage. Water comes out as tears with a sound like weeping. The phenomena is called **syneresis**.

15.16 COLLOID IN EVERYDAY LIFE

In our every day life and also in industry, colloid now plays a major role. In fact our day begins with colloid — toothpaste is a gel; tea (or cofee) is colloid and if we go to bed after taking a cup of milk (emulsion) then we end with colloid also. Many of our foods — milk; butter, protein etc. are colloid. Most of our medicines are either sols or emulsions — ointments are gels. Paints used are emulsions or gels.

We apply colloid chemistry to prepare smoke precipitator which uses electrical property of the colloid. Artificial dialysers are also application of colloid chemistry. Delta formation and purification of turbid water is also application of co-agulation property of colloids. In fine it can be said that colloid chemistry has a wide range of applications in every part of our life.

POINTS TO REMEMBER

· When a phase is dispersed in another phase in such a manner that its particle size is in between 10 to 1000 Å and the mixture is heterogeneous thau it is a colloids state. It may have several names such as sol; gel; emulsion etc. depending on nature of dispersed phase and dispersion medium some sols are lyophobic (solvent hating), while some others are lyophillic (solvent loving). They can be prepared by dispersion and condensation methods. Sols have many peculiar properties, notable among them are Kinetic, optical and electrical property. Because of kinetic property it can show Brownian movement; optical property leads to Tyndall effect and electrical property is responsible for electrophoresis; electro-osmosis; streaming potential etc, Electrical property is mainly responsible for the stability of colloid. Lyophillic sols are more stable and they can impart extrastability towards lyophillic sols. They are protective colloids protective capacity is measured by gold number. Co-agulation means destruction of colloids and it can be done by adding electrolytes. According to Schulze and Hardy co-agulating capacity increases with increase in valency of ions. Series showing the order of co-agulating capacities of different ions is called lyotropic series.

QUESTIONS

SHORT TYPE

1. Can a substance itself be called colloid ?
2. What are sol; gel and emulsions ?
3. Which one of starch and Au-sol is lyophobic and which one is lyophillic?
4. What is a molecular colloid ?
5. Why the sky is blue ?
6. Why a solution of $FeCl_3$ can stop bleeding from a wound ?
7. Whose co-agulation in difficult and why ?
   (i) As_2S_3 sol (ii) gelatin sol.
8. Out of SO_4^{2-} ; PO_4^{3-} and NO_3^- ions which one is most effective for co-agulating sol ?
9. What is thixotropy ?
10. The greater the zeta potential, the more is the stability — explain.
11. What is gold number ?
12. What do you understand by the statement — "Gold number of haemoglobin is 0.03."

ESSAY TYPE

1. What are colloids ? How are they classified ? How can you prepare a colloidal sol of (a) gold (b) As_2S_3 (c) Fe $(OH)_3$?
2. What are lyophobic and lyophillic colloids. Draw a comparison between them.
3. Explain optical and kinetic properties of a colloid.
4. What are the origin of electric charge on colloids ? What are double layer and zeta potential?
5. What do you mean by co-agulation. State and explain Schulze-Hardy rule.
6. Explain the terms :
   (i) Peptization (ii) Lyotropic series (iii) Floculation value (iv) Colloidal electrolyte (v) Critical micelle concentration (vi) Protective colloid (vii) Thixotropy (viii) Synerpsis (ix) Emulsion (x) Dialysis.

OBJECTIVE TYPE

1. Fill in the gaps.

(a) Colloid is a ——— of matter.

(b) Preparation of a colloid by striking a fresh ppt. by solvent is called ———.

(c) Process of separation of electrolytes from colloids is known as ———.

(d) Ultramicroscope was devised by ———.

(e) An example of colloidal electrolyte is ———.

(f) Milk is an example of ——— type emulsion.

(g) Sol $\rightleftharpoons$ Gel equilibrium is known as ———.

(h) The weeping sound during shrinkage of gels is called ———.

2. Write T against true and F against false statements.

(a) Sols of inorganic substances are lyphobic.
(b) NaCl is crystalloid.
(c) Electrodispersion method was devised by Brown.
(d) Osmotic pressure of a sol is more than that of true solution with same concentration.
(e) Zwitter ions carry both charges at two ends.
(f) Gelatin (Gold no : 0.005) is less effective in protective capacity than egg-albumin (Gold no : 0.08).

3. Match terms of column A with those of column B.

A	B
(a) Milk	(a) Brown
(b) Paste	(b) Dialyser
(c) Kinetic property	(c) Soap solution
(d) c.m.c. value	(d) Gel
(e) Kidney	(d) Emulsion

Answers :

1. (a) State (b) peptization (c) Dialysis (d) Zsigmondy (e) Soap/detergent (f) Oil-in-water (g) Thixotrophy (h) syneresis.

2. (a) T (b) F (c) F (d) F (e) T (f) F.

3. a – e; b – d; c → a; d – c; e – b;

Chapter 16

ELECTROLYTIC CONDUCTION

16.1 INTRODUCTION—CONDUCTORS — DIFFERENT TYPES

Substances through which electricity can pass are known as electrical conductors. On the contrary substances through which electricity can not pass are called insulators. Electrical conductors are of two types (i) electronic and (ii) electrolytic. Electronic conductors are those which have free electrons. All metals and few very very long chain conjugated dienes having completely delocalised free electrons are electronic conductors. They are generally solid with the exception of mercury, which is liquid. On the contrary electrolytic conductors are solutions of acids, bases and salts or fused salts. In solution they produce some charged particles (called ions) which carry electricity.

16.2 TERMINOLOGY

Since electrolytic conductors (or electrolytes) are liquids, they must be taken in a vessel. Two metal rods are dipped in the electrolyte and these are connected to an external source of electricity. The entire set is called **electrolytic cell**. The metal rods are called **electrodes**. The electrode attached to the negative end of the external cell is called **cathode** and the other which is attached to the positive end of the cell is **anode**. In this connection attention may be drawn to the case of **electrochemical cell** e.g., Daniel cell **where electricity is produced due to chemical reaction (in electrolytic cell electricity produces chemical change**). In electrochemical cell negative electrode is anode and positive electrode is cathode. No misunderstanding will arise if we keep in mind that at anode **oxidation (annation)** reaction occurs and at cathode **reduction (catation)** reaction occurs. This is applicable for any type of cell.

16.3 MECHANISM OF ELECTROLYTIC CONDUCTION

It was Grothus who first proposed a mechanism to explain electrolytic conduction. According to him electrolytes are dipoles. In solution they are most randomly distributed. As soon as electricity is applied they orient themselves in such a way that +ve part points towards –ve electrode and –ve part points towards +ve electrode. The +ve and –ve part just near the electrodes are then discharged leaving behind their partners. The remaining halves then combine with the opposite ends of the nearest neighbours and rotate through 180°. The process is then repeated (Fig. 16.1). As a working hypothesis this is very good but this lacks quantitative support. Because the theory demands application of a minimum potential to

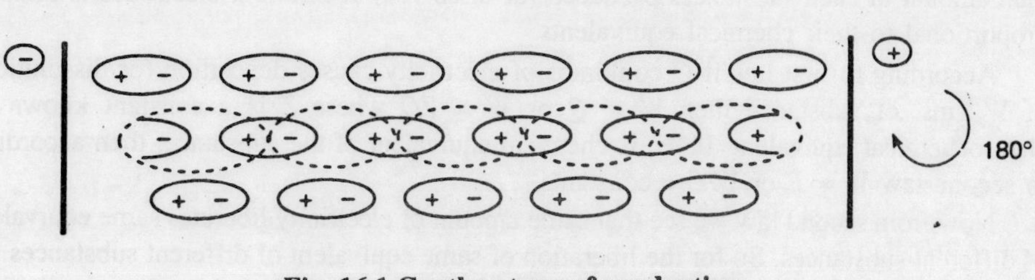

Fig. 16.1 Grothus type of conduction.

start electrolysis. But this is not found — any low potential can start electrolysis. So this theory was rejected. However, we shall see later that this idea has been used to explain high conductance value of H^+ ion.

16.4 ARRHENIUS THEORY OF ELECTROLYTIC DISSOCIATION

When Grothus theory failed to explain mechanism of electrolytic conduction, Arrhenius came in to picture. There were two problems before him, which were (i) Grothus theory had failed to explain mechanism of electrolytic conduction because it could not explain how the bond between charged particles was broken (ii) vant Hoff's equation for osmotic pressure was not obeyed by electrolytes because here number of particles appear to be more than dissolved. Arrhenius solved these two problems by assuming dissociation of electrolytes. He thought that electrolytes are those substances, which when dissolved breaks down considerably into charged particles. [Since breaking is spontaneous there is no question of minimum potential to break the bond and as breaking increases number of particles, so colligative property values become more than theoretically calculated values.]

These particles can move freely anywhere inside the solution. So they were called as ions (Greek : ion = Wanderer). Since positive ions move towards cathode they are cathode-seeking ions or cations. Similarly negative ions are anode-seeking or anions.

In any solution total number of cations and total number of anions may or may not be same. But total cationic charge must be equal to total anionic charge. (See table 16.1)

Table 16.1

Electrolyte	No. of Cation (n_c)	No. of anion (n_a)	Valency of Cation z_c	Valency of anion z_a	$n_c z_c$	$n_a z_a$
NaCl	1	1	+1	−1	+1	−1
Na_2SO_4	2	1	+1	−2	+2	−2
$Al_2(SO_4)_3$	2	3	+3	−2	+6	−6

So in every case $n_c z_c = n_a z_a$.

Arrhenius considered it as dissociation and not a decomposition.

16.5 FARADAY'S LAWS

Michael Faraday gave two laws connecting amount of product produced by electrolysis and quantity of electricity passed. The two laws are —

First law : The amount of a substance produced (or dissolved) at any electrode is directly proportional to quantity of electricity passed.

Second law : When same amount of electricity is passed through different solutions then amount of each substances produced (or dissolved) at different electrodes is directly proportional to their chemical equivalents.

According to first law if Q coulombs of electricity causes deposition (or dissolution) of W gms. of substance then $W \propto Q$ or $W = ZQ$ where Z is a constant known as electrochemical equivalent. If 'E' is chemical equivalent of the substance, then according to second law $W \propto E$ or W/E = constant.

Now from second law we see that same amount of electricity liberates same equivalent of different substances. So for the liberation of same equivalent of different substances we are to pass same electricity. Amount of electricity required for liberation of one equivalent of every substance will thus be same. This amount is called 'faraday'. If 'E' is equivalent wt. then for the liberation of 1 equivalent i.e., '*E*' gm. we require (*E/Z*) coulombs from first law. This (*E/Z*) is a constant and is called faraday. Its value is ~ 96,500 coulomb.

16.6 HITTORF'S RULE — CONCEPT OF TRANSPORT NUMBER

According to Faraday's second law if we pass one faraday of electricity through a solution of HCl by inserting Pt-electrodes we shall get one equivalent H-atoms and one equivalent Cl-atoms discharged at cathode and anode respectively. Before discharge they were ions and so same number of ions are discharged. This may lead us to the wrong idea that cations and anions are moving with same velocity. But because of a difference in size this is not possible. Hittorf considered an imaginary cell and showed that although cation and anion differ in their velocity, equivalent amount of each will be discharged during passage of same amount of electricity.

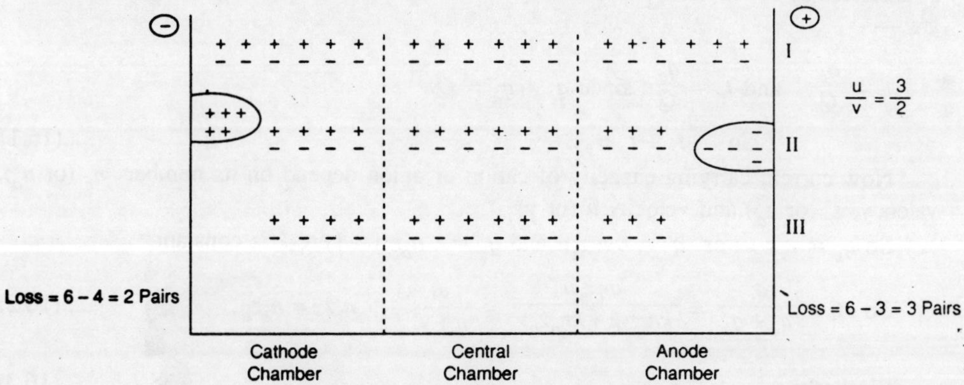

Fig. 16.2 Hittorf's imaginary cell.

Let us consider an imaginary cell consisting of three chambers cathode, central and anode. Let before electrolysis each chamber consisted of same number of cations and anions and let it be six cations and six anions (I, of above figure). Also let velocity ratio of cation and anion ($u : v$) be 3 : 2. So if in a given time three cations come towards cathode, two anions will go towards anode. Cathode chamber will have 9 cations, 4 anions and anode chamber will have 3 cations, 8 anions. Concentration of central chamber will not change. (State II). So cathode compartment has (9 – 4) = 5 excess cations and anode compartment has (8 – 3) = 5 excess anions. They will be discharged. Thus equivalent amounts are discharged though velocities are different. Actually number of ions discharged = Number of newcomers towards the electrode + Number, left partnerless due to movement of opposite ions. This total is same in both sides. Anyway after discharge cathode chamber will have 4 cations and 4 anions and anode will have 3 cations and 3 anions (state III). So loss at cathode = (6 – 4) = 2 which is proportional to velocity of anion and loss at anode = (6 – 3) = 3 which is proportional to velocity of cation. So the result can be summarised as —

Loss in conc. at anode compartment ∝ Velocity of cation (u).

Loss in conc. of cathode compartment ∝ Velocity of anion (v).

and total liberation ∝ ($u + v$).

$$\text{So} \quad \frac{\text{Loss of concentration of anode chamber}}{\text{Loss of concentration of cathode chamber}} = \frac{u}{v}.$$

This is known as *Hittorf's rule.*

Now in the just described imaginary cell 5 cations and 5 anions have been discharged i.e., cathode has lost 5 electrons and anode has gained 5 electrons i.e., 5 units of electricity has been passed. Since this is the result of movement of 3 cations and 2 anions, so contribution of cation is $\frac{3}{5}$th part and of anion is $\frac{2}{5}$th part. These fractional contributions are their transport number.

16.7 TRANSPORT NUMBER — TRANSFERENCE NUMBER OR HITTORF'S NUMBER

We can now define transport number, which is also known as transference number or Hittorf's number. It is the fraction of total amount of electricity carried by a type of ion (i.e., cation or anion). Thus if Q amount of electricity is passed through a solution, out of which q_c is carried by cation and q_a is carried by anion then cationic and anionic transport numbers are —

$$t_c = \frac{q_c}{Q} \quad \text{and} \quad t_a = \frac{q_a}{Q}; \text{ since } q_c + q_a = Q.$$

$$\text{So} \quad t_c + t_a = 1. \qquad .....(16.1)$$

Now current carrying capacity of cation or anion depend on its number n_c (or n_a), valency, z_c (or z_a) and velocity u (or v).

So $q_c \propto n_c z_c u$ or $q_c = k\, n_c z_c u$ and $q_a = k\, n_a z_a v$ where k is constant.

$$\text{So} \quad t_c = \frac{q_c}{q_c + q_a} = \frac{kn_c z_c u}{kn_c z_c u + kn_a z_a v} = \frac{u}{u+v}. \quad \because n_c z_c = n_a z_a. \qquad .....(16.2)$$

$$\text{Similarly} \quad t_a = \frac{v}{u+v}. \qquad .....(16.3)$$

Let we have sent such an amount of electricity through a solution that 't_c' equivalent of cation have moved towards cathode and 't_a' equivalent of anion has moved towards anode. Then amount discharged will be $(t_c + t_a)$. [See Hittorf's imaginary cell]. But $(t_c + t_a)$ equivalent is one equivalent. So one equivalent will be discharged which means that current sent is one faraday. So transport number of an ion can also be defined as its amount in equivalent unit moving through the solution during the passage of one faraday of electricity.

16.8 EXPERIMENTAL METHODS FOR THE DETERMINATION OF TRANSPORT NUMBER

There are three methods for experimental determination of transport number. These are —

(a) Hittorf's method (b) Moving boundary method and (c) e. m. f. method. Only the first two will be discussed.

(a) **Hittorf's method :** The principle of Hittorf's method is the result obtained from consideration of Hittorf's imaginary cell. There we have seen that —

$$\frac{\text{Loss of concentration of anode chamber}}{\text{Total amount liberated}} = \frac{u}{u+v}.$$

But $\frac{u}{u+v} = t_c$. If we express the amounts in equivalent units, then —

$$t_c = \frac{\text{Loss of concentration in equivalent of anode chamber}}{\text{Total equivalent liberated}}.$$

But total equivalent liberated is the total faraday passed. So —

$$t_c = \frac{\text{Loss of conc. in equivalent of anode chamber}}{\text{Total faraday passed}}.$$

The numerator can be known by quantitative analysis and the denominator can be determined by a coulometer (instrument for measuring flow of electricity). Thus t_c can be known. This is the working principle. Now let us see how Hittorf actually measured transport number of Ag^+ and NO_3^- in $AgNO_3$ solution.

The instrument used by Hittorf is shown in fig 16.3. It consisted of three limbs at the end of each there is a stop-cock which can be used to drain out solution of that limb. Initially the instrument is filled by a $AgNO_3$ solution of known concentration. Two electrodes are dipped at the two extreme limbs. They are connected to an external battery via a coulometer and key. Anode and cathode limbs are graduated. So volume of anode or cathode solution is known.

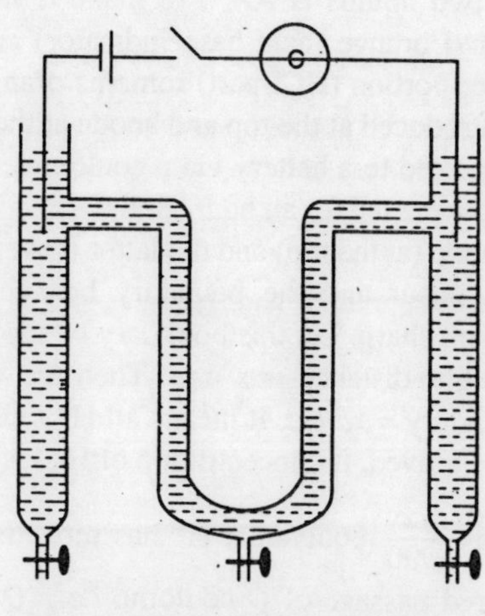

Fig. 16.3 Hittrof's instrument.

Hence from the known concentration of $AgNO_3$ solution and its known volume, equivalent of Ag present at anode is known. Let it be '*k*' equivalent. Now electricity is passed for about an hour and then circuit is disconnected and anode solution is completely drained out. Silver present in it is then determined by titration. Let it be '*l*' equivalent. Also let coulometer indicates that '*m*' faraday has been passed. Now transport number can be calculated. Type of calculation will depend on nature of electrode metal.

Pt-electrodes :

If Pt-electrodes have been used then loss in conc. of anode chamber = $(k - l)$ equivalent.

So $t_c = \left(\frac{k-l}{m}\right)$ and $t_a = 1 - t_c$.

Ag-electrodes :

If Ag-electrodes are used then at anode Ag metal will be attacked by NO_3^- ion and it will dissolve. For passage of '*m*' faraday '*m*' equivalent of Ag^+ will be dissolved. Initially its amount was '*k*' equivalent. Had there been no loss conc. would have been $(k + m)$ equivalent. But it is '*l*' equivalent. So loss = $(k + m) - l$. Hence here $t_c = \frac{(k+m)-l}{m}$ and $t_a = 1 - t_c$

The only precaution in this method is that there should not be any delay in draining out the liquid after electrolysis. Otherwise there will be diffusion.

(b) Moving boundary method : In moving boundary method transport number of ion is determined by noting directly the movement of ion under the influence of electric field. Here we take not only experimental electrolyte but also an auxilliary electrolyte. This secondary electrolyte is such that it has an ion common with the experimental electrolyte. The uncommon ion is such that it moves with less velocity. Thus if HCl be experimental electrolyte $CdCl_2$ may be the second electrolyte. Because Cd^{2+} moves with less velocity than H^+ and their boundary is always sharp.

The instrument used consists of a long tube of cross-sectional area 'a' cm.^{2}. Inside it we take solutions of both $CdCl_2$ and HCl. $CdCl_2$ being dense is kept at the lower part and HCl is above it. The initial position of boundary between the two liquids is AA' ; to make it sharp trace amount of methyl orange (acid-base indicator) can be added, so that upper portion (HCl-part) remains orange red. Now cathode is introduced at the top and anode at the bottom and they are connected to a battery via a coulometer and key. Electricity is now passed for an hour. Both Cd^{2+} and H^+ moves towards cathode (at the top) and the latter moves faster no overtaking will occur and the boundary between them will always remain sharp. Let this boundary be moved from AA′ to BB′ and this distance is x c.m. Then the volume swept by the boundary = xa c.c. It means all H^+ ions within this 'xa' c.c. have moved. If concentration of HCl is 'c' equivalent litre^{-1}, then $\frac{xac}{1000}$ equivalent H^+ has moved. Let coulometer registered passage of Q coulomb i.e., (Q/F) faraday. Then for one faraday equivalent of H^+ moved is $\frac{xacF}{1000Q}$. By definition this is the transport number of H^+. So $t_{H^+} = \frac{xacF}{1000Q}$ and

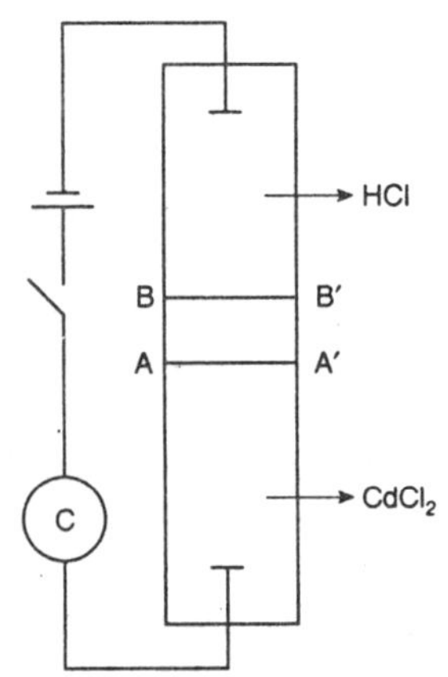

Fig. 16.4 Moving boundary method.

$$t_{Cl^-} = \left(1 - t_{H^+}\right).$$

16.9 TRUE AND APPARENT TRANSPORT NUMBER

The transport number that we measure by above mentioned methods is always apparent. This is because in polar solvent cations and anions are solvated. They attract those ends of solvent molecules whose charge is opposite to its. Thus they are covered by a solvent layer which they carry during movement. Hence the equivalent of cation or anion, found to move during the passage of one faraday (which is transport number) is always different from that which would have moved if there were no solvation and this figure is true transport number. Relation between true and apparent transport number and their measurements are beyond the scope of this book.

16.10 ABNORMAL TRANSPORT NUMBER

In dilute solution of CdI_2 transport number of Cd^{2+} shows normal value. But as concentration is increased $t_{Cd^{2+}}$ begins to decrease and at highly conc. solution it becomes negative. Negative transport number means Cd is being migrated towards anode. This can happen only when Cd forms a complex anion. Actually in concentrated solution CdI_2 dissociates as $2\,CdI_2 \rightleftharpoons Cd^{2+} + CdI_4^{2-}$. Because of larger size solvation of CdI_4^{2-} is less than that of Cd^{2+} and so its velocity is more. Hence more Cd is carried towards anode, as a result of which $t_{Cd^{2+}}$ becomes negative.

From Hittorf's rule $t_{Cd^{2+}} = \frac{\textbf{Loss in conc. of Cd at anode}}{\textbf{Current in faraday}}$.

Now loss = (Initial – final conc.). Since Cd is carried to anode so final conc. is more

and hence $t_{Cd^{2+}}$ is negative. So negative transport number is an indication of complex anion formation.

Q. Is transport number of Cl^- in NaCl and HCl same ?

Ans : In NaCl $t_{Cl^-} = \frac{V_{Cl^-}}{U_{Na^+} + V_{Cl^-}}$ and in HCl; $t_{Cl^-} = \frac{V_{Cl^-}}{U_{H^+} + V_{Cl^-}}$.

Since $U_{H^+} > U_{Na^+}$ so t_{Cl^-} in NaCl > t_{Cl^-} in HCl. Thus transport number of an ion in an electrolyte depends very much on its partner.

Q. How transport number change with temperature ?

Ans : In HCl, $t_{H^+} > t_{Cl^-}$; on heating t_{H^+} decreases.

In NaCl, $t_{Na^+} < t_{Cl^-}$; on heating t_{Na^+} increases.

In KCl, $t_{K^+} = t_{Cl^-}$; on heating t_{K^+} does not change.

So heating tends to equalise transport numbers of cation and anion..

16.11 CONDUCTANCE — SPECIFIC CONDUCTANCE OR CONDUCTIVITY

Ohm's law ; law regarding resistance etc. which are obeyed by electronic conductor are also obeyed by electrolytic conductor. However, for electronic conductor resistance is less and for electric conductors resistance is very high. For electrolytic conductors we generally measure conductance (C) which is reciprocal of resistance (R).

Thus $C = \frac{1}{R}$; its unit is ohm^{-1}.

For a wire of length l and uniform cross-sectional area A resistance R is given by

$$R = \rho \frac{l}{A}, \qquad .....(16.4)$$

where ρ is called specific risistance or resistivity. It is the resistance of a wire of unit length and unit cross-sectional area i.e., it is the resistance of a unit cube. Its unit is ohm. cm. So if we express conductance (C) of an electrolytic solution, then

$$C = \frac{1}{R} = \frac{1}{\rho} \cdot \frac{A}{l}$$

or

$$C = \kappa \frac{A}{l}. \qquad .....(16.5)$$

Where κ (kappa) is reciprocal of ρ and is called specific conductance or conductivity. For an electrolytic cell l = distance between two electrodes and A = cross-sectional area of each electrode. Thus specific conductance (κ) of a solution is the conductance of the solution placed between two electrodes which are unit distance apart (l = 1) and which are of unit area (A = 1 sq. unit). It can also be regarded as the conductance of a solution placed in a unit cube. **Its unit is ohm^{-1} cm^{-1}.** Dilution of a solution decreases number of ions present in the unit cube and since ions are carriers of electricity so conductance of the unit cube i.e., **specific conductance will decrease with dilution.**

16.12 EQUIVALENT CONDUCTANCE

In order to compare conductance of solution at different concentrations of the same electrolyte, a term equivalent conductance (Λ) has been defined. It is the conductance of a solution containing one gram equivalent of electrolyte placed between two sufficiently large electrodes which are exactly one centi-meter apart. Thus for a 0.1 (N) solution we are to pour 10 litres (to get one equivalent) for a 0.01 (N) solution we are to pour 100 litres etc. solution between two electrodes one centi-meter apart and measure their conductance. Direct measurement of equivalent conductance is almost impossible because it requires electrodes of enormous size. It is always measured indirectly as follows. If our solution is C(N) then $\frac{1000}{C}$ c.c. will contain one equivalent. If we are to pour such a volume of solution between two electrodes one c.m. apart (l = 1) then cross-sectional area of each should be $A = \frac{1000}{C}$ sq. cm. Its conductance (C) will be called equivalent conductance. So from equation 16.05

$$C = \Lambda = \kappa \times \frac{1000}{C}. \qquad .....(16.6)$$

This equation is utilised to measure equivalent conductance (Λ). Its unit is $ohm^{-1}\ cm^2$ $equivalent^{-1}$

Q. Do you find any inconsistency in the statement, " Equiv. conductance of a 0.1 N KCl solution is greater than its specific conductance?"

Ans : From the relation $\Lambda = \kappa \times \frac{10^3}{C}$ it may appear that $\Lambda > \kappa$ when $C = 0.1$ (N). However it should be remembered that unit of Λ is $ohm^{-1}\ cm^2$ $equivalent^{-1}$ whereas unit of κ is $ohm^{-1}\ cm^{-1}$. Since the two units are totally different, there should not be any comparison between them. So the statement is inconsistent.

16.13 MOLAR CONDUCTANCE

At present concentration is being expressed in molar units. That is why we use the term molar conductance instead of equivalent conductance. Molar conductance is the conductance of a solution containing one mole of the electrolyte placed between two sufficiently large electrodes which are exactly one centi-meter apart. Here also for a C(M) solution

$$\Lambda_{molar} = \kappa \times \frac{10^3}{C}.$$

For (1 – 1) electrolyte e.g. HCl, KNO_3 etc. equivalent mass = molar mass and $\Lambda_{molar} = \Lambda_{eq}$.

For (2 – 2); (2 – 1) ; (1 – 2) electrolytes e.g. $ZnSO_4$; $CaCl_2$; Na_2SO_4 etc. molar mass = 2 × equivalent mass. So $\Lambda_{molar} = 2 \times \Lambda_{eq}$.

16.14 MEASUREMENT OF CONDUCTANCE — CONDUCTIVITY CELL

All conductometric measurements are done by taking the solution inside conductivity cells. Conductivity cells are of many types — a simple type is shown below. (Fig 16.5).

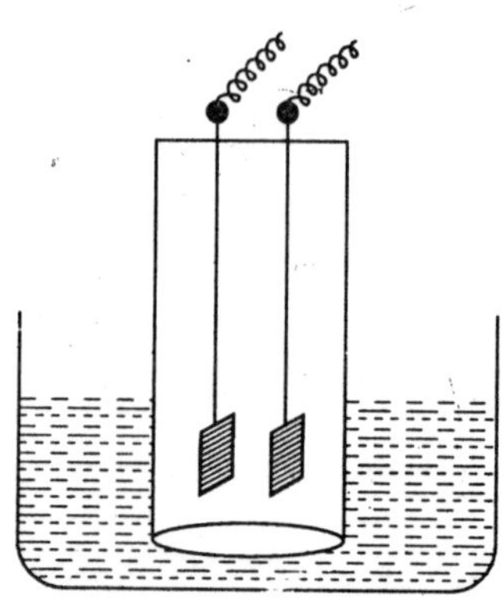

Fig. 16.5 Conductivity cell.

Two platinum discs of equal area are coated with Pt-black. They are fused at the ends of two wires. They are then kept inside a tube made of insulator.

The other two ends of the wire are attached to two screws fixed with the lid of the tube. They are connected to external source. The entire tube is then placed in a beaker where we take the solution whose conductance we want to measure. Its amount should be such that the electrodes are completely immersed in it. That can be seen by two windows. The windows are also used for washing the electrodes by a wash-bottle. Since for a given cell two electrodes are fixed in their own position, so the ratio $\frac{l}{A}$ is constant for a given cell. This ratio is called **cell constant.**

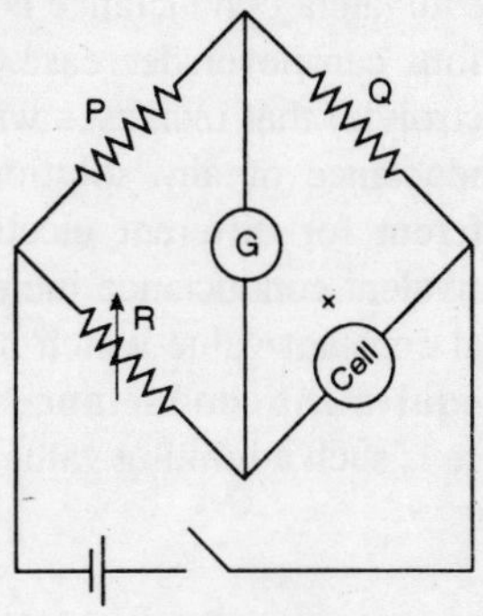

Fig. 16.6

Conductance of the solution taken in a given conductivity cell is measured by Wheatstone's bridge arrangement. In the adjacent figure P and Q are two fixed resistances, R is a variable resistance and X is resistance of cell. When null point is obtained $\frac{P}{Q} = \frac{R}{X}$. Thus '$X$' can be known. Its reciprocal is conductance.

16.15 MEASUREMENT OF EQUIVALENT CONDUCTANCE

We have already stated that equivalent conductance (Λ) of a solution is never measured directly. It is measured indirectly using the relation $\Lambda = \kappa \times \frac{10^3}{C}$. Now to measure specific conductance (κ) we use the realtion —

$$\frac{1}{R} = \kappa . \frac{A}{l}$$

or

$$\frac{l}{A} = \kappa \times R \qquad .....(16.7)$$

i.e. **Cell Constant = Specific Conductance × Resistance.**

Kohlrausch measured resistances of KCl solutions (by dissolving KCl in very very pure water, obtained by distilling thirty-two times !) at various concentration in conductivity cell of known cell constant. Then by utilising equation (16.7) their κ-values were calculated at various temperatures. We now use these values. First of all we take a cell, whose cell constant is not known. We fill it with a KCl solution of known concentration, whose κ is given in Kohlrausch's table. Then by measuring its resistance in a Wheatstone's bridge we can find out cell constant. Next the same cell is filled with the solution whose equivalent conductance is to be measured. We measure its resistance. Since cell constant is known we can calculate its specific conductance value (κ). Then by using the relation $\Lambda = \kappa \times \frac{10^3}{C}$ we can find out equivalent conductance Λ.

16.16 EQUIVALENT CONDUCTANCE AND DILUTION — DEBYE-HUCKEL-ONSAGAR EQUATION

We have seen that specific conductance of a solution decreases with increase in dilution, because number of ions present in the unit cube decreases with increase in dilution. However as equivalent conductance is the conductance of one equivalent electrolyte so here number of ions can never decrease. It is the volume of the solution (allotted for one equivalent electrolyte) that increases with dilution. Experimental measurements show that equivalent conductance of any solution increases with dilution. However, nature of increment is different for different electrolytes. For strong electrolytes (also known as **ionophore**) equivalent conductance increases with dilution but progressive dilution ultimately gives a final constant value which no longer changes with dilution. This constant value is known as **equivalent conductance at infinite dilution** (Λ^0). For weak electrolyte it appears that there is such a limiting value but simple dilution will not help to reach it. Thus we can write that

$$\Lambda^0 = \underset{C \to 0}{\mathrm{Lt}}(\Lambda). \qquad .....(16.8)$$

Equivalent conductance of a solution depends on many factor; of them only number of ions and velocity can increase with dilution. For weak electrolytes dilution increases degree of dissociation and hence number of ions increases with dilution. Since dissociation is never 100% due to existence of dissociation equilibrium we can never obtain Λ^0 - value by dilution.

For strong electrolytes there is no question of increase of number of ions because it is completely dissociated even in concentrated solution. The only thing that can increase is velocity. Debye-Huckel and Onsagar explained this in the following way.

In concentrated solution of strong electrolytes a large number of ions are present in a small volume and so there is strong interionic attraction. Each ion is surrounded by an ion atmosphere of opposite sign. This atmosphere was supposed to be spherically symmetrical. Now in polar solvent both central ion and ion-atmosphere are solvated i.e., surrounded by some loosely bound solvent molecules. During movement they carry these solvent molecules along with them. In concentrated solution due to proximity of the ions there will be a frictional force (between these two types of solvent layers) which will retard velocity of the ion. This effect is called **electrophoretic** effect.

As long as ion-atmosphere is spherically symmetrical the central ion feels no net attractive force. But during movement old atmosphere breaks down and new atmosphere is created. Destruction of old and formation of new atmosphere is not instantaneous. There is a time gap between the two. During this time, known as relaxation time the ion-atmosphere becomes elliptical (assymetric) and so more charge accumulates on the side opposite to that towards which central ion is moving. This will exert a net backward force, which will reduce its velocity. This effect is known as **asymmetric** effect or **relaxation** effect. Both electrophoretic and asymmetric effect will gradually decrease with increase in interionic distance and will disappear at infinite dilution. Considering these two effects Debye-Huckel and Onsagar derived the following equation which is known as Debye-Huckel-Onsagar equation. The equation is —

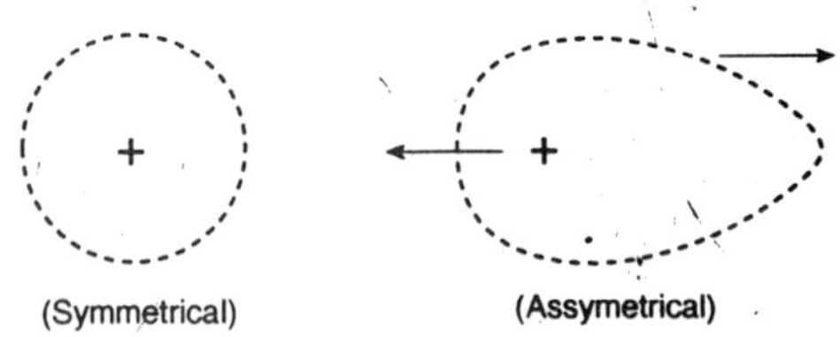

Fig. 16.7 Ion-atmosphere.

$$\Lambda = \Lambda^0 - \left(A + B\Lambda^0\right)\sqrt{C}. \qquad .....(16.8)$$

Here Λ is equivalent conductance at some finite concentration 'C', Λ equivalent conductance at infinite dilution and 'A' and 'B' represents electrophoretic and asymmetry effect.

Table 16.2

Variation of specific and equivalent conductance at 18°C

Conc.	Sp. conductance κ		Equivalent conductance Λ	
	KCl	CH_3COOH	KCl	CH_3COOH
N	0.0982	0.00132	98.3	1.32
N/10	0.01119	0.00024	112	2.30
N/100	0.001225	0.000143	122.4	14.3
N/500	—	—	126.3	30.2

16.17 DETERMINATION OF EQUIVALENT CONDUCTANCE AT INFINITE DILUTION

Equivalent conductance at infinite dilution (Λ^0) for strong electrolytes can be determined graphically using Debye-Huckel-Onsagar equation. $\Lambda = \Lambda^0 - \left(A + B\Lambda^0\right)\sqrt{C}$. What we are to do is to measure equivalent conductance (Λ) of the electrolyte at three or four different concentrations. We then plot Λ vs $\sqrt{C}$. A straight line is obtained. Its intercept on Λ-axis will give us Λ^0.

This method of determining Λ^0 is not valid for weak electrolytes, because weak electrolytes do not obey Debye-Huckel-Onsagar equation. In such cases Λ^0 is determined indirectly applying Kohlrausch's law. This is discussed in section (16.18).

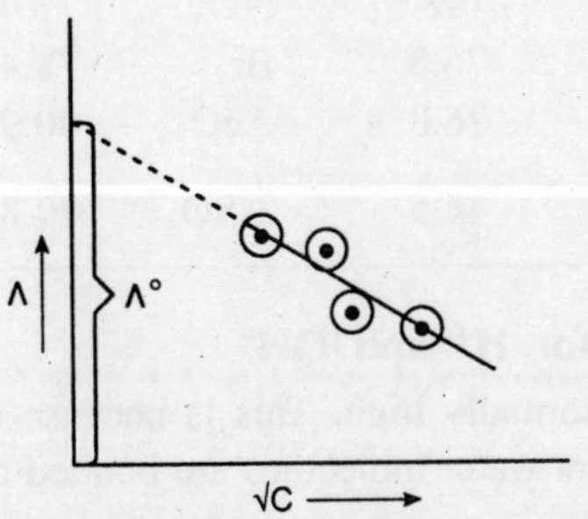

Fig. 16.8 Λ vs $\sqrt{C}$ plot.

16.18 KOHLRAUSCH'S LAW OF INDEPENDENT ION CONDUCTANCE

Kohlrausch measured equivalent conductance at infinite dilution for several strong electrolytes and found an interesting relation, which is —

$$\left(\Lambda^0_{KCl} - \Lambda^0_{NaCl}\right) = \left(\Lambda^0_{KNO_3} - \Lambda^0_{NaNO_3}\right) = \left(\Lambda^0_{KOH} - \Lambda^0_{NaOH}\right).$$

Also — $$\left(\Lambda^0_{HCl} - \Lambda^0_{HNO_3}\right) = \left(\Lambda^0_{KCl} - \Lambda^0_{KNO_3}\right) = \left(\Lambda^0_{NaCl} - \Lambda^0_{NaNO_3}\right).$$

To explain such a regularity Kohlrausch stated following law. It states that equivalent conductance at infinite dilution of any electrolyte is the summation of equivalent cationic and anionic conductances and each ion conductance term is independent of its partner. Thus

$$\Lambda^0 = l^0_c + l^0_a$$

[Λ = Equivalent conductance at infinite dilution and l^0 = equivalent ion conductance].

So $$\Lambda^0_{KCl} = l^0_{K^+} + l^0_{Cl^-} \text{ and } \Lambda^0_{NaCl} = l^0_{Na^+} + l^0_{Cl^-}.$$

Hence $$\Lambda^0_{KCl} - \Lambda^0_{NaCl} = l^0_{K^+} - l^0_{Na^+}$$

$$\Lambda^0_{KNO_3} - \Lambda^0_{NaNO_3} = l^0_{K^+} - l^0_{Na^+}$$

and $$\Lambda^0_{KOH} - \Lambda^0_{NaOH} = l^0_{K^+} - l^0_{Na^+}.$$

By applying this law we can compute Λ^0 — values for weak electrolytes. Thus —

$$\Lambda^0_{CH_3COOH} = l^0_{H^+} + l^0_{CH_3COO^-}$$
$$= \left(l^0_{H^+} + l^0_{Cl^-}\right) + \left(l^0_{Na^+} + l^0_{CH_3COO^-}\right) - \left(l^0_{Na^+} + l^0_{Cl^-}\right)$$
$$= \Lambda^0_{HCl} + \Lambda^0_{CH_3COONa} - \Lambda^0_{NaCl}$$

and $\Lambda^0_{NH_4OH} = \Lambda^0_{NH_4Cl} + \Lambda^0_{NaOH} - \Lambda^0_{NaCl}$.

Since HCl ; CH_3COONa ; NaCl ; NH_4Cl ; NaOH etc. all are strong electrolytes - their Λ^0 — values can be known graphically. Hence $\Lambda^0_{CH_3COOH}$ or $\Lambda^0_{NH_4OH}$ become known.

Table 16.3

Equivalent ion conductances at infinite dilution (25°C)

IONS	l^0_c	IONS	l^0_c	IONS	l^0_a	IONS	l^0_a
H^+	349.8	Na^+	50.1	OH^-	198.5	NO_3^-	71.4
Li^+	38.69	K^+	73.5	Cl^-	76.3	Br	78.4
NH_4^+	73.4	Ag^+	61.9	I^-	76.8	AcO	40.9
$\frac{1}{2}Ba^{2+}$	63.6	$\frac{1}{2}Mg^{2+}$	53.1	HCO_3^-	44.5	$\frac{1}{2}SO_4^{2-}$	79.8

Q. Explain abnormally high conductance values for H^+ and OH^-

Ans : Conductance values of H^+ and OH^- are abnormally high. This is because of Grothus type of conductance. In aqueous solution adjacent water molecules are bonded by

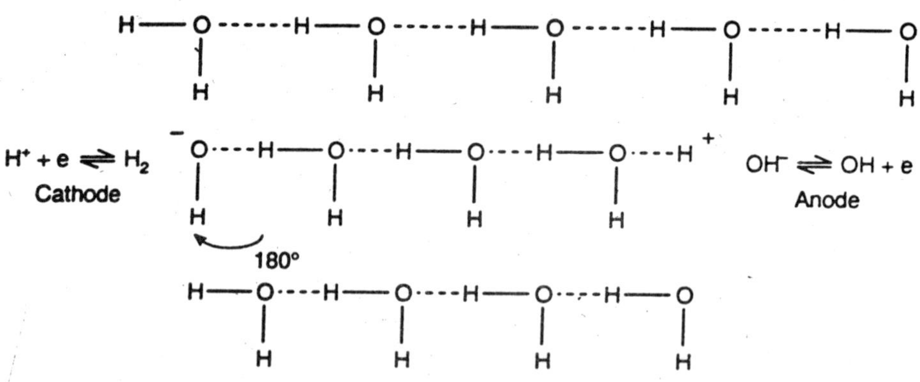

Fig. 16.9 High conductance of H^+ and OH^- ion.

H - bond. When one H^+ is released at cathode the hydrogen bonds convert to ordinary bonds releasing one OH^- at anode. They then turn through 180°. The discharge process becomes quick and consequently their conductance values become very high.

Q. All conductance measurements are done by A. C. — why ?

Ans : If conductance measurement is done by *D. C.* then there will be electrolysis and so concentration of solution will continuously change with time; we shall not get any constant value of Λ. Furthermore there will be concentration polarisation due to a difference in concentration at cathode and anode. On the contrary if we use *A. C.* electrolysis will not occur and these difficulties will not come.

16.19 IONIC MOBILITY AND ABSOLUTE VELOCITY

Velocity of an ion is the distance through which it moves in one second. Its traditional unit is cm. sec^{-1}. In order to compare velocities of different ions ionic mobility and absolute velocity have been defined. Ionic mobility is velocity at unit potential gradient. And absolute velocity is mobility at infinite dilution. Thus if '*E*' is applied potential, '*l*' is distance between two electrodes i.e., (*E/l*) is potential gradient, and under these circumstances if '*x*' is velocity of an ion at infinite dilution then absolute velocity = $\frac{x \times l}{E}$. Its unit is cm^2 sec^{-1} volt^{-1}. If '*u*' and '*v*' are absolute velocities of cation and anion then it can be shown that

$u = \frac{l^0_c}{F}$ and $v = \frac{l^0_a}{F}$ or $(u + v).\ F = l^0_c + l^0_a = \Lambda^0$.

16.20 APPLICATION OF CONDUCTANCE MEASUREMENTS

Measurement of conductance of solution helps us to know (a) degree of dissociation of weak electrolytes (b) solubility and solubility product of spairingly soluble salt (c) ionic product of water (d) hydrolysis constant of salts and (e) conductometic titration. Topics 'c'and 'd' will be discussed in next chapter. Here the remainings are being discussed.

(a) Degree of dissociation : The term degree of dissociation means fraction of the total amount of electrolyte undergoing dissociation. Let Λ be the equivalent conductance of a weak electrolyte at finite concentration '*c*' when its degree of dissociation = α. Also let Λ^0 be its equivalent conductance at infinite dilution when dissociation is supposed to be complete. So at finite conc. '*c*' fraction dissociated = (Λ/Λ^0). This is degree of dissociation (α.). 'Λ' can be experimentally measured. Λ^0 can be known from Kohlrausch's table. So 'α' is known.

(b) Solubility and solubility product : Solubility and solubility product can be determined conductometrically. We are to prepare a saturated solution of the spairingly soluble salt and measure its conductance in a cell of known cell constant. We can then know its specific conductance ($\kappa_{Sol^n.}$).

Now $\qquad \kappa_{Sol^n.} = \kappa_{Salt} + \kappa_{H_2O}$

Knowing κ_{H_2O} we can calculate κ_{salt}. Since the salt is spairingly soluble, the solution is infinitely dilute, so its equivalent conductance (Λ) is Λ^0.

But $\qquad \Lambda = \kappa \times \frac{10^3}{C}$.

So $\qquad \Lambda^0 = \kappa_{salt} \times \frac{10^3}{C}$.

Again $\Lambda^0 = l^0_c + l^0_a$. So Λ^0 can be known from Kohlrausch's table. As κ_{salt} is determined, we can calculate solubility of the salt '*c*'.

(c) Conductometric titration : End points of titration reaction between two electrolytes can be determined conductometrically if natures of variation of conductance (during reaction) before end point and after end point are different. We discuss it with some examples —

(i) Titration between a strong acid and strong base :

Let we have taken a strong acid (say HCl) solution in the conductivity cell. Its conductance value is high due to presence of high conducting H^+ ion. If we now add NaOH slowly then due to reaction H^+ ions are replaced by low conducting Na^+ ions. So overall conductance will decrease. This will continue up to end point, where conductance will be minimum because of total absence of H^+ ions. If we still go on adding NaOH, conductance will now increase due to fast moving OH^- ion and excess Na^+ ion. A rough plot of conductance vs volume of alkali added is shown in figure 16.10. The point of intersection between two straight lines correspond to end point.

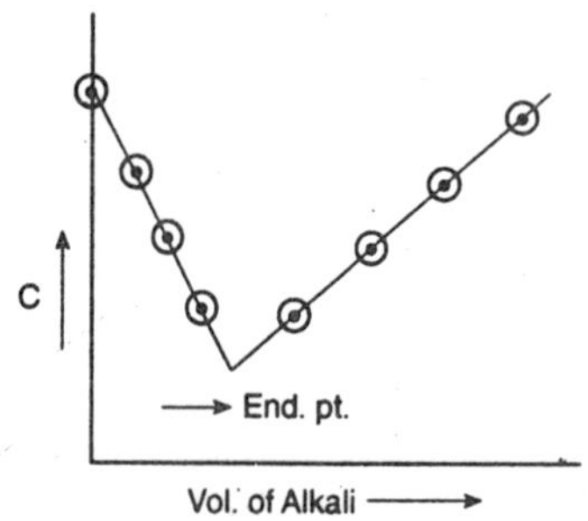

Fig. 16.10 Strong acid vs strong base.

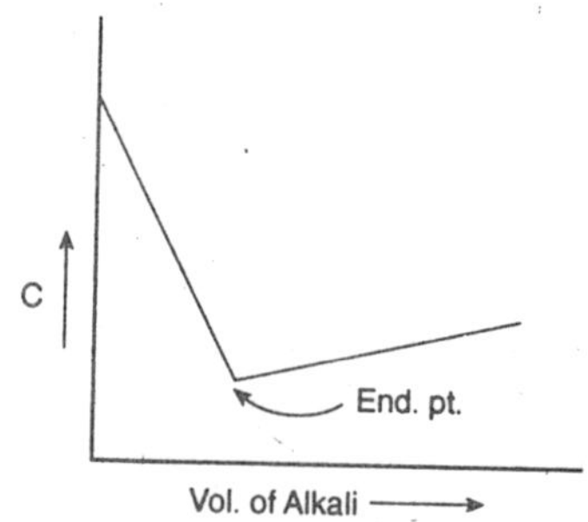

Fig. 16.11 Strong acid vs weak base.

(ii) Strong acid vs weak base :

If the conductometric titration is between a strong acid (HCl) and a weak base (NH_4OH) then at the initial stages conductance of the solution continuously decreases due to replacement of high conducting H^+ by low conducting NH_4^+ ion. It becomes lowest at end point and if addition of NH_4OH is still continued conductance of the solution remains more or less constant because dissociation of weak base is negligible in presence of common-ion. A rough sketch showing variation of conductance is shown in figure 16.11.

(iii) Weak acid vs strong base :

During conductometric titration of a weak acid (CH_3COOH) by a strong base (NaOH), a momentary drop in conductance is noted when first few drops of NaOH are added. This is due to replacement of high conducting H^+ ion by low conducting Na^+ ion. But very soon the equilibrium

$CH_3COOH \rightleftharpoons H^+ + CH_3COO^-$ shifts towards right and the loss of H^+ ion is compensated. Due to excess NH_4^+ and CH_3COO^- i.e. the salt, conductance slowly increases up to end point. After this if we still add NaOH conductance will increase sharply because of the high conducting OH^- ion and also excess Na^+ ions. A rough sketch (Fig. 16.12) showing variation of conductance with vol. of alkali is drawn.

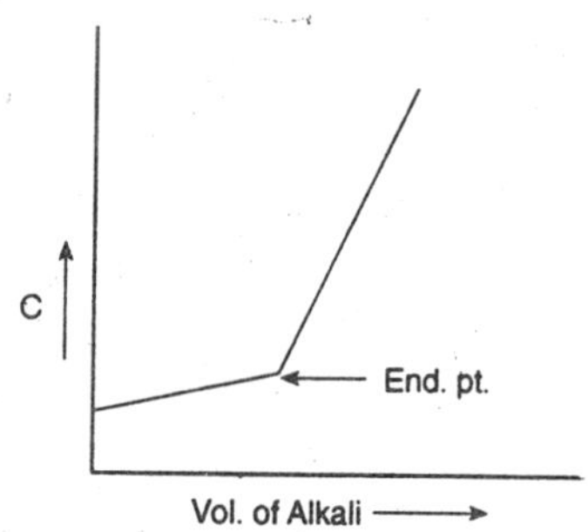

Fig. 16.12 Weak acid vs strong base.

(iv) Weak acid vs weak base :

Such titrations can not be carried out by using indicators. However they can be done conductometrically. At the beginning conductance is quite low due to poor dissociation of the acid. But as we go on adding base salt is produced which is completely dissociated. Due to increase in number of ions conductance begins to increase slowly up to end point. After end point if we still add base conductance will remain more or less constant due to negligible amount of the added base in presence of common ion. Variation of conductance during titration is shown in figure 16.12.

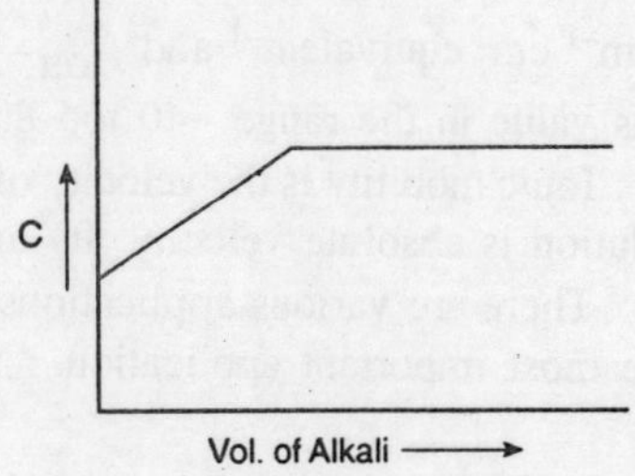

Fig. 16.13 Weak acid vs weak base.

(v) Precipitation titration :

Precipitation titrations e.g. titration between KCl and $AgNO_3$ can be done conductometrically. During addition of $AgNO_3$ to KCl, AgCl is ppted, which means that Cl^- in KCl solution are replaced by NO_3^- ions. Since their ion conductance values are quite close [$l^0_{Cl^-} = 76.3$; $l^0_{NO_3^-} = 71.4$] overall conductance of the solution does not show any remarkable change up to end point. After end point conductance will increase due to excess Ag^+ ions and NO_3^- ions. A rough sketch showing variation of conductance with vol. of $AgNO_3$ added is shown in figure 16.13.

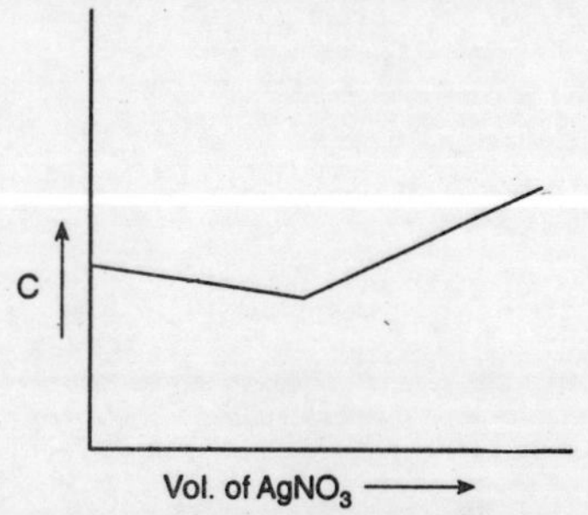

Fig. 16.14 $AgNO_3$ vs KCl titration.

Q. In all conductometric titration curves we plot conductance vs volume of titre and not resistance vs volume. Can we plot resistance ? $\left[R = \frac{1}{C}\right]$

Ans : Conductance of a solution varies directly with number of ions present which again depends on volume of titre added. So conductance and volume of titre are linearly related. But resistance is reciprocal of conductance. So it will not have linear relation with volume. Hence plot of R vs. Vol. will not be straight line. That is why resistance is not plotted.

POINTS TO REMEMBER

Those liquid conductors, where electricity is carried by some charge particles (ions) are electrolyte. Particles bearing +ve charge are cations and those having –ve charge are anions. In any electrolytic solution $n_c z_c = n_a z_a$. [n = number; z = valency]. Passage of one faraday (= 96,500 coulombs) electricity causes deposition (or dissolution) of one equivalent of a substance Transport number (t_c or t_a) [of cation or anion] is the fraction of total current carried by an ion. It is also its amount in equivalent, moving through the solution when one faraday is passed. Experimental methods for its measurement are (a) Hittorff's method (b) Moving boundary method and (c) E.M.F. method.

Specific conductance (κ) of a solution is conductance of a solution enclosed in a unit cube. Its unit is *ohm*$^{-1}$ *cm*$^{-1}$. ($\kappa \times v$), where v is the volume of solution containing one equivalent is equivalent conductance (Λ). [*unit = ohm*$^{-1}$ *cm*^{2} *equivalent*$^{-1}$]. For a C(N) solution $\Lambda = \kappa \times \frac{10^3}{C}$. Dilution decreases κ ; but increases Λ. Successive dilution leads to a limiting value of Λ, known as equivalent conductance at infinite dilution (Λ^0). According to Kohlrausch $\Lambda^0 = l_c^0 + l_a^0$; where l^0 is equivalent ion-conductance. At 25°C $l^0_{H^+} \approx 350$ ohm^{-1} cm^2 $equivalent^{-1}$ and $l^0_{OH^-} \approx 198$ ohm^{-1} cm^2 $equivalent^{-1}$ while all other ions have this value in the range ~40 to ~80.

Ionic mobility is the velocity of ion under unit potential gradient and mobility at infinite dilution is absolute velocity. Its unit is cm^2 sec^{-1} $equivalent^{-1}$.

There are various applications of conductance measurement. Conductance titration is the most important application.

QUESTIONS

SHORT TYPE

1. What is a faraday ?
2. 'The sentence transport number of Ag^+ is 0.47' is incomplete. Why ?
3. What is the unit of equivalent ion conductance ?
4. Will (a) conductance (b) specific conductance and (c) equivalent conductance of a solution change if the cell is changed ?
5. What is cell constant ? What is its unit ?
6. Define equivalent conductance at infinite dilution ?
7. Why conductance is measured by *A. C.*?
8. Can we plot resistance vs volume in conductometric titration curves ?
9. Why H^+ has so high equivalent ion-conductance value ?
10. What are true and apparent transport numbers. ?
11. Distinguish between velocity and absolute velocity of an ion.
12. "In 1 (M) CdI_2 solution transport number of I^- ion 1.12" — explain.

ESSAY TYPE

1. State and explain the term 'transport number'. Describe the principle of Hittorf's method for its determination.
2. What is Hittorf's rule — establish it from consideration of his imaginary cell.
3. Describe the moving boundary method for the determination of transport number.
4. Explain the terms specific and equivalent conductance. How do they vary with dilution.
5. What is equivalent conductance at infinite dilution ? Discuss the factors which are responsible for the fact that "equivalent conductance of an electrolyte at finite conc. is always less than that at infinite dilution."
6. Describe experimental procedures for the determination of Λ^0 for (i) HCl and (ii) NH_4OH.
7. State and explain Kohlrausch's law. Discuss one of its practical application.
8. How can you determine degree of dissociation of CH_3COOH conductometrically.
9. Discuss the method of determining solubility product of $BaSO_4$ conductometrically.
10. Draw and explain conductometric titration curves for the titration of (i) HNO_3 vs KOH (ii) NH_4OH vs HCl (iii) Na_2SO_4 vs $BaCl_2$.

PROBLEMS

1. Find the equivalent conductivity of $\frac{N}{10}$KCl solution having a specific conductivity of 0.01119 mhos.

   [*Hints* :
   $$\Lambda = \kappa \times \frac{10^3}{C}$$
   $\kappa = 0.01119$; $C = 10^{-1}$; find Λ].

2. The ionic conductance of H^+ is 315 mhos and that of Na^+ is 43.2 mhos. If the equivalent conductivity at infinite dilution of sodium acetate is 78.2 mhos, find the equivalent conductivity at infinite dilution of acetic acid.

[*Hints* : $\Lambda^0_{CH_3COOH} = \Lambda^0_{CH_3COONa} - l^0_{Na^+} + l^0_{H^+}$
$= 78.2 - 43.2 + 315$].

3. If the equivalent conductance at infinite dilution of NaCl, HCl and CH_3COONa are 126.45, 426.16 and 91.0 respectively, what will be that of CH_3COOH ?

[*Hints* : $\Lambda^0_{CH_3COOH} = \Lambda^0_{HCl} + \Lambda^0_{CH_3COONa} - \Lambda^0_{NaCl}$
$= 426.16 + 91 - 126.45.$]

4. At 25°C the specific conductivity of butyric acid at a dilution of 64 litres (i.e., 1 mol in 64 litres) is 1.812×10^{-4} mhos. The equivalent conductivity at infinite dilution is 380. What is the degree of dissociation of the acid, the concentration of hydrogen ions in the solution and the dissociation constant of the acid.

[*Hints* : $\Lambda = \kappa \times \frac{10^3}{C} = 1.812 \times 10^{-4} \times \frac{10^3}{1/64}$

$\alpha = \frac{\Lambda}{\Lambda_0}$; $\Lambda_0 = 380$].

5. The specific conductivity of 0.1 molar solution of a 1 - 1 salt is 0.00920. Calculate the equivalent conductance and the degree of dissociation. Given that the equivalent conductance is 108.9 at infinite dilution.

[*Hints* : Find Λ using $\Lambda = \kappa \times \frac{10^3}{C}$.

$\Lambda^0 = 108.9$; so $\alpha = \frac{\Lambda}{\Lambda_0}$].

6. The speed ratio of silver and nitrate ions in a solution of $AgNO_3$ electrolysed between silver electrodes is 0.916. Find the transport number of the two ions.

[*Hints* : $\frac{u}{v} = 0.916$;

$t_{Ag^+} = \frac{u}{u+v} = \frac{0.916}{1.916}$; $t_{NO_3^-} = 1 - t_{Ag^+}$].

7. Find out the velocity of K^+ ion in aqueous solution at 18°C under a potential gradient of 6 volt. per cm. from the following data :

Λ_0 for LCl at 10°C =130. Transport number of Cl^- ion KCl = 0.495.

[*Hints* : Absolute velocity $U_{K^+} = \frac{t_c l_o}{F}$

$= \frac{0.495 \times 130}{96{,}500}$ cm^2 sec^{-1} volt^{-1}.

Here pot. gradient = 6 volt cm^{-1}.

So velocity = $\frac{0.495 \times 130}{96{,}500} \times 6$ cm. sec^{-1}].

OBJECTIVE TYPE

1. Write T against true and F against false statements.

(a) In an electrolytic solution number of cations and number of anions must be same.

(b) In an electrolytic solution total cationic and anionic charges must be same.

(c) Transport number of an ion is a dimensionless quantity.

(d) Equivalent conductance depends on cell constant.

(e) Conductance depends on cell constant.

(f) Specific conductance depends on cell constant.

(g) Ionic mobility and absolute velocity are same.

2. Fill in the gaps.

(a) Unit of specific conductance is ——— but that of equivalent conductance is ———.

(b) In NaCl solution t_{Cl^-} is ——— than that in HCl solution.

(c) In conc. CdI_2 solution transport number of Cd^{2+} is ———.

(d) Increase in tempareture tends to ——— the transport numbers of cation and anion.

(e) Specific conductance ——— but equivalent conductance ——— with increase in dilution.

(f) In a conductivity cell electrodes are made of ——— platinum.

(g) For weak electrolytes dilution increase number of ——— but for strong electrolytes dilution increases ionic ———.

3. Match terms of column A with column B .

A	B
(a) Faraday	(a) Conductance
(b) Arrhenius	(b) Law of ion conductance
(c) Hittroff	(c) Absolute velocity
(d) Kohlrausch	(d) Law of electrolytic dissociation
(e) $ohm^{-1}\ cm^{-1}$	(e) Transport number measurement
(f) $ohm^{-1}\ cm^{2}$ equivalent^{-1}	(f) Law of electrolysis
(g) ohm^{-1}	(g) Equivalent conductance
(h) $cm^2\ sec^{-1}\ volt^{-1}$	(h) Specific conductance.

Answers :

1. (a) F (b) T (c) T (d) F (e) T (f) F (g) F

2. (a) $ohm^{-1}\ cm^{-1}$; $ohm^{-1}\ cm^{2}$ equivalent^{-1} (b) more (c) negative (d) equalise (e) decrease, increase (f) platinised (g) ions; velocity.

3. a – f; b – d; c – e; d – b; e – h; f – g; g – a; h – c.

Chapter 17

IONIC EQUILIBRIUM

17.1 OSTWALDS DILUTION LAW

Weak electrolytes are those, for which dissociation is always partial. In aqueous solution, all organic acids, bases and few inorganic acids and base behave as weak electrolytes. In their solution an ionic equilibrium between dissociated and undissociated part exists. Let the weak electrolyte be a weak organic acid HA. Then the equilibrium is

$$HA \rightleftharpoons H^+ + A^-.$$

Applying law of mass action, the equilibrium constant (K_a) is —

$$K_a = \frac{a_{H^+} \cdot a_{A^-}}{a_{HA}}, \text{ where } a = \text{activity.}$$

Since it is a dissociation equilibrium, K_a is called as dissociation constant. In dilute solution activity $\simeq$ concentration. So

$$K_a = \frac{C_{H^+} \cdot C_{A^-}}{C_{HA}}$$

Let C(M) be the concentration of acid HA prior to dissociation and α is its degree of dissociation, then at equilibrium $C_{H^+} = \alpha C$; $C_{A^-} = \alpha C$; $C_{HA} = C - \alpha C = (1 - \alpha)C$.

$$\text{So} \qquad K_a = \frac{\alpha^2 C}{1-\alpha}. \qquad .....(17.1)$$

If V is molar volume i.e., volume of solution in litre containing one mole of solute, then $C = \frac{1}{V}$

$$\text{or} \qquad K_a = \frac{\alpha^2}{(1-\alpha)V}. \qquad .....(17.2)$$

Dilution of a solution decreases C (or increases V). So to keep K_a constant α must increase i.e., there will be **increase in dissociation with increase in dilution. This is Ostwald's dilution law.**

17.2 DISSOCIATION CONSTANT — STRENGTHS OF WEAK ACIDS OR BASES

For a weak acid HA, existing in equilibrium as $HA \rightleftharpoons H^+ + A^-$, dissociation constant $K_a = \frac{C_{H^+} \cdot C_{OH^-}}{C_{HA}}$. Similarly for a weak base BOH existing as $BOH \rightleftharpoons B^+ + OH^-$, the dissociation constant is $K_b = \frac{C_{B^+} \cdot C_{OH^-}}{C_{BOH}}$.

If α is degree of dissociation for a C (M) acid solution then from equation (17.1)

$$K_a = \frac{\alpha^2 C}{1-\alpha}$$

when dissociation is 50% ; $\alpha = \frac{50}{100} = 0.5.$

So $\quad K_a = \frac{(0.5)^2 C}{1-0.5} = 0.5\ C$

or $\quad C = 2K_a.$

So dissociation constant of an acid (or base) is half the concentration at which it is 50% dissociated. Now for two acids HA_1, and HA_2 at same concentration C if degrees of dissociation are α_1 and α_2 then C_{H^+} in $HA_1 = \alpha_1 C$ and C_{H^+} in $HA_2 = \alpha_2 C$.

So $\quad \frac{[C_{H^+}]_{HA_1}}{[C_{H^+}]_{HA2}} = \frac{\alpha_1}{\alpha_2}.$

Thus strengths can be compared from α values at same concentration. Again $K_a = \frac{\alpha^2 C}{1-\alpha}$; generally α is very small at moderate dilution i.e., $\alpha << 1$.

So $\quad K_a \approx \alpha^2 C \quad$ or $\quad \alpha = \sqrt{\frac{K_a}{C}}.$

Thus $\quad \frac{\alpha_1}{\alpha_2} = \sqrt{\frac{K_{a_1}}{K_{a_2}}}.$

Hence strengths of two weak acids can be compared by taking the ratio of square root of their dissociation constants. In fact K_a - measures the strength. The greater the K_a -value the more is the strength. Sometimes we use pK_a value. The term 'p' symbolises negative logarithm (base = 10). Thus $pK_a = -\log_{10} K_a$. So the less the pK_a - value the more is the strength. As an example let us see the following —

Table 17.1 (25°C)

HCOOH	CH_3COOH
$K_a = 1.77 \times 10^{-4}$	$K_a = 1.8 \times 10^{-5}$
$pK_a = 3.6$	$pK_a = 4.7.$

So HCOOH is stronger.

17.3 DISSOCIATION CONSTANT—TRUE AND APPARENT—MEASUREMENT

For the dissociation equilibrium $HA \rightleftharpoons H^+ + A^-$ the true dissociation constant is

$$K_a = \frac{a_{H^+} \cdot a_{A^-}}{a_{HA}}.$$

Here 'a = activity'. K_a is truely constant at a given temperature. So 'K_a' is called true or thermodynamic dissociation constant. Now $a = f.C$ where 'f' is called *activity co-efficient* and C is molar concentration.

So $$K_a = \frac{C_{H^+} \cdot C_{A^-}}{C_{HA}} \cdot \frac{f_{H^+} \cdot f_{A^-}}{f_{HA}}.$$

The term $\dfrac{C_{H^+} \cdot C_{A^-}}{C_{HA}}$ is called apparent dissociation constant $\left(K'_a\right)$.

So $$K_a = K'_a \frac{f_{H^+} \cdot f_{A^-}}{f_{HA}}.$$

In very dilute solution f-values are generally unity. So $K_a \simeq K'_a$. But with change in ionic strength of the medium f-values may change. Then to keep K_a constant K'_a must change slightly. Thus apparent dissociation constant $\left(K'_a\right)$ is not truely constant. Its value can be determined conductometrically as follows.

$$K'_a = \frac{C_{H^+} \cdot C_{A^-}}{C_{HA}} = \frac{\alpha^2 C}{1-\alpha}.$$

But $$\alpha = \frac{\Lambda}{\Lambda_0}$$

[See chapter-16. applications of conductometric measurement].

So $$K'_a = \frac{\Lambda^2 C}{\Lambda_0\left(\Lambda_0 - \Lambda\right)}. \qquad .....(17.3)$$

'Λ' can be experimentally measured. 'Λ_0' can be known from Kohlrausch's table. Thus K'_a is determined.

17.4 COMMON-ION EFFECT

Let us consider a weak acid (HA) solution remaining in equilibrium as $HA \rightleftharpoons H^+ + A^-$.

Dissociation constant $K_a = \dfrac{C_{H^+} \cdot C_{A^-}}{C_{HA}}$. If to this solution common ion H^+ (or A^-) is added from outside then to keep 'K_a' constant, some of the added ions will combine with opposite ion to form undissociated HA. Thus amount of undissociated acid will increase i.e., degree of dissociation will decrease. Thus addition of common-ion decrease degree of dissociation. This is known as common-ion-effect.

17.5 ISOHYDRIC SOLUTION

If to a solution of HCOOH, CH_3COOH solution is added then as both are getting common-ion (H^+) so degree of dissociation of both of them should decrease. However it can be shown that if concentration of common ion i.e., H^+ is same before mixing then though they are getting common-ion, yet their degrees of dissociation will remain unchanged. Such solutions are called isohydric.

17.6 IONIC PRODUCT OF WATER

Pure water is a very poor electrolyte. However it dissociates as —

$$2H_2O \rightleftharpoons H_3O^+ + OH^-.$$

For all practical purposes dissociation is written as $H_2O \rightleftharpoons H^+ + OH^-$; though it is a wrong representation. Because H^+ is a bare proton, which is non-existent. However still we write —

$$H_2O = H^+ + OH^-$$

The dissociation constant $K = \dfrac{a_{H^+} \cdot a_{OH^-}}{a_{H_2O}}$

or $\qquad K \times a_{H_2O} = a_{H^+} \cdot a_{OH^-}$.

As in water ionic strength is negligible, so activity $\simeq$ concentration. Hence

$$K \times C_{H_2O} = C_{H^+} \cdot C_{OH^-}.$$

In pure water $\qquad C_{H_2O} = \dfrac{1000}{18}$ moles/litre.

[1 litre water $\equiv$ 1000 gm. water $\equiv \dfrac{1000}{18}$ mole].

So $K \times C_{H_2O}$ = constant. This constant is called ionic product of water and is written as K_w.

So $\qquad K_w = C_{H^+} \cdot C_{OH^-}$

And $\qquad K_w = \dfrac{1000}{18} \times K = 55.5\ K.$

Value of K_w can be determined conductometrically. If 'α' is degree of dissociation of water then in water $C_{H^+} = C_{OH^-} = \alpha C$.

So $\qquad K_w = [\alpha\ C]^2.$

But $\qquad \alpha = \dfrac{\Lambda}{\Lambda^0}$. $\qquad$ So $\qquad K_w = \left[\dfrac{\Lambda C}{\Lambda^0}\right]^2$

Again $\qquad \Lambda = \kappa \times \dfrac{10^3}{C}$ or $\Lambda C = \kappa \times 10^3$

Hence $\qquad K_w = \left[\dfrac{\kappa \times 10^3}{\Lambda^0}\right]^2 \qquad$ .....(17.4)

κ can be measured by measuring resistance of water in a cell of known cell constant.

$$\text{and } \Lambda^0 = l^0_{H^+} + l^0_{OH^-}.$$

It can be known from Kohlrausch's table. Thus K_w can be calculated. Its value depends on temperature.

$$\text{At } 25°C \; K_w = 10^{-14}.$$

17.7 pH — pOH — SIGNIFICANCE

In pure water $C_{H^+} = C_{OH^-}$; water is said to be neutral. As $K_w = C_{H^+}.C_{OH^-} = 10^{-14}$ at 25°C; so in water $C_{H^+} = C_{OH^-} = 10^{-7}$ (M).

In aqueous HCl solution C_{H^+} is larger than C_{OH^-} because H^+ is coming from HCl and also H_2O and dissociation of H_2O has been depressed by common-ion-effect so C_{OH^-} is negligible. However as $C_{H^+}.C_{OH^-} = 10^{-14}$; so C_{H^+} must be greater than 10^{-7} (M).

Similarly in NaOH solution due to common-ion effect $C_{OH^-} >> C_{H^+}$. Here $C_{OH^-} > 10^{-7}$ (M) and C_{H^+} is $< 10^{-7}$ (M).

So we see that at 25°C neutrality means $C_{H^+} = C_{OH^-} = 10^{-7}$ (M) ; acidic solution means $C_{H^+} > 10^{-7}$ (M) or $C_{OH^-} < 10^{-7}$ (M) and alkaline solution means $C_{H^+} < 10^{-7}$ (M) or $C_{OH^-} > 10^{-7}$ (M).

However expressing concentration as 'ten to the power minus something' is a rather cumber-some process. Sorensen simplified it by expressing acidity or alkalinity in terms of pH. ['pH' → Puissance de 'H' i.e. concentration of H^+]. He defined pH as — $pH = -\log_{10} a_{H^+}$ or taking activity = concentration, $pH = -\log_{10} C_{H^+}$. Similarly pOH can be defined, $pOH = -\log_{10} a_{OH^-} \simeq -\log_{10} C_{OH^-}$.

So at 25°C

for neutral sol^n $pH = -\log 10^{-7} = 7$

for acids : $C_{H^+} > 10^{-7}$ i.e., $-\log C_{H^+} < -\log 10^{-7}$ i.e., $pH < 7$

for alkalis : $C_{H^+} < 10^{-7}$ i.e., $-\log C_{H^+} > -\log 10^{-7}$ i.e., $pH > 7$

Let us make a table.

Table 17.2 ; t = 25°C

Substance	C_{H^+}	C_{OH^-}	pH	pOH	pH + pOH
Water	10^{-7}	10^{-7}	7	7	14
0.1 M HCl	10^{-1}	10^{-13}	1	13	14
0.01 M NaOH	10^{-12}	10^{-2}	12	2	14

So in each case pH + pOH = 14 at 25°C.

The general relation can be obtained as —

$$C_{H^+} \, C_{OH^-} = K_w$$

or $$-\log C_{H^+} - \log C_{OH^-} = -\log K_w$$

or $$\text{pH} + \text{pOH} = \text{p}K_w. \qquad .....(17.5)$$

As $\text{pH} = -\log a_{H^+}$ — and highest activity of any ion is generally one, so lowest pH value = 0. Similarly $\text{pOH} = -\log a_{H^+}$; so lowest pOH value = 0. But pH + pOH = 14 (at 25°C) ; so highest pH value = 14. Hence the pH scale at 25°C is 0 → 14.

$$0 \xleftarrow{\text{Acid}} \overset{(N)}{7} \xrightarrow{\text{Alkali}} 14 \quad \text{(at 25°C)}.$$

The less the pH is the more is acidity and the more the pH is the more is alkalinity.

Q. Can pH be negative ?

Since $\text{pH} = -\log a_{H^+}$ and highest activity is **generally** 1 lowest pH value is **generally** zero. However in too much concentrated solution, strong repulsive force comes in to force and activity of ions suddenly becomes greater than one. In such cases pH may be negative.

Q. Will pH of water be 7 at 70°C ?

In water pH = pOH ; so from equation (17.5), pH of water = $\frac{1}{2}\text{p}K_w$. Now $\text{p}K_w = 14$ at 25°C as $K_w = 10^{-14}$. Since dissociation of water is endothermic in nature so equilibrium constant will increase with increase in temperature. Hence K_w (at 70°C) > K_w (at 25°C) i.e., K_w (at 70°C) > 10^{-14}. Hence $\text{p}K_w$ (at 70°C) < 14. Thus pH of water (at 70°C) < 7. This does not mean that water is behaving as an acid. The pH⁻ scale is

$$0 \xrightarrow{\text{Acid}} \overset{(N)}{\tfrac{1}{2}\text{p}K_w} \xrightarrow{\text{Alkali}} \text{p}K_w$$

As $\text{p}K_w$ value change with temperature — pH scale will also change with temperature.

Q. What is the pH of 10^{-9} (M) HCl ?

$\text{pH} = -\log C_{H^+}$; or $C_{H^+} = 10^{-\text{pH}}$

Generally during calculation of pH of acid solutions we neglect the hydrogen ion conc. coming from water as it is a very weak electrolyte. But when $C_{HCl} = 10^{-9}$ (M) i.e., abnormally small we can not neglect conc. of H^+ coming from water ; let it be x then total H^+ conc. = $(10^{-9} + x)$ and $C_{OH^-} = x$.

So $(10^{-9} + x)\,x = 10^{-14}$

or, $x^2 + 10^{-9}x - 10^{-14} = 0$

or, $x = 9.95 \times 10^{-8} = 99.5 \times 10^{-9}$.

So $C_{H^+} = 10^{-9} + 99.5 \times 10^{-9} = 1.005 \times 10^{-7}$

or, $\text{pH} = 6.997.$

17.8 BUFFER SOLUTION

Let we have one litre very pure water. Its pH at 25°C is '7'. To it let us add 2 drops (0.1 c.c.) very dilute (N/10) HCl. The purity is lost. It became a HCl solution of concentration nearly 10^{-5} (M). So its pH is now 5! Thus we see that pH of ordinary solutions are highly sensitive towards addition of little amount of acid or alkali. However, during analytical work we often require solutions, whose pH should remain reasonably constant during addition of little amount of acid or alkali. Such solutions whose pH remain more or less same during addition of little amount of acid or base are called buffer solutions. They are solutions of reserved acidity or reserved alkalinity.

Buffer solutions can be prepared by mixing —

(i) a weak acid (HA) with any of its salts (BA)

(ii) a weak base (BOH) with its salt (BA)

(iii) two acid salts of same acid e.g. $NaH_2PO_4 + Na_2HPO_4$.

17.9 MECHANISM OF BUFFER ACTION

To understand the mechanism of buffer action let us consider a buffer made by mixing a weak acid HA with its salt BA. In solution BA is completely dissociated to B^+ and A^- but dissociation of HA is depressed due to common-ion-effect. So there are plenty of undissociated HA and A^- remaining in equilibrium $HA \rightleftharpoons H^+ + A^-$; if we add H^+ it will combine with A^- to give HA and if we add OH^- it will react with H^+ but loss of H^+ will be compensated by dissociation of HA. In both cases total H^+ and hence pH will remain same. So here HA and its anion A^- are responsible for buffering action.

If the buffer is made by mixing a weak base BOH with its salt BA then there are plenty of B^+ (from salt) and undissociated BOH. They are in equilibrium $BOH \rightleftharpoons B^+ + OH^-$. If we add H^+ , BOH will dissociate to compensate loss of OH^- and if we add OH^- , it will combine with B^+ to give BOH. In each case total OH^- and hence pOH or pH will remain same. This is virtually a buffer of weak base and its cation BOH and B^+.

If we mix two acid salts NaH_2PO_4 and Na_2HPO_4, then there are large amounts of $H_2PO_4^-$ and HPO_4^{2-} existing in equilibrium $H_2PO_4^- \rightleftharpoons H^+ + HPO_4^{2-}$. So this is a buffer of weak acid and its anion.

pH of buffer :

Let us consider a buffer made by mixing a weak acid (HA) with its salt (BA). In mixture concentration of acid = C_a (M) and concentration of salt = C_s (M). We consider the equilibrium

$$HA \rightleftharpoons H^+ + A^-.$$

Applying law of mass action —

$$\text{dissociation constant } K_a = \frac{C_{H^+} \cdot C_{A^-}}{C_{HA}} \quad \text{[Activity co-efficient are taken as unity].}$$

So $$C_{H^+} = K_a \cdot \frac{C_{HA}}{C_{A^-}}$$

or, $$-\log_{10} C_{H^+} = -\log_{10} K_a - \log_{10} \frac{C_{HA}}{C_{A^-}}$$

or, $$pH = pK_a - \log_{10}\frac{C_{HA}}{C_{A^-}}$$

or, $$pH = pK_a - \log_{10}\frac{C_{A^-}}{C_{HA}}$$

In this mixture if x (M) is concentration of H^+ produced by dissociation of weak acid HA. Then $C_{HA} = C_a - x$; and $C_{A^-} = C_s + x$ [∵ A^- is coming mainly from complete dissociation of salt].

So $$pH = pK_a + \log \frac{C_s + x}{C_a - x}.$$

But x is generally much small than either C_a or C_s. So

$$pH = pK_a + \log \frac{C_s}{C_a}. \quad .....(17.6)$$

Similarly it can be shown that for a buffer of weak base and its salt

$$pOH = pK_b + \log \frac{C_s}{C_b}. \quad .....(17.7)$$

Above two equations (17.6) and (17.7) are called Henderson's equation.

Q. What will be the pH of the mixture when 500 c.c. of 0.2M CH_3COOH is mixed with 500 c.c., 0.2 M CH_3COONa ? If to this mixture 10 c.c. (N) HCl is added — what will be new pH ? [pK_a = 4.74]

The mixture is buffer where $C_a = \frac{0.2}{2} = 0.1$ and $C_s = \frac{0.2}{2} = 0.1$. So $C_s/C_a = 1$.

Hence $pH = pK_a + \log \frac{C_s}{C_a} = 4.74$.

When 10 c.c. (N) HCl is added if there was no reaction C_{HCl} would be $10 \times 1 = 1010 \times x$

or $x = \frac{10}{1010} = 0.009$ (M). But the reaction $H^+ + CH_3COO^- = CH_3COOH$ occurs.

So $C_{CH_3COO^-}$ becomes = 0.1 – 0.009 = 0.091 and C_{CH_3COOH} = 0.1 + 0.009 = 0.109.

$$pH = pK_a - \log_{10}\frac{C_{A^-}}{C_{HA}} \; ,$$

$$pH = 4.74 + \log \frac{0.091}{0.109},$$

$$pH = 4.65.$$

Thus pH remains almost same. [cf. problems at the beginning of section 17.7].

17.10 HYDROLYSIS OF SALT SOLUTION

As a salt is produced during neutralisation reaction of equivalent amount of acid and base, so it is expected that pH of a salt solution should be 7 at 25°C. Let us now see what experiment tells.

Table 17.3 (t = 25°C)

Name of salt solution		pH (exptl. value)
NaCl	solution	7
NH_4Cl	,,	< 7
CH_3COONa	,,	> 7
CH_3COONH_4	,,	7 (accidental)
NH_4CN	,,	> 7

If we critically analyse the above table we shall see that pH of salt solutions of strong acid and strong base is 7. But when acid is strong, base is weak, pH is < 7 and when acid is weak, base strong pH > 7. But if both are weak then pH $\lesseqgtr$ 7 depending on relative weakness. So in such case where pH ≠ 7 we are to assume that salt has undergone some interaction with solvent to produce excess H^+ (pH < 7) or OH^- (pH > 7). Since the process is interaction with solvent, its general name is solvolysis. When the solvent is water, it is called hydrolysis. We shall now discuss each case separately.

(a) Salt of strong acid (HA) and weak base (BOH) : Let us consider a salt BA made of strong acid (HA) and weak base (BOH). In solution it will dissociate to B^+ and A^- ; the former is the conjugate of a weak base and so is a Bronsted acid. It will accept OH^- from water as —

$$B^+ + H_2O \rightleftharpoons BOH + H^+ \qquad \text{........ (i)}$$

[cf. $NH_4Cl \rightarrow NH_4^+ + Cl^-$; NH_4^+ is conjugate of NH_3 and so is an acid $NH_4^+ + H_2O \rightleftharpoons NH_4OH + H^+$].

The resultant solution is acidic due to excess H^+. Applying law of mass action to above equilibrium (i) we get equilibrium constant (K) as —

$$K = \frac{C_{BOH} \cdot C_{H^+}}{C_{B^+} \cdot C_{H_2O}} \qquad [\text{assumption} : a \approx c]$$

or $$K \times C_{H_2O} = K_h = \frac{C_{BOH} \cdot C_{H^+}}{C_{B^+}}.$$

As C_{H_2O} is a constant $\left[\simeq \frac{10^3}{18} \text{ M/L}\right]$ so K_h is a new constant ; we call it hydrolysis constant. So

Hydrolysis constant $$K_h = \frac{C_{BOH} \cdot C_{H^+}}{C_{B^+}}.$$

But $$K_b = \frac{C_{B^+} . C_{OH^-}}{C_{BOH}}$$ [For weak base K_b = dissoc^n constant]

or, $$\frac{C_{BOH}}{C_{B^+}} = \frac{C_{OH^-}}{K_b}.$$

Hence $$K_h = \frac{C_{H^+} . C_{OH^-}}{K_b} = \frac{K_w}{K_b}. \quad .....(17.8)$$

Again if C (M) be the concentration of salt before hydrolysis and 'h' is degree of hydrolysis, then for equilibrium (i)

$$C_{B^+} = (C - hC) = (1 - h) C$$

$$C_{BOH} = C_{H^+} = hC.$$

So $$K_h = \frac{hC \times hC}{(1-h)\,C} = \frac{h^2 C}{1-h}.$$

As $h \ll 1$ $\quad K_h \simeq h^2 C \quad$ or $h = \sqrt{\frac{K_h}{C}}$.

To find the pH of this solution, we see

$$pH = -\log_{10} C_{H^+} = -\log_{10} (hC) = -\log_{10} \left[\sqrt{\frac{K_h}{C}} . C\right]$$

or, $$pH = -\log_{10} \left(K_h . C\right)^{1/2}; \quad \text{but } K_h = \frac{K_w}{K_b}.$$

So $$pH = -\log_{10} K_h = \left(\frac{K_w . C}{K_b}\right)^{1/2}$$

or, $$pH = -\frac{1}{2} \log_{10} K_w + \frac{1}{2} \log_{10} K_b - \frac{1}{2} \log_{10} C$$

or, $$pH = \frac{1}{2} pK_w - \frac{1}{2} pK_w - \frac{1}{2} \log_{10} C. \quad .....(17.9)$$

At 25°C, $$\frac{1}{2} pK_w = 7.$$

So $$pH < 7.$$

(b) Salt of weak acid and strong base : If the salt BA is of weak acid (HA) and strong base BOH, then anion of the salt A^- is a conjugate base of weak acid HA and so it will act as a strong Bronsted base. It will hydrolyse as proton acceptor

$$A^- + H_2O \rightleftharpoons HA + OH^-. \quad .....(ii)$$

[Example : CH_3COONa is a salt of weak acid CH_3COOH and strong base NaOH, the anion hydrolyses as $CH_3COO^- + H_2O \rightleftharpoons CH_3COOH + OH^-$].

The hydrolysis constant $K_h = \dfrac{C_{HA} \cdot C_{OH^-}}{C_{A^-}}$.

But dissociation constant of weak acid — $K_a = \dfrac{C_{H^+} \cdot C_{A^-}}{C_{HA}}$

so $$\frac{C_{HA}}{C_{A^-}} = \frac{C_{H^+}}{K_a}.$$

Hence $$K_h = \frac{C_{H^+} \cdot C_{OH^-}}{K_a} = \frac{K_w}{K_a}. \qquad .....(17.10)$$

If 'C' is starting concentration of salt and 'h' is its degree of hydrolysis, then at eqm.

$$C_{A^-} = C - hC \; ; \; C_{HA} = C_{OH^-} = hC.$$

So $$K_h = \frac{hC \times hC}{(1-h)\,C} = \frac{h^2 C}{1-h}.$$

Once again assuming $h << 1$ we write

$$K_h \simeq h^2 C \qquad \text{or } h = \sqrt{\frac{K_h}{C}} = \sqrt{\frac{K_w}{K_a C}}$$

Now $$\text{pOH} = -\log_{10} C_{OH^-} = -\log_{10}(hC) = -\log_{10}\left[\frac{K_w C}{K_a}\right]^{1/2}$$

or, $$\text{pOH} = -\frac{1}{2}\log_{10} K_w + \frac{1}{2} K_a C - \frac{1}{2}\log_{10} C$$

or, $$\text{pOH} = \frac{1}{2}\,\text{p}K_w - \frac{1}{2}\,\text{p}K_a - \frac{1}{2}\log_{10} C.$$

So $$\text{pH} = \text{p}K_w - \text{pOH}$$

$$= \frac{1}{2}\,\text{p}K_w + \frac{1}{2}\,\text{p}K_a + \frac{1}{2}\log_{10} C. \qquad .....(17.11)$$

Here pH > 7 (at 25°C).

(c) Salt of weak acid and weak base : If the salt BA is of weak acid (HA) and weak base (BOH), then in solution both cation and anion of salt will hydrolyse as —

$$B^+ + H_2O \rightleftharpoons BOH + H^+$$

and $$A^- + H_2O \rightleftharpoons HA + OH^-$$

Adding $$B^+ + A^- + H_2O \rightleftharpoons BOH + HA \qquad .....(iii$$

Applying law of mass action, the hydrolysis constant (K_h) is

$$K_h = \frac{C_{BOH} \cdot C_{HA}}{C_{B^+} \cdot C_{A^-}} = \frac{C_{H^+} \cdot C_{OH^-}}{\frac{C_{H^+} \cdot C_{A^-}}{C_{HA}} \cdot \frac{C_{B^+} \cdot C_{OH^-}}{C_{BOH}}}$$

or, $$K_h = \frac{K_w}{K_a \cdot K_b} \quad .....(17.12)$$

If before hydrolysis conc. of salt was C (M) and 'h' is degree of hydrolysis, then at hydrolysis equilibrium —

$$C_{B^+} = C_{A^-} = (1-h)\,C$$

and $$C_{BOH} = C_{HA} = hC$$

So $$K_h = \frac{hC \times hC}{(1-h)\,C \times (1-h)\,C} = \left[\frac{h}{1-h}\right]^2 \quad .....(17.13)$$

To find pH we consider the expression

$$K_a = \frac{C_{H^+} \cdot C_{A^-}}{C_{HA}}$$

or, $$C_{H^+} = K_a \cdot \frac{C_{HA}}{C_{A^-}} = K_a \cdot \frac{hC}{(1-h)\,C} = K_a \cdot \left[\frac{h}{1-h}\right]$$

So from equation (17.13)

$$C_{H^+} = K_a \cdot \sqrt{K_h}$$

Again taking the help of equation (17.12)

$$C_{H^+} = K_a \sqrt{\frac{K_w}{K_a\,K_b}} = \left[\frac{K_w\,K_a}{K_b}\right]^{1/2}$$

So $$-\log_{10} C_{H^+} = -\frac{1}{2}\log_{10} K_w - \frac{1}{2}\log_{10} K_a + \frac{1}{2}\log_{10} K_b$$

or, $$\text{pH} = \frac{1}{2}\text{p}K_w + \frac{1}{2}\text{p}K_a - \frac{1}{2}\text{p}K_b \quad .....(17.14)$$

Following cases are to be noted —

(i) If both acid and base are equally weak then $K_a = K_b$ or $\text{p}K_a = \text{p}K_b$ or $\text{pH} = \frac{1}{2}\text{p}K_w = 7$ (at 25ºC). This is the case with NH_4OOCCH_3. $\text{p}K_a$ (for CH_3COOH) = $\text{p}K_b$(for NH_4OH).

(ii) If acid is weaker i.e., $K_a < K_b$ or $\text{p}K_a > \text{p}K_b$ then $\text{pH} > \frac{1}{2}\text{p}K_w$ i.e., pH > 7 (at 25ºC) — solution will be alkaline.

(iii) If base is weaker i.e., $K_a > K_b$ or $\text{p}K_a < \text{p}K_b$ then $\text{pH} < \frac{1}{2}\text{p}K_w$ i.e., pH < 7 (at 25ºC) — solution will be acidic.

17.11 SOLUBILITY PRODUCT OF SALTS

In a saturated solution of solid salt if we add more salt it will not dissolve any more. The solid salt (let $A_x B_y$) will remain in equilibrium with the dissolved salt. But dissolved salt exists in the completely dissociated form. So the equilibrium is —

$$A_x B_y (S) \rightleftharpoons x\, A^{+y} + y\, B^{-x}.$$

Applying law of mass action, the equilibrium constant (= let K_s) is —

$$K_s = \left(a_{A^{+y}}\right)^x \left(a_{B^{-x}}\right)^y. \qquad .....(17.15)$$

This constant K_s is known as solubility product or more accurately as activity solubility product. Its value depends on temperature. Thus solubility product can be defined as the maximum product of ionic activities (in a saturated solution of the salt) each activity term being raised to a power which is its co-efficient in the stoichiometric equation showing the dissociation of one mole salt. **Any salt whether spairingly soluble (like Bi_2S_3) or highly soluble (like NaCl) will have solubility product value.** But solubility product values of highly soluble salts do not find any analytical application.

Concentration solubility product :

We know that in dilute solution activity $\simeq$ concentration. So equation (17.15) can be written as $K_s = \left(a_{A^{+y}}\right)^x \left(a_{B^{-x}}\right)^y$. This is concentration solubility product. This value slightly changes with ionic strength. Let us examine the following table.

Table 17.4

Name of Eqm.	Solubility	C_{cation}	C_{Anion}	K_s
$AgCl \rightleftharpoons Ag^+ + Cl^-$	S (M)	$C_{Ag^+} = S$	$C_{Cl^-} = S$	$K_S = S \times S = S^2$
$CaF_2 \rightleftharpoons Ca^{2+} + 2F^-$	S (M)	$C_{Ca^{2+}} = S$	$C_{F^-} = 2S$	$K_S = S \times (2S)^2 = 4S^3$
$Ca_3(PO_4)_2 \rightleftharpoons 3Ca^{2+} + 2PO_4^{3-}$	S (M)	$C_{Ca^{2+}} = 3S$	$C_{PO^{3-}} = 2S$	$K_S = (3S)^3 \times (2S)^2$ $= 108\,S^5$

Variation of solubility :

Let us consider the case of a spairingly soluble salt (say AgCl). Its equilibrium is —

$$AgCl\ (s) \rightleftharpoons Ag^+ + Cl^-$$

and $\quad K_s = a_{Ag^+} \cdot a_{Cl^-} = C_{Ag^+} \cdot C_{Cl^-} \cdot f_{Ag^+} \cdot f_{Cl^-}$

where C = concentration in moles/litre

and f = activity co-efficient.

If we now want to dissolve AgCl in $AgNO_3$ solution, then since there are already Ag^+ ions so C_{Ag^+} is high. To keep 'K_s' constant C_{Cl^-} should be decreased. But Cl^- ions are coming from AgCl alone. So solubility of AgCl will decrease. Similar will be the case if we try to dissolve AgCl in KCl solution, here Cl^- ion is common. So we see that solubility decreases in presence of common ion. But if we try to dissolve it in KNO_3 then as ionic strength has been increased, so 'f' terms will decrease. To keep K_s constant, C_{Ag^+} and

C_{Cl^-} should increase, i.e., solubility will rise. So **solubility of the salt decreases in presence of common ion but increases in presence of uncommon ion.**

Q. Why pptⁿ occurs :

For the equilibrium AgCl (s) $\rightleftharpoons$ $Ag^+ + Cl^-$, solubility product $K_s = C_{Ag^+} . C_{Cl^-}$. This value is 10^{-10} at 25°C. Now let we have a 0.1 (M) $AgNO_3$ solution. So, here $C_{Ag^+} = 10^{-1}$ (M). Hence maximum amount of Cl^- ion that can exist in this solution is 10^{-9} (M). If we add Cl^- from outside by large amount then excess Cl^- ions will be removed. According to Le-chatelier principle the equilibrium AgCl (s) $\rightleftharpoons$ $Ag^+ + Cl^-$, will shift towards left, and thus solid AgCl (s) will appear. We say that precipitation has occurred. Thus precipitation is nothing but shifting of a dissociation equilibrium; it will occur only when product of ionic activities with appropriate power exceed the solubility product.

Applications :

Solubility product principle is widely applied in analytical chemistry. The entire group analysis for basic radicals is based on solubility product principle. We know that during group analysis solution is kept acidic in group-II. Let us try to know the reason. Metal sulphide of group - II (HgS; CuS etc.) have very low solubility product [K_s for Hgs is 3 $\times 10^{-54}$ and for CuS is 3×10^{-42} at 25°C]. Now metal sulphides of group III B [e.g. ZnS; MnS etc.] have also low solubility product [K_s for ZnS is 1×10^{-23} for MnS is 2×10^{-15} etc.]. If in a mixture containing metal ions of both group-II and group-III B large amount of S^{2-} ions enter then all metal ions will be precipitated. However, if the mixture is acidified and H_2S is passed then the weak dibasic acid H_2S will be dissociate feebly due to common ion effect and thus negligible amount of S^{2-} ions will be produced. This amount is sufficient to exceed K_s values of group-II metal ions but not sufficient to exceed K_s -values of group-III B metal ions. Thus only group-II metal ions will be precipitated. For this reason we acidify the solution in group-II.

Q. Why NH_4Cl is added in group-III A ?

During qualitative analysis metal ions of group-III A are precipitated as their hydroxides. These hydroxides have very small solubility product values. [e.g. K_s for Fe $(OH)_3 = 1 \times 10^{-38}$ for $Al(OH)_3 = 1 \times 10^{-33}$ at 25°C]. Now some metal hydroxides of IIIB, IV, V etc. have also small K_s values [e.g. K_s for Co $(OH)_2 = 2 \times 10^{-16}$ for $Mg(OH)_2 = 6 \times 10^{-12}$ at 25°C]. In order to prevent their precipitation in group-IIIA we add NH_4OH along with NH_4Cl. NH_4OH is a weak base and supplies small amount of OH^- ions. In presence of NH_4Cl, there is common-ion-effect which depresses the dissociation further. Thus very small amount of OH^- ion is produced. This is sufficient to exceed solubility product values of only group-IIIA metal hydroxides. That is why NH_4OH is added.

Purification of common salt

In a saturated solution of impure common salt if we pass HCl gas ther C_{Cl^-} goes on increasing and the product $C_{Na^+} . C_{Cl^-}$ exceeds the solubility product of NaCl Thus pure NaCl is precipitated. In this way purification of common salt can be done.

17.12 ACID BASE (OR NEUTRALISATION) INDICATOR

Indicators are those substances, which indicate end points of titrations, by means of colour change. Depending on nature of titrations indicators are classified in to three classes.

These are —

(i) Acid-base indicators used in acid-base titrations.

(ii) Redox indicators used in redox titrations and

(iii) Adsorption indicators used in precipitation titrations.

Presently we shall discuss only type (i).

Acid base indicators show one colour in low pH range and another in high pH range. Colour shown at low pH range is acid-colour and that at high pH range is alkali colour. Some common indicators with the nature of their acid colour and alkali colour are listed below.

Table 17.5

Name	Colour		pH range	
	Acid	Alkali	Acid colour	Alkali colour
Methyl orange	Red	Orange yellow	≤ 3.1	≥ 4.4
Methyl red	Red	Yellow	≤ 4.2	≥ 6.3
Litmus	Red	Blue	≤ 5	≥ 8
Phenolphthalien	Colour less	Pink	≤ 8.3	≥ 10
Phenol red	Yellow	Red	≤ 6.8	≥ 8.4
Bromothymol blue	Yellow	Violet	≤ 6	≥ 7.6

Theory of indicators :

The theory of acid base indicators was proposed by Oslwald. According to him acid-base indicators are generally organic acids or bases. Their structures are such that they can not dissociate. However, they will dissociate after some tautomeric change. Thus structures of dissociated and undissociated forms are different. Hence their colour become different. Generally the dissociated form has quinonoid structure which has brilliant colour. To explain this let our indicator is HIn. If can dissociate via its tautomer HIn* as —

$$HIn \rightleftharpoons HIn^* \qquad .....(17.16)$$

$$HIn^* \rightleftharpoons H^+ + In^{*-} \qquad .....(17.17)$$

In acid solution HIn form will be predominant due to common-ion-effect and in alkaline solution In^{*-} will be predominant. Their colours are different. Two examples are given —

Phenolphthalien (Organic acid indicator)

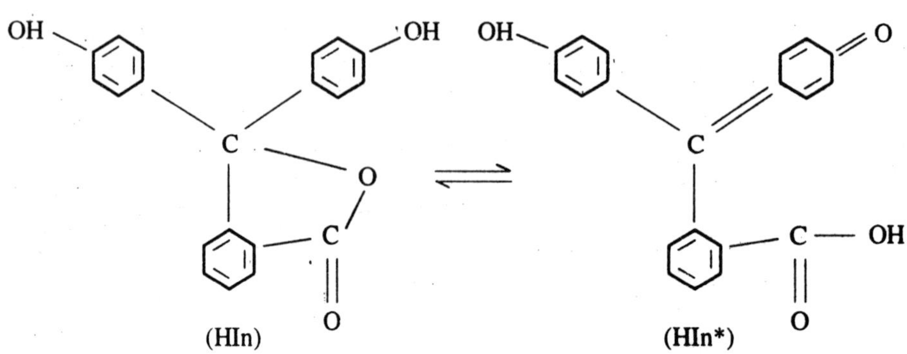

(HIn) (Colourless) **(HIn*) (Red)**

Methyle Orange (Organic base indicator)

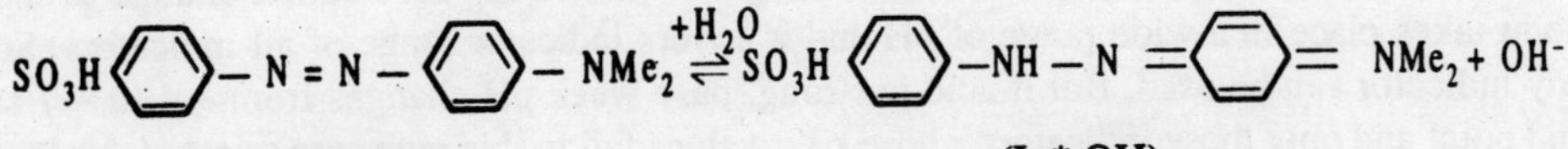

(InOH) (Yellow-orange) (In* OH) (Orange-red)

Now let us consider equilibrium (17.16) and (17.17); the respective equilibrium constants are $K_1 = \frac{[HIn^*]}{[HIn]}$ and $K_2 = \frac{[H^+][In^{*-}]}{[HIn^*]}$.

Here square brackets symbolises molar concentration.

Neither K_1 nor K_2 is experimentally determinable, because of the existence of intermediate [HIn*] concentration term. But $(K_1 \times K_2)$ can be determined. This $(K_1 \times K_2)$ is a new constant K_{In}, known as indicator constant.

Thus $$K_{In} = \frac{[H^+][In^{*-}]}{[HIn]}$$

or $$[H^+] = K_{In} \cdot \frac{[HIn]}{[In^{*-}]}$$

or $$-\log_{10}[H^+] = -\log_{10} K_{In} - \log_{10} \frac{[HIn]}{[In^{*-}]}$$

or $$pH = pK_{In} + \log_{10} \frac{[In^{*-}]}{[HIn]}$$

Since colours of $[In^{*-}]$ and [HIn] i.e., dissociated quinonoid form and undissociated forms are different, the colour that we shall see will depend on the ratio $[In^{*-}]/[HIn]$.

Generally if $\frac{[In^{*-}]}{[HIn]} \geq 10$, colour of $[In^{*-}]$ will be seen

When $\frac{[In^*]}{[HIn]} \leq \frac{1}{10}$, colour of [HIn] will be seen.

But if $\frac{[In^{*-}]}{[HIn]} \geq 10$, then $pH \geq pK_{In} + 1$ and

if $\frac{[In^{*-}]}{[HIn]} \leq \frac{1}{10}$, then $pH \leq pK_{In} - 1$

Thus acid colour is seen at $pH \leq pK_{In} - 1$ and alkali colour at $pH \geq pK_{In} + 1$. The range $pK_{In} \pm 1$ is the range where change of colour generally occurs. It is called indicator range.

Choice of indicators :

As pK_{In} values of different indicators are different, they will show colour change at different pH ranges. During titration of a strong acid and strong base colour change at end point takes place in a wide range of pH and it covers indicator range of all indicators. So any indicator can be used. But if acid is strong, base weak pH changes from ~ 4 to ~ 7 at end point and only those indicators whose pK_{In} values fall in this range are selected. Methyl orange, methyl red are suitable. Similarly when acid is weak and base is strong then pH range at end point is ~ 7 to ~ 11. So those indicators whose pK_{In} values are in this range are selected. Phenolphthalien is such an indicator.

Q. In the titration of Na_2CO_3 by H_2SO_4 titre-value with methyl orange is twice the titra-value with phenolphthalien — why ?

Ans : The titration of Na_2CO_3 by H_2SO_4 occurs via two steps.

(i) $CO_3^{2-} + H^+ \rightleftharpoons HCO_3^-$

and (ii) $HCO_3^- + H^+ \rightleftharpoons H_2CO_3$.

At half neutralisation point i.e., after completion of step (i) pH becomes 8.35. This falls in the indicator range of phenolphthalien. So phenolphthalien will show end point at half-neutralisation point. After completion of step (ii) pH becomes ~ 4, which is in the indicator range of methyl orange. Thus methyl orange will show end point after complete neutralisation. Hence titre-value with methyl-orange is twice the titre-value with phenolphthalien.

POINTS TO REMEMBER

Weak electrolytes are those, which never dissociate fully. Ostwald's dilution law states that degree of dissociation will increase with increase in dilution. The equilibrium constant for dissociation equilibrium is called dissociation constant (K_a or K_b). It is a measurement of its strength. **pK_a (or pK_b) is $-\log_{10} K_a$ (or $-\log_{10} K_b$).** The less the pK_a the more is the strength. Ionic product of water **(K_w) is $C_{H^+} \cdot C_{OH^-}$** ; its value is **10^{-14} at 25°C.** Acidity or alkalinity of a solution is measured by pH (or pOH) values. **$pH = -\log_{10}C_{H^+}$** and **$pOH = -\log_{10}C_{OH^-}$. At 25°C, pH = 7 means neutral nature; pH < 7 is acidic and pH > 7 is alkaline.** The scale will change is temperature is changed. Mixtures of weak acid and its salt or weak base and its salt have the capacity of preserving its pH-value. These are **'buffer'** solution. For them **$pH = pK_a + \log \frac{C_{salt}}{C_{base}}$** or **$pOH = pK_b + \log \frac{C_{salt}}{C_{base}}$**. Though pH of salt solutions are expected to be 7 at 25°C yet excepting salts of strong acid and strong base; almost all salt solution are either acidic or alkaline. This is due to **hydrolysis** of salts which produces excess H^+ or OH^- Solubility product is the maximum product of ionic activities of a salt; where each activity term has a power equal to its co-efficient in the stoichiometric equation showing dissociation of one mole salt. When product of ionic conc. of a salt with appropriate power exceed solubility product ppt. occurs. Solubility of a salt decreases in presences of common ion but increases in presence of uncommon ion. Indicators are those which show one colour at low pH and another at high pH. They show colour change in the range $pK_{In} \pm 1$. For titration of strong acid by strong base any indicater can be used.

QUESTIONS

SHORT TYPE

1. Whose degree of dissociation is more and why ?
(a) 0.1 (M) CH_3COOH (b) 0.05 (M) CH_3COOH
2. 'K_a for CH_3COOH is 1.8×10^{-5} at 25°C' — at what conc. it will be 50% dissociated ?
3. pK_a values for acids HA_1 and HA_2 are 5 and 3 respectively, — which one is weaker ?
4. What is the condition for becoming isohydric solution ?
5. pH of two solutions are 2 and 4 — which one is more acidic ?
6. pH of two solutions are 10 and 13 — which one is more alkaline ?
7. Can pH be negative ?
8. pH of water at 70°C is less than 7. Does it mean that water becomes acidic at 70°C ?
9. 'x' equivalent of a weak acid is mixed with 'y' equivalent of a base. Under what condition the mixture will be a buffer ?
10. pH of CH_3COONa solution is > 7 — why ?
11. Though CH_3COONH_4 hydrolyses pH of its solution is 7 — why ?
12. Discuss the solubility of PbI_2 in (i) Pb $(NO_3)_2$ and (ii) KNO_3 solution.
13. Why solution is kept acidic in group-II and why NH_4Cl is added in group- IIIA ?
14. For the titrations of (i) NH_4OH by HCl and (ii) NaOH by CH_3COOH — which are suitable indicators ?
15. In the titration of Na_2CO_3 by H_2SO_4, titre value with phenolphthalien is half the titre value with methyl orange — why ?

ESSAY TYPE

1. State and explain Ostwald's dilution law ? What are true and apparent dissociation constant; how is latter determined ?
2. What is ionic product of water ? How is it determined ? What is the relation between ionic product and dissociation constant of water ?
3. What is a buffer solution ? How are they made ? Discuss the mechanism of buffer action.
4. Explain the term hydrolysis. Derive an expression for hydrolysis constant and hence p_H for a solution of salt of weak acid and weak base.
5. Explain the term solubility product and also discuss when precipitation will occur.
6. What are acid-base indicators ? Discuss the mechanism of colour change. What is indicator range ?

PROBLEMS

1. Formic acid is 4.5% dissociated in a 0.1N solution at 20°C. Calculate the dissociation constant of the acid.

[*Hints :* $\alpha = 0.45$; $C = 0.1$; $K_a = \dfrac{\alpha^2 C}{1-\alpha}$].

2. The equivalent conductivity of $\frac{N}{32}$ solution of a weak acid is 9.2 mhos. If the conductivity at infinite dilution is 389 mhos, what is the dissociation constant of the acid ?

[*Hints* :

$$K_a = \frac{\Lambda^2 C}{\Lambda_0(\Lambda_0 - \Lambda)}\ ; \quad \Lambda = 9.2; \quad C = \frac{1}{32}; \quad \Lambda^0 = 389; \quad \text{Find } K_a].$$

3. The specific conductivity of a 0.5% solution of acetic acid is 0.0004525 mhos at 18°C. Calculate its dissociation constant. Eq. conductivity at infinite dilution is 352.

[*Hints* :

$$0.5\% \equiv 5\text{gm./litre} \equiv \frac{5}{60} \text{ eq./litre.}$$

$$\Lambda = \kappa \times \frac{10^3}{C} = 0.0004525 \times \frac{10^3}{(5/60)},$$

$$K_a = \frac{\Lambda^2 C}{\Lambda_0(\Lambda_0 - \Lambda)}\ ; \quad \Lambda_0 = 352].$$

4. What is the percentage of hydrolysis of NaCN in $\frac{N}{80}$ solution, when the dissociation constant of HCN is 1.3×10^{-9} and the ionic product of water is 10^{-14}.

[*Hints* :

$$K_h = \frac{K_w}{K_a} = \frac{10^{-14}}{13 \times 10^{-9}} = \frac{h^2 C}{1 - C};$$

$$C = \frac{1}{80}\ ; \text{ find '}h\text{'.]}$$

5. Knowing that the solubility of AgCl is 0.0015 gm. per litre at 18°C.
(a) Calculate the solubility product of AgCl
(b) Calculate the quantity of AgCl which remains dissolved if one adds a solution of NaCl such that the total concentration of it is 0.4585 gm. of NaCl per litre.

[*Hints* :

$$0.0015 \text{ gm./litre} \equiv \frac{0.0015}{143.5} \text{ M/L} = \text{let } C$$

$K_s = C^2$. If Conc. of NaCl = x moles/litre.
Then $K_s = C'(C' + x)$; find C'].

6. The solubility of lead chloride is 1.1×10^{-3} mole per litre. Calculate the solubility product.

[*Hints* : For $PbCl_2$ if S is solubility. $K_s = 4S^3$].

OBJECTIVE TYPE

1. **Write √ mark on the correct answer.**
(a) Addition of common ion will (increase / decrease) degree of dissociation.
(b) Ionic product of water is (dependent / independent) on temperature.

(c) A solution having pH = 7 at 70°C is (neutral / alkaline).
(d) A mixture of benzoic acid and Na-benzoate is (isohydric / buffer) solution.
(e) pH of CH_3COONa solution is (greater/less) than 7 at 25°C.
(f) Solubility of CaF_2 will (increase / decrease) in $Ca(NO_3)_2$ sol^n.
(g) Solubility of PbI_2 will (increase / decrease) in KNO_3 solution.
(h) For the titration of HCl vs NH_4OH (methyl orange / phenolphthalien) is suitable indicator.

2. Write T against true and F against false statements.

(a) At 25°C ionic product of water is 10^{-14}.
(b) For any solution at 25°C pH + pOH = 14.
(c) pH of 10^{-9} (M) HCl is 9.
(d) A mixture of 30 c.c I(M) CH_3COOH and 25 c.c. I(M) NaOH will be a buffer.
(e) Highly soluble salts like NaCl have no solubility product.
(f) pH of CH_3COONH_4 is 7 (at 25°C) because the salt does not hydrolyse.
(g) Phenolphthalien is a suitable indicator for titration of oxalic acid by sodium hydroxide.

3. Match terms of column A with those of column B.

A	B
(a) Ionic Product of water.	(a) >7
(b) C_{H_2O} in water.	(b) 14
(c) pK_w value	(c) 55.5
(d) pH of water	(d) < 7
(e) pH of NH_4Cl	(e) 7
(f) pH of CH_3COONa	(f) 10^{-14}

Answers :

1. (a) decrease (b) dependent (c) alkaline (d) buffer (e) greater (f) decrease (g) increase (h) methyl orange.

2. (a) T ; (b) T ; (c) F ; (d) T ; (e) F ; (f) F ; (g) T.

3. a – f ; b – c ; c – b ; d – e ; e – d ; f – a .

Chapter 18

ELECTROMOTIVE FORCE

18.1 CELLS CLASSIFICATIONS

Cells are devices where interconversion of chemical and electrical energy takes place. If electrical energy is converted to chemical energy (i.e. electrolysis occurs) then the cell is called **electrolytic cell.** On the contrary if chemical energy is converted to electrical energy (e.g. Daniel cell) then the cell is an **electrochemical cell.** In the present chapter we shall discuss electrochemical cell.

18.2 REACTION IN A CELL

In every electrochemical cell, the reaction that occurs is nothing but a **'redox'** reaction. Two metal rods are present on whose surface oxidation and reduction reaction occurs. They serve as sink and source for electrons during oxidation and reduction. They are called *electrodes*. The metal rod on whose surface oxidation occurs becomes negatively charged (due to collection of accumulated electron) and is called anode (annation = oxidation). And the metal rod on whose surface reduction occurs becomes positively charged due to shortage of electron; it is called *cathode*. When they are joined electrons flow from –ve electrode (anode) to +ve electrode (cathode). And we say that electricity is flowing in opposite direction.

18.3 CONVENTIONS DURING WRITING A CELL

While writing a cell symbolically we follow following conventions —

(a) Cathode (i.e. +ve electrode) should be written in right hand side and anode (i.e. –ve electrode) in the left hand side.

(b) A vertical line is to be drawn to indicate phase difference (i.e. solid /liquid; solid/ gas; liquid./gas etc.) between electrode material.

(c) When there are two electrolytes, single vertical line between them means diffusion is allowed but double vertical lines mean diffusion is prohibited.

Some examples are —

Table 18.1

Cells	Reaction	Symbolic representation
1.	$Zn \rightleftharpoons Zn^{2+} + 2e$ $2H^+ + 2e \rightleftharpoons H_2$ (on Cu-surface)	$\overset{(-)}{Zn}\,\vert H_2SO_4\vert\,\overset{(+)}{Cu}$

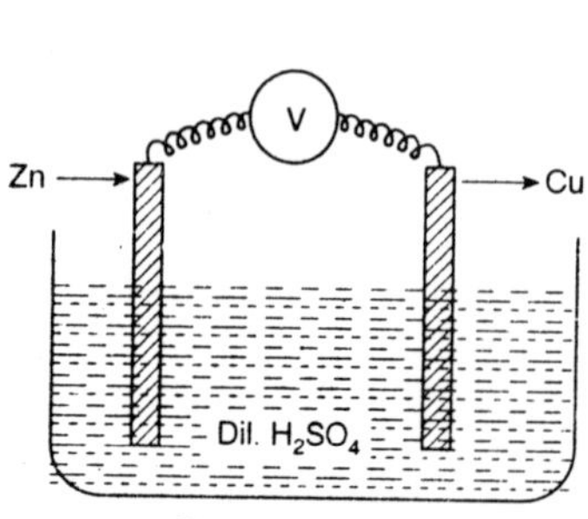

Simple Voltaic Cell

Fig. 18.1

2.

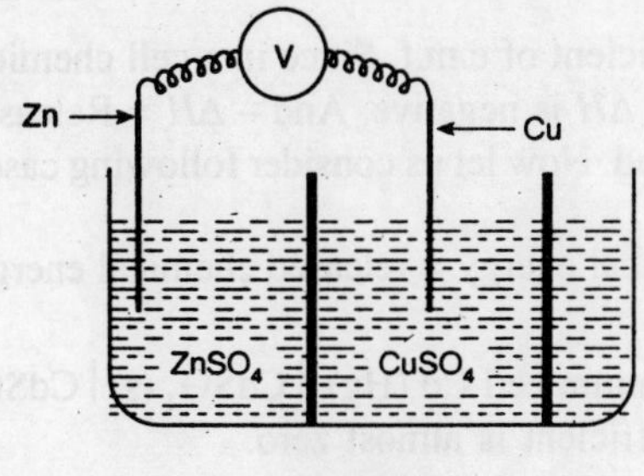

$Zn \rightleftharpoons Zn^{2+} + 2e$

$Cu^{2+} + 2e \rightleftharpoons Cu$

$Zn\,|ZnSO_4|\,CuSO_4|\,Cu$

Fig. 18.2

3. Same as '2' $Zn|ZnSO_4||CuSO_4|Cu$

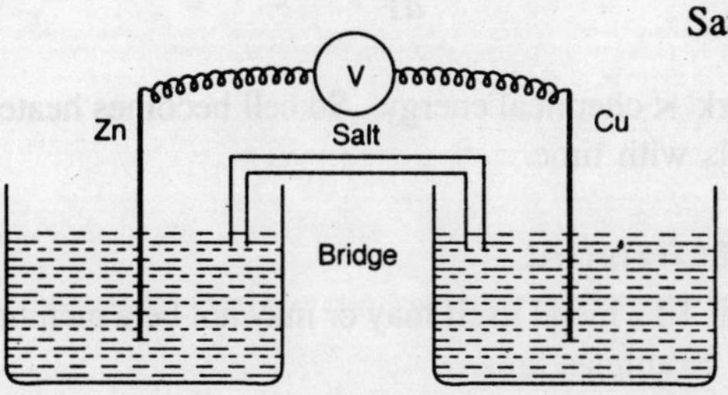

Fig. 18.3

When polarity of the electrodes are not previously known we can write whatever we like i.e. any one electrode can be written in R.H.S. or L.H.S. If writing is wrong then in course of calculation e.m.f. will come as negative.

18.4 E.M.F. AND ELECTRICAL WORK

Due to the redox reaction occurring inside a cell one electrode gets excessive negative charge and the other becomes electron-defficient. So there becomes a difference in electronic pressure between the two electrodes. And this pressure difference is the driving force for flow of electrons. E.M.F. (electro motive force) of a cell is a measurement of this electronic pressure difference. If 'E' volt is electromotive force and in a given time Q coulomb electricity flows through the external circuit then electrical work done = EQ volt-coulomb = EQ Joule. If during this time one mole of metal (M) of valency 'n' in oxidised, then as here 1 mole = n equivalent and n equivalent carries n faraday i.e. nF coulombs so $Q = nF$. Hence electrical work obtained = nFE Joule. Thermodynamics tells us that electrical work is net work. And net work obtained in any process is equal to decrease in Gibbs free energy (ΔG). So here —

$$-\Delta G = nFE$$

or $$\Delta G = -nFE. \quad .....(18.1)$$

18.5 TEMPERATURE CO-EFFICIENT OF E.M.F.

From Gibbs-Helmholtz equation we know —

$$\Delta G = \Delta H + T \cdot \frac{d}{dT}(\Delta G)_p .$$

Now putting $\Delta G = -nFE$, we get

$$-nFE = \Delta H + T \cdot \frac{d}{dT}(-nFE)$$

or $$E = -\frac{\Delta H}{nF} + T \cdot \left(\frac{dE}{dT}\right)_p \quad .....(18.2)$$

The term $(dE/dT)_p$ is called temperature co-efficient of e.m.f. Since in a cell chemical energy released is converted to electrical energy so ΔH is negative. And $-\Delta H$ = Released chemical energy. Also nFE is electrical work obtained. Now let us consider following cases.

(i) If $\left(\frac{dE}{dT}\right)_p \simeq 0$ then $nFE = -\Delta H$ i.e. electrical energy = released chemical energy.

Such cells are called *standard cell* e.g. Weston cadmium cell Cd (Hg) | $CdSO_4$ (s) | $CdSO_4$ satd. soln. | Hg_2SO_4(s) | Hg ; its temperature co-efficient is almost zero.

(ii) If $(dE/dT)_p > 0$ then electrical work > chemical energy. So excess energy is absorbed from cells surface – it becomes cooled (dT is –ve) $\frac{dE}{dT}$ = + ve; so 'E' will fall with time.

(iii) If $(dE/dT)_p < 0$, then electrical work < chemical energy . So cell becomes heated (dT is +ve) but $(dE/dT)_p$ is –ve, so 'E' falls with time.

18.6 ELECTRODE AND ELECTRODE POTENTIAL

Any metal in contact with a redox system, (the metal itself may or may not be a member of the system) can behave as an electrode. Examples are —

Zn | Zn^{2+} ; (Pt) H_2 | H^+ ; (Pt) Cl_2 | Cl^- ; Pt | Fe^{2+} ; Fe^{3+} etc.

Let us consider a metal M dipped in a solution containing its own ion M^{n+} [n = valency].

Three cases may arise —

(i) $M^{n+} + ne \rightarrow M$;surface becomes +ve.

(ii) $M \rightarrow M^{n+} + ne$; surface becomes –ve.

(iii) Nothing happens ; electrode is null electrode.

So in case of reduction (i) or oxidation (ii) electrode surface gets some charge. It attracts oppositely charged ions from solution towards it and an electrical double layer is produced. Potential of this double layer is called *electrode potential.* Its value can be calculated. If we write reduction reaction

$M^{n+} + ne \rightarrow M$, then from Vant Hoff's reaction isotherm

$$\Delta G = \Delta G^0 + RT \ln \frac{a_M}{a_{M^{n+}}}.$$

But (M) is solid so $a_M = 1$ and $\Delta G = -nFE$ and $\Delta G^0 = -nFE^0$.

$$\text{So} \qquad -nFE_{red} = -nFE^0_{red} - RT \ln a_{M^{n+}}$$

$$\text{or} \qquad E_{red} = E^0_{red} + \frac{RT}{nF} \ln a_{M^{n+}}$$

It can also be written as —

$$E_{M^{n+}/M} = E^0_{M^{n+}/M} + \frac{RT}{nF} \ln a_{M^{n+}} \qquad .....(18.3)$$

Similarly if we write oxidation reaction

$M \rightarrow M^{n+}$ + ne, then

$$E_{ox} = E^0_{ox} - \frac{RT}{nF} \ln a_{M^{n+}}$$

$$\text{or} \qquad E_{M/M^{n+}} = E^0_{M/M^{n+}} - \frac{RT}{nF} \ln a_{M^{n+}}. \qquad .....(18.4)$$

It can be shown that $E_{M^{n+}/M} = -E^0_{M/M^{n+}}$.

So $E_{M^{n+}/M} = -E_{M/M^{n+}}$

or $E_{red} = -E_{ox}$ .....(18.5)

Thus reduction potential and oxidation potential are equal in magnitude but opposite in sign. It should be kept in mind that when we use the word potential' only we mean reduction potential.

18.7 TYPES OF ELECTRODES

There is no uniformity in classifying electrodes into different types. We are here writing the easiest way of recollecting any electrode. We classify electrode into three kinds.

Table 18.2

Type	Description	Example	Reaction/Electrode Pot.
First kind	Metal or non metal and its ion. For non-metal (Pt) is conductor.	1. $Zn\|Zn^{2+}$ 2. $(Pt)H_2(P)/H^+$ 3. $(Pt)Cl_2(P)/H^+$	1. $Zn^{2+} + 2e \rightleftharpoons Zn$ $E_R = E_R^0 + \frac{RT}{2F} \ln a_{Zn^{2+}}$ 2. $H^+ + e \rightleftharpoons \frac{1}{2} H_2(P)$ $E_R = E_R^0 + \frac{RT}{F} \ln \frac{a_{H^+}}{P^{1/2}}$ 3. $\frac{1}{2}Cl_2(P) + e \rightleftharpoons Cl^-$ $E_R = E_R^0 + \frac{RT}{F} \ln \frac{P^{1/2}}{a_{Cl^-}}$
Second kind	Metal, its insoluble salt and anion of insoluble salt	1. Ag, AgCl (s)| Cl^- 2. Pb, $PbSO_4$ (s)| SO_4^{2-}	1. $AgCl + e \rightleftharpoons Ag + Cl^-$ $E_R = E_R^0 - \frac{RT}{F} \ln a_{Cl^-}$ 2. $PbSO_4 + 2e \rightleftharpoons Pb + SO_4^{2-}$ $E_R = E_R^0 - \frac{RT}{2F} \ln a_{SO_4^{2-}}$
Third kind	An inert metal in contact with a redox mixture in solution.	1. Pt| Fe^{2+}, Fe^{3+} 2. Pt| $Cr_2O_7^{2-}, Cr^{3+}$	1. $Fe^{3+} + e \rightleftharpoons Fe^{2+}$ $E_R = E_R^0 + \frac{RT}{F} \ln \frac{a_{Fe^{3+}}}{a_{Fe^{2+}}}$ 2. $Cr_2O_7^{2-} + 14H^+ + 6e \rightleftharpoons 2Cr^{3+} + 7H_2O$ $E_R = E_R^0 + \frac{RT}{6F} \ln \frac{a_{Cr_2O_7^{2-}}\left(a_{H^+}\right)^{14}}{\left(a_{Cr^{3+}}\right)^2}$

Remember that general formula of electrode potential is $E = E^0 + \frac{RT}{nF} \ln X$,

where $X = \frac{\text{Product of activities of reactants with appropriate power}}{\text{Same for products}}$.

18.8 REFERENCE ELECTRODES

Reference electrodes are those with the help of which potentials of other electrodes can be known. For this purpose potential of reference electrodes are to be known. If potential of reference electrode is arbitrarily fixed then it is called a *primary reference electrode* e.g., *a standard hydrogen electrode* (SHE) i.e., (Pt) H_2 (1 atm) | H^+ (a = 1). Its potential at 25°C is arbitrarily taken as zero. If however potential of an electrode is determined first by some primary reference electrode and then it is used to determine potential of other electrodes then the electrode is secondary reference electrode. Examples being calomel electrode, glass electrode quinhydrone electrode etc. We are now describing them one by one.

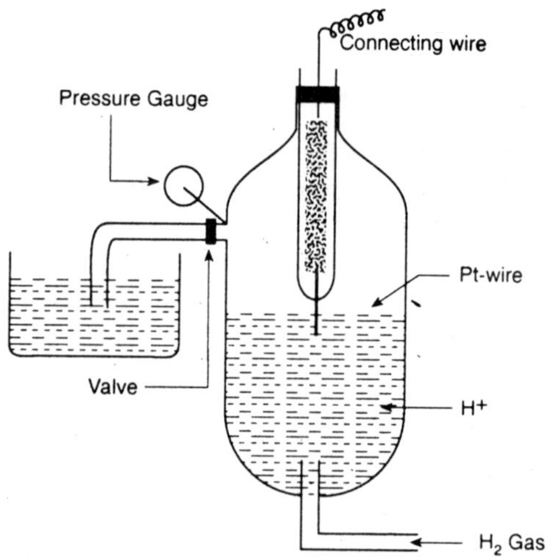

Fig. 18.4 Hydrogen electrode.

18.9 HYDROGEN ELECTRODE

Hydrogen electrode consists of a cylindrical bulb inside which a solution containing H^+ is taken. A Pt-wire is dipped in it and H_2-gas is slowly bubbled. The gas accumulates above the solution. Its pressure can be fixed at any desired value with the help of a pressure gauge and valve. The electrode when expressed symbolically is — (Pt) H_2 (P) | H^+ (a), when P = 1 atm and a = 1 then it will be called a standard hydrogen electrode (SHE). Its potential at 25°C is arbitrarily fixed at zero. Now the reaction is —

$$H^+ (a) + e \rightleftharpoons \frac{1}{2} H_2 (P).$$

So $$E = E^0 + \frac{RT}{F} \ln \frac{a_{H^+}}{P^{1/2}}.$$

For S. H. E. $E = E^0$ [as $a_{H^+} = 1$ & P = 1 atm].

But arbitrarily $E = 0$ so $E^0 = 0$.

Hence for ordinary hydrogen electrode

$$E = \frac{RT}{F} \ln \frac{a_{H^+}}{P^{1/2}}.$$

If the electrode is (Pt) H_2 (1atm) | H^+ (a), then

$$E = \frac{RT}{F} \ln a_{H^+}.$$

At 25°C $$E = \frac{8.313 \times 298}{96500} \times 2.303 \log a_{H^+}$$

or $\quad E = 0.059 \log a_{H^+}$

or $\quad E = -0.059\ \text{pH}. \quad$ .....(18.6)

Thus we see that hydrogen electrode can be used to measure pH of solution. However it fails to measure pH above 8 and in presence of some metal ions.

18.10 QUINHYDRONE ELECTRODE

This is a very simple electrode used for measuring pH of a solution. Quinhydrone is an equimolecular mixture of quinone (Q) and hydroquinone (H_2Q). If trace amount of quinhydrone is added to a solution having H^+ ion and then a Pt–metal is dipped in it, the total set us behaves as an electrode and is known as quinhydrone electrode.

The electrode is (Pt), $|\ H_2Q\ ;\ Q\ ;\ H^+$ and the reaction is

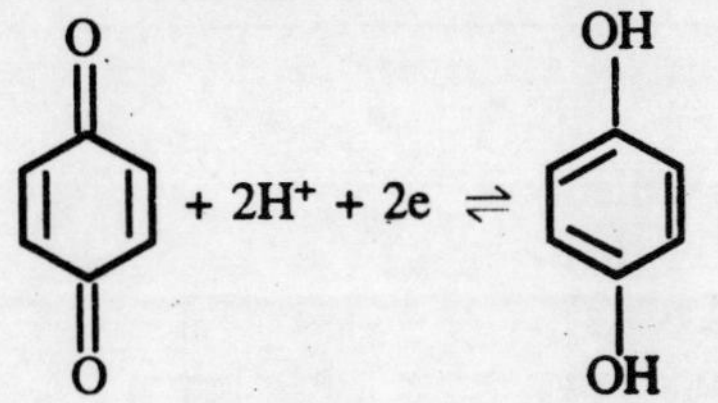

i.e., $\quad Q + 2H^+ + 2e \rightleftharpoons H_2Q.$

The potential is $\quad E = E^0 + \dfrac{RT}{2F} \ln \dfrac{a_Q \cdot \left(a_{H^+}\right)^2}{a_{H_2O}}.$

In solution quinone and hydroquinone are present in equimolar amount.

So $\quad a_Q = a_{H_2O}$ and hence

$$E = E^0 + \frac{RT}{F} \ln a_{H^+}.$$

At 25°C $\quad E = E^0 - 0.059$ pH (see H–electrode).

The value of E^0 for quinhydrone electrode is 0.6994 volt at 25°C. So

$$\boxed{E = 0.6994 - 0.059\ \text{pH}} \quad .....(18.7)$$

The electrode is simple and can be prepared very easily. However, it also fails to measure pH above 8. For measuring pH of alkaline solutions, we use glass electrode.

18.11 CALOMEL ELECTRODE

The most extensively used secondary reference electrode is calomel electrode. Calomel is the commercial name of Hg_2Cl_2. In a glass vessel we take shinning Hg at the bottom – it is covered by a paste of calomel and then the vessel along with its side tube which is dipped in a beaker is filled with KCl solution. The KCl solutions chosen are (N/10); (N) and saturated. The electrode is Hg, Hg_2Cl_2 (s) $|$ Cl^- and the reaction is

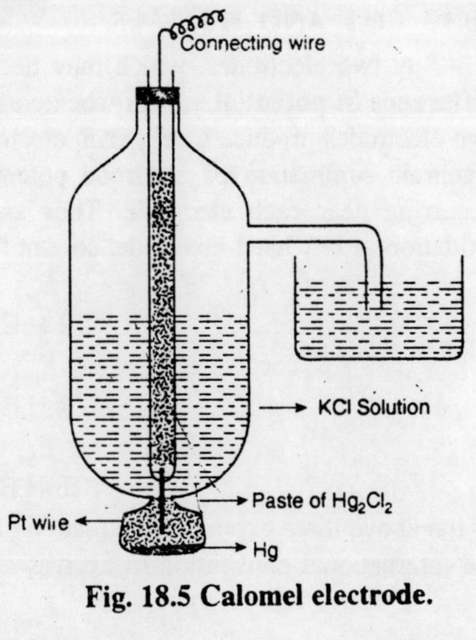

Fig. 18.5 Calomel electrode.

Hg_2Cl_2 (s) + 2e $\rightleftharpoons$ 2Hg + $2Cl^-$. So

$$E = E^0 - \frac{RT}{F} \ln a_{Cl^-} \quad .....(18.8)$$

For three types of calomel electrode potential values are different. These are shown below.

Table 18.3

Electrode	E (red. pot.) at 25°C.
Decinormal calomel	0.334 volt
Normal calomel	0.280 volt
Saturated calomel	0.242 volt

18.12 GLASS ELECTRODE

Glass electrode consists of a thin-walled glass bulb inside which a solution of 0.1 (M) HCl is taken. A silver-wire coated with AgCl(s) is dipped in it. If now this set up is dipped in a solution containing H^+ ion, then a potential is found to be developed on the wall of the glass vessel. The reason for the development of potential is not quite clear. It may be due to a difference in H^+ ion concentration between the inner and outer side of glass membrane. However, the expression for potential has been found to be $E_G = E^0_G - 0.059$ pH; here E^0_G is a constant dependent on quality of glass. This is such an electrode, which can measure pH with supreme accuracy in the range 0 – 12; above pH 12 the accuracy is little less.

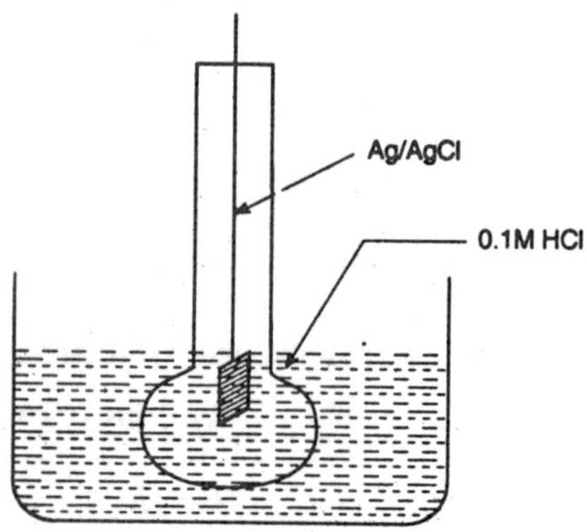

Fig. 18.6 Glass electrode.

18.13 CELL AND ITS E.M.F.

Any two electrodes, which may be made of same or different materials but have a **difference in potential values** when connected will produce a cell. Since combination of two electrodes produce a cell, each electrode is called a *half cell. E. M. F.* of a cell is the algebraic summation of electrode potentials taking into consideration actual processes occurring near each electrode. Thus as reduction occurs at right hand electrode and oxidation at left hand electrode, so e.m.f. is

$$E_{Cell} = \underset{\text{(R.H.E.)}}{E_{red}} + \underset{\text{(L.H.E)}}{E_{ox}} \quad .....(18.9)$$

$$\text{or} \quad E_{Cell} = \underset{\text{(R.H.E.)}}{E_{red}} - \underset{\text{(L.H.E.)}}{E_{red}} \quad .....(18.10)$$

$$\text{or} \quad E_{Cell} = \underset{\text{(L.H.E)}}{E_{ox}} - \underset{\text{(R.H.E.)}}{E_{ox}} \quad .....(18.11)$$

of the above three expressions equation (18.10) is widely used to calculate e.m.f. And it is the international convention to express e.m.f. of a cell as

$$E_{Cell} = \underset{\text{(R.H.E.)}}{E_{red}} - \underset{\text{(L.H.E.)}}{E_{red}}$$

18.14 STANDARD ELECTRODE POTENTIAL

For an arbitrary electrode reaction — Reactant + ne = Product, expression for electrode potential is —

$$E_{red} = E^0_{red} + \frac{RT}{nF} \ln \frac{a_{reactant}}{a_{product}}.$$

Now if all constituents are in their standard states then each activity term = 1. So $E_{red} = E^0_{red}$. This is called *standard electrode potential.* Thus standard electrode potential of an electrode is its potential when all the members participating in electrode reaction are in their standard state i.e. with unit activity. It is numerically equal to the e.m.f. of a cell whose LHE is a SHE and RHE is the given electrode, where activities of all participating in electrode reaction are unity. To explain let, our electrode be M / M^{n+}, then its standard potential is equal to e.m.f. of following cell —

Std. H-electrode || M^{n+} ($a = 1$) | M.

Because its e.m.f. is

$$E = E_{M^{n+}/M} - E_{SHE} = E_{M^{n+}/M} \qquad [\because E_{SHE} = 0]$$

$$\text{or} \qquad E = E^0_{M^{n+}/M} \qquad [\text{ as } a_{M^{n+}} = 1].$$

Q. Standard electrode potentials of Zn |Zn^{2+} and Cu |Cu^{2+} are written with opposite signs why? Which one should be positive ?

Ans : By definition standard potentials of Zn |Zn^{2+} and Cu |Cu^{2+} are equal to e.m.f. of following cells (i) and (ii) respectively.

(i) SHE || Zn^{2+} ($a = 1$) | Zn and

(ii) SHE || Cu^{2+} ($a = 1$) | Cu

Now Zn can replace H^+ from acid, but Cu can not. So out of H_2 and Zn tendency towards oxidation is more for Zn. So it should be written as LHE. But as it is written as RHE, so E_{Cell} will be negative (section 18.3). Hence standard potential of Zn | Zn^{2+} electrode is negative. For Cu its tendency towards oxidation is less than that of H_2. So cell (ii) is written correctly ; hence its 'E' will be +ve. Thus standard potential of Cu | Cu^{2+} electrode is +ve.

N.B. Following international convention we always mean reduction reaction when saying about standard potential.

The following is the table showing standard potential values of some common electrode at 25°C.

Table 18.4

Name of electrode	Reaction	E^0 (in volt at 25°C)
Li | Li^+	$Li^+ + e \rightleftharpoons Li$	– 2.959
Na | Na^+	$Na^+ + e \rightleftharpoons Na$	– 2.715
Mg | Mg^{2+}	$Mg^{2+} + 2e \rightleftharpoons Mg$	– 1.866
Zn | Zn^{2+}	$Zn^{2+} + 2e \rightleftharpoons Zn$	– 0.762
Sn | Sn^{2+}	$Sn^{2+} + 2e \rightleftharpoons Sn$	– 0.136
Pb | Pb^{2+}	$Pb^{2+} + 2e \rightleftharpoons Pb$	– 0.122

Name of electrode	Reaction	E^0 (in volt at 25°C)
(Pt) H_2 | H^+	$H^+ + e \rightleftharpoons \frac{1}{2} H_2$	ZERO
Cu | Cu^{2+}	$Cu^{2+} + 2e \rightleftharpoons Cu$	+ 0.344
(Pt) | Fe^{2+}, Fe^{3+}	$Fe^{3+} + e \rightleftharpoons Fe^{2+}$	+ 0.771
Ag | Ag^+	$Ag^+ + e \rightleftharpoons Ag$	+ 0.798
Hg | Hg_2^{2+}	$Hg_2^{2+} + 2e \rightleftharpoons 2Hg$	+ 0.799
(Pt) Cl_2 | Cl^-	$\frac{1}{2} Cl_2 + e \rightleftharpoons Cl^-$	+ 1.360

Notes-

(1) When any two electrodes are joined to produce a cell, a first hand idea about which one will be oxidised and which one will be reduced can be obtained from their E^0 values, though final decision should be taken from E-values, which is $E = E^0 + \frac{RT}{nF} \ln a_{M^{n+}}$. The electrode whose electrode potential is more will be reduced i.e. the more the reduction potential, the more is tendency for reduction and the more the oxidation potential, the more is tendency for oxidation.

(2) Let us take an example. Let we are to connect a saturated calomel electrode and a quinhydrone electrode to construct a cell. At 25°C

$$E_{\text{Satd. Cal}} = 0.242 \text{ volt} \quad \text{and} \quad E_{(Pt)Q/H_2Q} = 0.6994 - 0.059 \text{ pH}$$

if pH = 7
$$E_{Q/H_2Q} = 0.6994 - 0.413 = 0.2864 \text{ volt.}$$

Since its E is more so quinhydrone electrode will be reduced and the cell will be —

$$\overset{(-)}{\text{Satd. Calomel}} \,\|\, H^+, H_2Q, Q \,|\, \overset{(+)}{(Pt)}.$$

But if pH of the solution becomes '8' then

$$E_{Q/H_2Q} = 0.6994 - 0.472 = 0.2274 \text{ volt.}$$

Now $E_{\text{satd. Cal}}$ is more so calomel electrode will be reduced and the cell will be —

$$\overset{(-)}{(Pt)} \,|\, Q, H_2Q, H^+ \,\|\, \overset{(+)}{\text{Satd. Calomel}}.$$

So we see that cell reactions are perfectly reversible.

18.15 CLASSIFICATION OF CELLS

We have seen that any two electrodes, which may be of same or different materials, but have a difference in potential values when connected will produce a cell. Depending on types of electrodes cells can be classified as follows —

Electrochemical cells
- Chemical Cells
  - Without Transference
  - With Transference
- Concentration cells
  - Electrode conc. cell
  - Electrolyte conc. cell
    - With Transference
    - Without transference

We shall now give a very brief account of all these types.

18.16 CHEMICAL CELLS

In chemical cells two electrodes are made of different materials, and chemical energy released during reaction is converted to electrical energy. Depending on nature of electrodes we may have one or more electrolytes. When two electrolytes are taken, then if there is diffusion it is a cell without transference and if diffusion is prohibited it is a cell without transference.

Following examples are to be noted

(a) Chemical cell with one electrolyte :

(Pt) H_2 (1atm) | $H^+ Cl^-$ | AgCl (s), Ag.

Reaction :

$$\text{R H E} : AgCl\,(s) + e \rightleftharpoons Ag + Cl^-$$

$$\text{L.H.E.} \quad \frac{1}{2} H_2\,(1atm) \rightleftharpoons H^+ + e$$

$$\frac{1}{2} H_2\,(1atm) + AgCl\,(s) \rightleftharpoons Ag + HCl$$

$$E_{Cell} = E^0_{Cell} - \frac{RT}{F} \ln a_{HCl}. \quad .....(18.12)$$

[activities of solid = 1 of gas = P = 1 = 1atm]

(b) Chemical cell with two electrolytes :

Zn | $ZnSO_4$ (a_1) || $CuSO_4$ (a_2) | Cu.

Reaction :

$$\text{R.H.E.} \quad Cu^{2+} + 2e \rightleftharpoons Cu$$

$$\text{L.H.E.} \quad Zn \rightleftharpoons Zn^{2+} + 2e$$

$$Zn + Cu^{2+} \rightleftharpoons Zn^{2+} + Cu$$

$$E_{Cell} = E^0_{Cell} + \frac{RT}{2F} \ln \frac{a_{Cu^{2+}}}{a_{Zn^{2+}}}. \quad .....(18.13)$$

18.17 CONCENTRATION CELL

In concentration cells the two electrodes are of same materials but their potentials are made different by making a difference in concentration either at electrode surface or in electrolytic solution. In the former case, it is an electrode concentration cell and in the later it is electrolyte concentration cell. In such case no chemical reaction in electrolytic solution occurs but there is a diffusion of charge. Examples are given below —

(a) Electrode concentration cell : Examples of such a cell is

(Pt) H_2 (P_1) | H^+ | H_2 (P_2) (Pt)

(Pt) Cl_2 (P_1) | Cl^- | Cl_2 (P_2) (Pt)

Process at different electrodes for the first are —

$$\text{R.H.E.} \quad H^+ + e \rightleftharpoons \frac{1}{2} H_2\,(P_2)$$

$$\text{L.H.E.} \quad \frac{1}{2} H_2\,(P_1) \rightleftharpoons H^+ + e$$

$$\frac{1}{2} H_2(P_1) \rightleftharpoons \frac{1}{2} H_2\,(P_2)$$

$$E_{Cell} = \frac{RT}{F} \ln\left(\frac{P_1}{P_2}\right)^{1/2} = \frac{RT}{2F} \ln \frac{P_1}{P_2} \qquad .....(18.14)$$

and for the second —

$$\text{R.H.E.} \quad \frac{1}{2} Cl_2 (P_2) + e \rightleftharpoons Cl^-$$

$$\text{L.H.E.} \quad Cl^- \rightleftharpoons \frac{1}{2} Cl_2 (P_1) + e$$

$$\frac{1}{2} Cl_2 (P_2) \rightleftharpoons \frac{1}{2} Cl_2 (P_1)$$

or $$E_{Cell} = \frac{RT}{F} \ln\left(\frac{P_2}{P_1}\right)^{1/2} = \frac{RT}{2F} \ln \frac{P_2}{P_1}. \qquad .....(18.15)$$

Expressions for E_{Cell} do not contain E^0 terms. Because $E_{Cell} = E^0_{RHE} - E^0_{LHE}$; and in concentration cell both electrodes are of same material, so $E^0_{RHE} = E^0_{LHE}$ and so $E^0_{Cell} = 0$.

(b) Electrolyte concentration cells :

In such concentration cells potentials of two electrodes of same materials are made different by bringing a difference in concentration in electrolytic solution. If these two electrolytes are well separated by (X) dipping a common electrode or (Y) by a salt bridge, then it is an electrolyte concentration cell without transference and if they are remaining side by side allowing diffusion, then it is an electrolyte concentration cell with transference. Examples are —

(i) Electrolyte conc. cell without transference

(X) By dipping common electrode

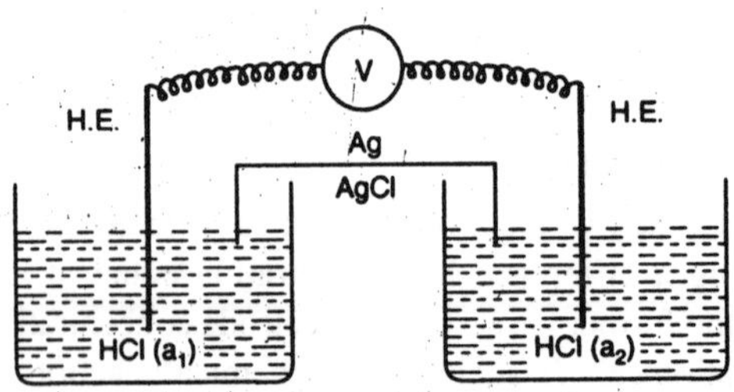

Fig. 18.7 Conc. cell without transference.

i.e., (Pt) H_2 (1 atm) | HCl (a_1) | Ag – AgCl | HCl (a_2) | H_2 (1atm) (Pt)

cell process is HCl (a_2) = HCl (a_1)

$$E = \frac{RT}{F} \ln \frac{a_2}{a_1} = \frac{2RT}{F} \ln \frac{a_{\pm 2}}{a_{\pm 1}}. \qquad .....(18.16)$$

[$a\pm$ = Mean ionic activity $= \sqrt{a}$]

(*Y*) By a salt bridge :

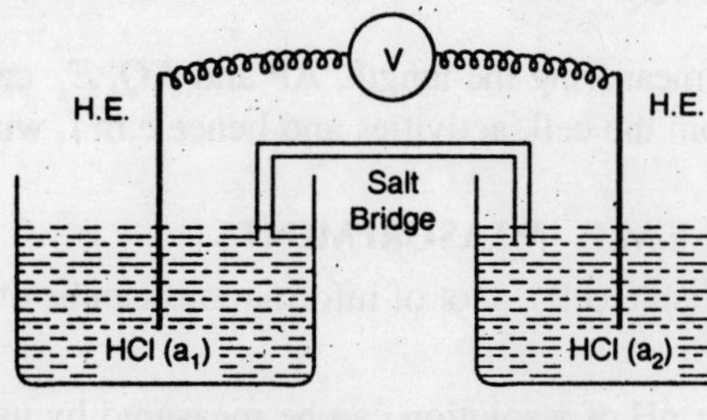

Fig. 18.8 Use of salt bridge.

i.e., (Pt) H_2 (1atm) | HCl (a_1) || HCl (a_2) | H_2 (1atm) (Pt)

cell process is $H^+ (a_{\pm 2}) \rightleftharpoons H^+ (a_{\pm 1})$

$$a_{\pm} = \text{mean ionic activity} = \sqrt{a}$$

$$E_{\text{Cell}} = \frac{RT}{F} \ln \frac{a_{\pm 2}}{a_{\pm 1}}. \quad .....(18.17)$$

N.B. Salt bridge is a U-tube inside which an agar-agar gel along with either KCl or NH_4NO_3 is taken. It acts as a conductor.

(ii) Electrolyte conc. cell with transference :

Example is

(Pt) H_2 (1atm) | HCl (a_1) | HCl (a_2) | H_2 (1atm) (Pt)

Here diffusion of electrolyte occurs at junction. Details are beyond the scope of this book. Its e.m.f. is

$$E_{\text{Cell}} = 2t - \frac{RT}{F} \ln \frac{a_{\pm 2}}{a_{\pm 1}}. \quad .....(18.18)$$

18.18 MEASUREMENT OF E.M.F. OF A CELL — POTENTIOMETER

E.M.F. of a cell is always measured by a potentiometer and not by a voltmeter. Because if a voltmeter is used, some electricity is drawn from the cell and so the cell reaction occurs. Hence activities of cell materials will alter. This will cause a change in e.m.f. In potentiometer, e.m.f. is measured by Poggendroff's compensation method in such a manner that no electrical energy is drawn from the cell. The method is as follows —

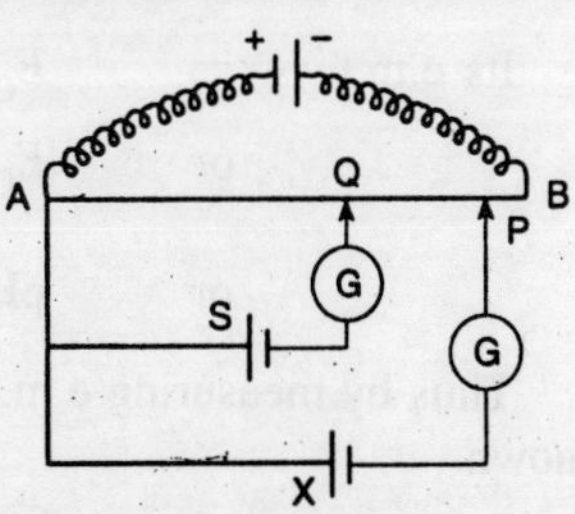

Fig. 18.9 Cercuit diagram of potentiometer.

AB is a wire with high resistance having uniform cross section. It is connected to the terminals of a storage cell, 'A' end is connected to the +ve end of the cell. '*X*' is the cell whose potential is to be measured. Its +ve end is also connected with 'A'. Its negative end is attached to a sliding contact through a galvanometer 'G'. The contact is moved over the wire AB to find that point where galvanometer will not show any deflection (null point). Let this point is 'P'. Then E.M.F. of the cell = Potential drop between A and P. Now the same experiment is done with a standard cell 'S' of known e.m.f. Let the new null point is Q. Then —

$$\frac{E_x}{E} = \frac{\text{Potential drop between A and P}}{\text{Potential drop between A and Q}} = \frac{\text{Resistance of AP}}{\text{Resistance of AQ}}$$

or $\frac{E_x}{E_s} = \frac{\text{Length of AP}}{\text{Length of AQ}}$. $[R = \rho\frac{l}{A}$, here ρ and A are same].

Thus if we know E_s then measuring the length, AP and AQ, E_x can be known. Since here no electricity is drawn from the cell, activities and hence e.m.f. will remain the same.

18.19 APPLICATIONS OF E.M.F. MEASUREMENTS

By measuring e.m.f. of suitable cells, a lot of informations can be obtained. Only a few of them are discussed here.

(a) Measurement of pH : pH of a solution can be measured by using (i) H-electrode (ii) quinhydrone electrode and (iii) glass electrode. We are discussing the method with quinhydrone electrode. This method is extensively used by students in practical class. We construct the cell —

Satd. Cal | Salt Bridge | H^+, H_2Q, Q | (Pt)

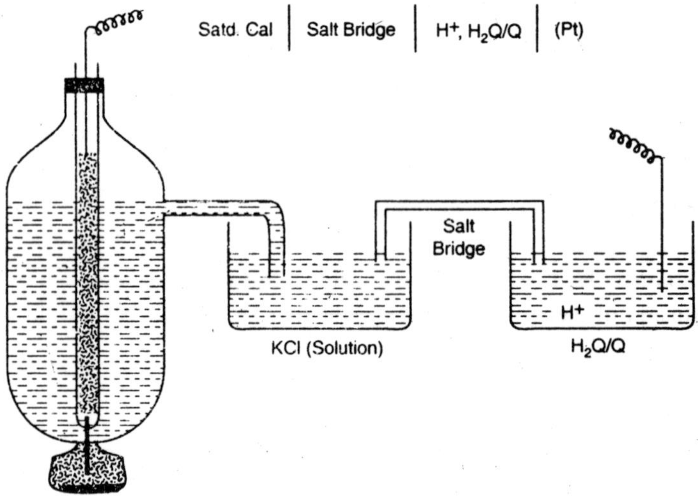

Fig. 18.10 Cell for pH-measurement.

Its e.m.f. is $E_{Cell} = E_{Q/H_2O} - E_{Satd.\ Cal.}$

or $E_{Cell} = 0.6994 - 0.059\ pH - 0.242$

or $pH = \frac{0.4574 - E_{Cell}}{0.059}$.

Thus by measuring e.m.f. of this cell with a potentiometer, pH of the solution can be known.

(b) Standard free energy change and equilibrium constant : Standard free energy change

$$\Delta G^0 = -nFE^0_{Cell} = -RT \ln K.$$

or $$E^0_{Cell} = \frac{RT}{nF} \ln K.$$

Thus if E^0_{Cell} is known then equilibrium constant 'K' and also standard free energy change ΔG^0 can be known.

(c) Solubility product calculation : Let us consider a cell

$$Ag \mid Ag^+ \parallel Cl^-, AgCl\,(s) \mid Ag.$$

Cell reaction is R. $AgCl\,(s) + e \rightleftharpoons Ag + Cl^-$

L. $Ag \rightleftharpoons Ag^+ + e$

$$AgCl\,(s) \rightleftharpoons Ag^+ + Cl^-$$

Its equilibrium constant $K = a_{Ag^+} \cdot a_{Cl^-} = K_s$ (solubility product).

So $E^0_{Cell} = \frac{RT}{F} \ln K_s.$

But $E^0_{Cell} = E^0_{AgCl\,.\,Ag} - E^0_{Ag^+/Ag}$ (red).

Thus by knowing standard potential values of two electrodes, E^0_{Cell} can be known; from this solubility product of AgCl can be calculated.

18.20 SECONDARY CELL/STORAGE CELL

Theoretically all cell reactions are reversible. The process of drawing current from the cell is called *discharging* and the process of sending current in the cell is charging. *Charging* rejuvenates the cell. Theoretically all cells can both be discharged and charged. But in practice there are some cells which can not be charged after being discharged. This is due to some side reactions. Such cells are called primary cells e.g. Lechlance's cell. Cells which can both be discharged and charged are secondary cell or storage cell. They are also known as accumulators. Examples being —

(i) Pb – accumulater :- $Pb \mid H_2SO_4 \mid PbO_2, Pb$

(ii) Alkali – accumulater :- $Fe \mid FeO(s), KOH, Ni_2O_3\,(s), \mid NiO\,(s) \mid Ni$

In case (i) the over all reaction is —

$$Pb + PbO_2 + 2H_2SO_4 = 2PbSO_4 + 2H_2O$$

Its e.m.f. is ~ 2 volts.

and in case (ii) the reaction is

$$Fe + Ni_2O_3 = FeO + 2NiO$$

Its e.m.f. is ~ 1.35 volts.

POINTS TO REMEMBER

Electrochemical cell are those where chemical energy is converted to electrical energy. The chemical reaction is a redox reaction. Oxidation and reduction occur on the surface of two metal electrodes which become electron-rich and electron-defficient. An electrical double layer is produced around each of them whose potential is electrode potential. According to Nernst expression for electrode potential for a M/M^{n+} electrode is —

$E_{M^{n+}/M} = E^0_{M^{n+}/M} + \frac{RT}{nF} \ln a_{M^{n+}}$. Any metal in contact with a redox system is an electrode or half-cell. When two such half-cells with a difference in potential values are connected we get a cell. If the two electrodes are of different materials, then the cell is chemical but when the electrodes are of same material, it is a concentration cell. For any cell $E_{Cell} = \underset{(R.H.E.)}{E_{red}} - \underset{(L.H.E.)}{E_{red}}$ [E_{red} = reduction pot; R.H.E. and L.H.E. are right hand and left hand electrodes]. The convention is to write negative electrode as L.H.E. and positive electrode as R.H.E. (Pt) H_2 (P) | H^+ (a) is a hydrogen electrode. When P = 1 atm & a = 1 it is a **standard hydrogen electrode (SHE).** Its potential is **arbitrarily taken as zero at 25°C.** Besides hydrogen electrode, there are other reference electrodes such as **quinhydrone; glass** and **calomel** electrode. **Hydrogen; quinhydrone and glass electrodes are used to measure pH.**

QUESTIONS

SHORT TYPE

1. Distinguish between electrolytic and electrochemical cells.
2. What are anodes and cathodes and what are their signs in electrolytic and electrochemical cells?
3. What is an electrode and electrode potential ?
4. Prove that $E_{red} = - E_{ox}$.
5. What is a SHE and what is its potential ?
6. What thing is common among hydrogen, quinhydrone and glass electrode ?
7. What is standard electrode potential ?
8. Why standard potentials of Zn | Zn^{2+} and Cu | Cu^{2+} electrodes are opposite in sign ?
9. What is the peculiarity of concentration cell ?
10. Why a voltmeter is not used for e.m.f. measurement ?
11. Expression for potential of a H-electrode is – 0.059 pH – how the factor 0.059 comes ?
12. What is a storage cell ? Name some of them ?

ESSAY TYPE

1. What is an electrode and electrode potential ? Derive an expression for electrode potential of a M/M^{n+} electrode ?
2. Describe the construction, reaction of a H-electrode. What is the expression for its potential ?
3. What are primary and secondary reference electrodes ? Describe the construction, and reaction of calomel electrode.
4. What is the necessary condition for construction of a cell ? Distinguish between a chemical and concentration cell.
5. Write down the electrode reactions and overall cell reactions for the following cells. Also write the expressions for their e.m.f.

   (i) Zn | $ZnSO_4$ || $CuSO_4$ | Cu

   (ii) (Pt) H_2 (1atm) | HCl | AgCl(s), Ag.
6. Describe the potentiometric methods for —

   (i) determination of pH of a solution.

   (ii) calculation of solubility product of AgCl.

PROBLEMS

1. Calculate the equilibrium constants for the reactions at 25°C

   (a) $Cu^{2+} + Zn \rightleftharpoons Cu + Zn^{2+}$

   (b) $Fe + 2Fe^{3+} \rightleftharpoons 3Fe^{2+}$ and

   (c) $PbSO_4$ (s) $\rightleftharpoons Pb^{+2} + SO_4^{2-}$

Given the following E^0 values.

$$E^0_{Cu^{+2}.Cu} = 0.337;\ E^0_{Zn^{2+}.Zn} = -0.763;\ E^0_{Fe^{2+}.Fe} = -0.44;$$

$$E^0_{Fe^{3+}.Fe^{2+}} = 0.771;\ E^0_{PbSO_4.Pb} = -0.356;\ E^0_{Pb^{2+}.Pb} = -0.126.$$

[*Hints* :

(a) The reaction $Cu^{2+} + Zn \rightleftharpoons Cu + Zn^{2+}$ occurs in a cell $Zn\ |Zn^{2+}\ ||\ Cu^{2+}\ |Cu$.

$$E^0 = E^0_{Cu^{2+}/Cu} - E^0_{Zn^{2+}/Zn} = 0.337 + 0.763 = 1.000V.$$

$$E^0 = \frac{RT}{2F} \ln K; \text{ find } K.$$

(b) $E^0 = E^0_{Fe^{3+}.Fe^{2+}} - E^0_{Fe^{2+}.Fe} = \frac{RT}{2F} \ln K.$

(c) $E^0 = E^0_{PbSO_4.Pb} - E^0_{Pb^{2+}.Pb} = \frac{RT}{2F} \ln K$].

2. The standard potentials of Cu/Cu^{+2} and Cu/Cu^+ are 0.337 and 0.530 V respectively. What is the equilibrium constant for the reaction

$$2Cu^+ \rightleftharpoons Cu^2 + Cu\ ?$$

[*Hints* :

The reaction $2Cu^+ \rightleftharpoons Cu^2 + Cu$ occurs in cell **Cu/Cu^{2+}** ; Cu^+/Cu

$$E^0_{Cell} = E_{Cu^+/Cu} - E_{Cu^2/Cu} = \frac{RT}{2F} \ln K].$$

3. The reaction $H_2 + Cl_2 + 2H_2O \rightleftharpoons 2H_3O^+ + 2Cl$ has an equilibrium constant $K = 10^{46}$ at 25°C. Calculate E^0 at 25°C for cell (Pt) H_2 | HCl (aq.) | Cl_2 (Pt).

[*Hints* : $E^0_{Cell} = \frac{RT}{2F} \ln K$].

4. A hydrogen electrode and a calomel half cell are used to determine the pH of a salution on a mountain top where the pressure is 500 m.m. of Hg. The hydrogen is allowed to bubble out of the electrode at the atmospheric pressure prevailing there. If the pH is found to be 4.00 by a visiting scientist who wrongly assumed pH = 1 atm., what is the correct pH. Take temperature as 25°C.

[*Hints* :

For H-electrode $E_{H^+/H_2} = \frac{RT}{F}\left(\ln P^{1/2} - \ln a_{H^+}\right)$

$= -0.0295 \log P - 0.059$ pH at 25°C

When P = 1atm pH = 4

if P = 500 m.m pH = ?]

OBJECTIVE TYPE

1. **Write √ mark on correct answer.**

(a) Daniel cell is an (electrolytic/electrochemical) cell.

(b) In an electrochemical cell nFE is equal to $(-\Delta H\ /\ -\Delta G)$.

(c) Example of primary reference electrode is (S.H.E./calomel electrode).

(d) pH can be measured by (glass/calomel) electrode.

(e) In a salt bridge we use (KCl/KNO_3).

2. Write T against true and F against false statements.

(a) For weston cadmium cell temperature co-efficient of e.m.f. is zero.

(b) Glass electrode fails to measure pH above 8.

(c) In concentration cell the electrodes are of different metals.

(d) E.M.F. of a cell is measured by a voltmeter.

(e) Storage cells are also called accumulators.

3. Match terms of column A with those of column B.

A	B
(a) Weston	(a) Potentiometer.
(b) Calomel	(b) Salt-bridge.
(c) Lead-accumulator	(c) Electrode potential.
(d) KCl	(d) Standard cell.
(e) Poggendroff	(e) Storage cell
(f) Nernst	(f) Secondary reference

Answers :

1. (a) electrochemical (b) $-\Delta G$ (c) SHE (d) glass (e) KCl.

2. (a) T (b) F (c) F (d) F (e) T.

3. a – d; b – f; c – e; d – b; e – a; f – c.

Chapter 19

PHYSICAL PROPERTIES AND MOLECULAR STRUCTURE

19.1 MOLECULAR STRUCTURE — HOW IT IS KNOWN ?

The term 'structure of a molecule' means knowledge of (i) molecular formula and (ii) spatial distribution of constituent atoms. Structures of molecules are of eternal curiosity. Many methods (both direct and indirect) are available to peep into the inner world of molecules. These methods include study of both physical and chemical properties of the substance. We shall here discuss some of the physical properties of a molecule whose study helps us to have informations regarding its structure.

19.2 ADDITIVE AND CONSTITUTIVE PROPERTIES

Physical properties of a substance may be of two types (a) additive and (b) constitutive.

(a) Additive properties : Properties, which depend upon number of atoms present in a molecule and whose magnitude is equal to the sum of the corresponding properties of atoms are additive. A truely additive property is 'mass'. Molar volume is mostly additive in nature.

(b) Constitutive properties : Properties, which depend entirely upon the constitution of molecules, and are independent of the amount are constitutive properties. Optical activity; absorption frequency of molecules are constitutive in nature.

Besides these, there are some properties, which are partly additive and partly constitutive in nature. Parachor (See section 4.4) is such a porperty.

19.3 STRUCTURAL INFORMATIONS

Study of the physical properties which give us information regarding molecular structure are (i) Parachor (ii) Molar refraction (iii) Optical rotation (iv) Molar-polarisation and dipole moment and (v) Molecular spectroscopy. Properties from (i) to (iv) help us to select the correct structure from a number of proposed structures, but molecular spectroscopy directly gives us the structure. It can be regarded as the finger-print of the molecules. Discussion about parachor has already been done in the chapter on liquids (section 4.4). Here we shall discuss only molar polarisation, –dipole moment and molecular spectroscopy.

19.4 MOLAR POLARISATION — CLAUSIUS — MASSOTTI EQUATION

We know that the force of attraction (F) between two parallel plate condensers in vacuum is $F = \frac{q_1 q_2}{r^2}$ where q_1 and q_2 are charges on the plates and r is the distance between them. If however some substance is present between the two plates then the attractive interaction between them is reduced by a factor say D i.e., $F = \frac{1}{D}\frac{q_1 q_2}{r^2}$.

This factor 'D' is called dielectric constant of the medium. The reason for this decrease is that the nuclei and electron cloud of the molecule are attracted by that plate whose charge is opposite to theirs. Thus the geometry of the molecule is distorted. We say that the molecule has been polarised. The polarisation (P_D) is said to be distortion or induced polarisation and is the summation of atomic (i.e., due to shifting of nuclei) and electronic (distortion of molecular orbital) polarisation i.e. $P_D = P_A + P_E$. Generally nuclei are much heavier and so their distortion is much less than that of electron cloud. P_A is generally only one percent of P_E so that

$$P_D = 1.01\ P_E \qquad .....(19.1)$$

Molar polarisation should therefore depend on the field strength (F) (reason behind distortion)

$$\text{So} \qquad P_D \propto F \text{ or } P_D = \alpha\ F. \qquad .....(19.2)$$

Here 'α' is a constant, known as polarisability of the medium. It is a measurement of the ease with which the medium can be polarised. When polarisation of a molecule takes place between two parallel plate condensers, its two ends with opposite charges will be attracted by the two plates and consequently the attractive force between twp plates will be reduced. The factor by which attraction is reduced is called $D.E.$ constant (D). So polarisation and $D.E.$ constant must be related. Clausius and Massotti deduced an equation (19.3) connecting molar polarisation and $D.E.$ constant. The equation is —

$$P_M = \frac{D-1}{D+2} \cdot \frac{M}{\rho}. \qquad .....(19.3)$$

where M is molar mass and ρ is density. Thus molar polarisation has the dimension of volume. Actually it is the total distortion of volume among molecules of one mole substance.

19.5 MOLAR POLARISATION AND TEMPERATURE - POLAR MOLECULE

Debye found that molar polarisation values of some molecules is independent of temperature but for some other molecules it is dependent on temperature or more appropriately it decreases with increase in temperature. Closer examination revealed that molecules for which $P_M \propto \frac{1}{T}$ are permanent dipole, so that they are polarised even in absence of any field. Such molecules when placed in an electric field will try to orient themselves in such a manner that positive end of the dipole will point towards negative plate and negative end towards positive plate. This gives an additional polarisation, known as orientation polarisation. This is in excess of the polarisation caused by the distortion of molecular geometry. Thus while for a non polar molecule $P_M = P_D$ alone, for a polar molecule $P_M = P_D + P_O$. Distortion polarisation is independent of temperature, since we can not change position of nuclei or electron cloud by simple heating the molecule. However heating of a gaseous substance brings randomness and chaotic motion; so orientation polarisation will decrease with increase in temperature i.e., $P_O \propto \frac{1}{T}$. Hence for non-polar molecules polarisation does not depend on temperature – but for polar molecules it decreases with increase in temperature.

19.6 BOND MOMENT AND DIPOLE MOMENT

In case of a heteronuclear bond (i.e., bond between two dissimilar atoms) the electron cloud of the molecular orbital surrounding the nuclei will be shifted towards the more electronegative atom. As a result of this the centre of gravity of negative charges will not coincide with that of nuclear charges (+ve). The bond will therefore act as a tiny bar magnet, moment of which will be called bond moment. In a molecule a number of heteronuclear links may be present; each will have such bond moment. These are vector quantities. Vector summation of all such bond moments is called dipole moment (μ) of the molecule. Thus dipole moment (μ) = Σ bond moments.

Generally bond moment = $e \times l$ where 'l' is the distance between centre of gravities of +ve or −ve charge. Bond moment can also be written as = $\partial e \times l_1$ here l_1 = bond length and ∂e is the small amount of negative (or positive) charge that accumulates on the more electronegative(or less electronegative) atom. Since in e.s.u. unit electronic charge is of 10^{-10} order and distance is of 10^{-8}cm. order a factor 10^{-18} comes in the e.s.u. magnitude of dipole moment. This 10^{-18} e.s.u. is called 1 debye. So

$$1 \text{ debye} = 10^{-18} \text{ e.s.u.}$$

In the following table, some bond moments are given.

Table 19.1

Bond	Moment (Debye)	Bond	Moment (Debye)
H-F	1.9	C-F	1.4
H-Cl	1.1	C-Cl	1.5
H-Br	0.8	C-Br	1.4
H-I	0.4	C-I	1.2
H-O	1.5	C-O	0.7
H-N	1.3	C-O	2.3
H-C	0.4	C-N	0.2

19.7 APPLICATIONS OF DIPOLE MOMENT MEASUREMENT

Knowledge of dipole moment of molecules helps us in getting following informations —

(a) Percentage of ionic character : Dipole moment of H-Cl is 1.07 debye. Its bond length = 1.275 Å = 1.275×10^{-8} cm. Had it been 100% ionic, charges on H-atom or Cl-atom [$H^+ Cl^-$] would have been one unit electronic charge i.e., 4.8×10^{-10} e.s.u. so dipole moment should be $4.8 \times 10^{-10} \times 1.275 \times 10^{-8} = 6.12$ debye. But it is 1.07 debye. So percentage of ionic character of H–Cl bond is $\frac{1.07}{6.12} \times 100 = 17\%$. Hence dipole moment helps us to know ionic character of bond.

(b) Bond angle calculation : Bond moment of O – H bond is 1.5 debye and dipole moment of H_2O is 1.84 debye. From these two data $\widehat{H–O–H}$ bond angle in water can be calculated. If O – H bond moment is μ_{O-H} and bond angle is 'α' then from principle of vector addition, dipole moment (μ) for water is —

$$\mu = \left[\mu_{O-H}^2 + \mu_{O-H}^2 + 2\mu_{OH} \cdot \mu_{O-H} \cos \alpha\right]^{1/2}$$

or, $$1.84 = [2(1.5)^2 + 2(1.5)^2 \cos \alpha]$$

Solving $\alpha = 104°24'7''$.

Thus dipole moment gives us knowledge about bond angle.

(c) Molecular Geometry : Molecular geometry i.e., structures of molecules can be established from dipole moment study. Some examples are given.

(i) Molecule *AB* :

Such a molecule is always polar and will have dipole moment; its structure is linear.

(ii) Molecule AB_2 :

If a molecule AB_2 is non-polar then it must have a linear structure B – A – B (e.g. CO_2 ; O = C = O) because only then vector addition of two bond moments is zero. But if it is polar, then the structure must be angular e.g. H_2O [H–O–H] so that vector summation of two bond moments has some value.

(iii) Molecule AB_3 :

If a molecule AB_3 is non-polar then it will have a triangular structure, because only then vector summation of bond moments will be zero. BF_3 has this structure. But if it is polar, it may have pyramidal structure.

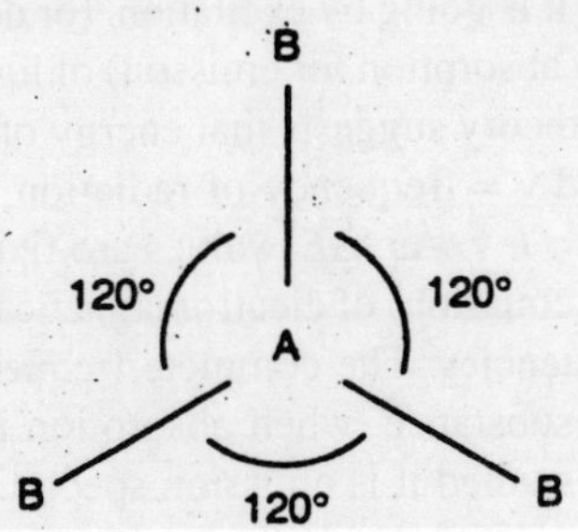

Fig. 19.1 AB_3 non-polar.

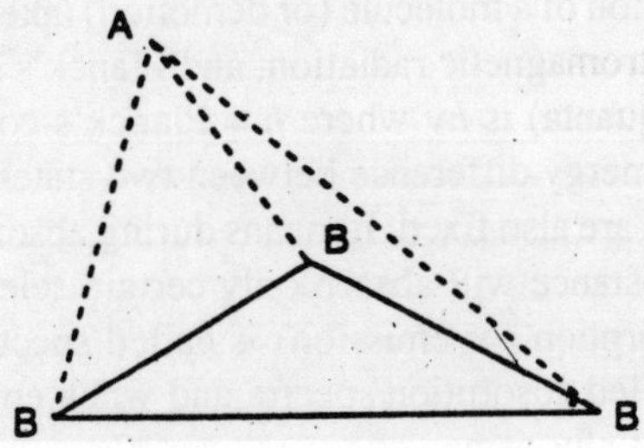

Fig. 19.2 AB_3 Polar.

(iv) Molecule AB_4 :

For non polar AB_4 molecules possible structures are tetrahedral or square plannar but for polar structure is square pyramid.

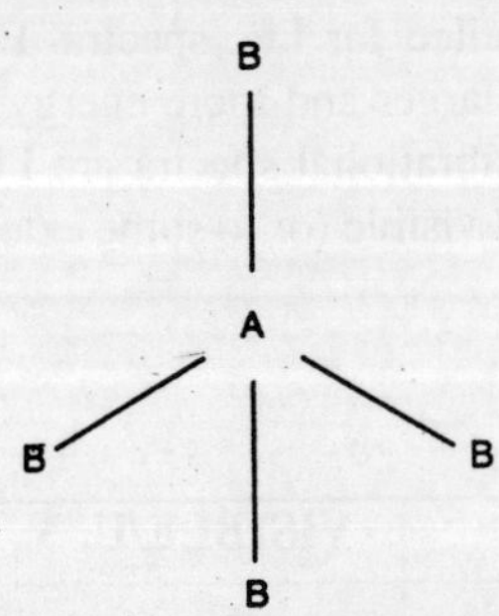

Fig. 19.3 Non-polar AB_4.

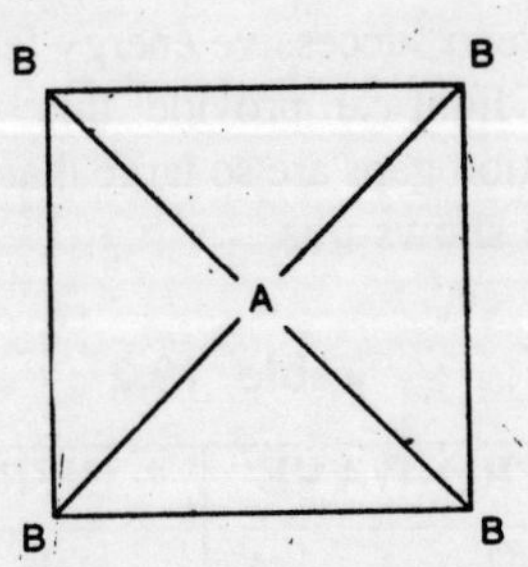

Fig. 19.4 Non-polar AB_4.

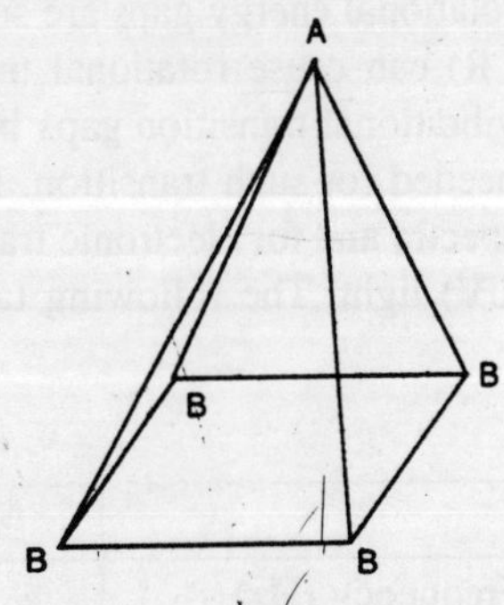

Fig. 19.5 Polar AB_4.

(d) Some more informations regarding geometry : Though C – H bond moment is 0.4 debye, dipole moment of benzene is zero because of its symmetrical structure which makes vector summation zero. But mono substituted benzene compounds will have dipole moment. For para disubstituted benzene compounds dipole moment is generally zero because of cancellation of two bond moments. However, for p-dihydroxy benzene and p-diamino benzene, dipole moment is not zero, probably because the bond moments do not act on the plane of the ring.

Anyway the conclusion that is drawn is - Dipole moment is one of the important tools for elucidation of structures of molecules.

MOLECULAR SPECTROSCOPY

19.8 SPECTRA-ABSORPTION EMISSION

Every substance possesses certain amount of energy inside it. This may be due to (i) translational motion (in case of gases and to some extent liquids) (ii) rotational motion of the nuclei (if the substance is polyatomic molecule) (iii) vibrational motion of the nuclei (again for polyatomic molecule) and (iv) electronic arrangement. Modern quantum theory tells that all these energies are quantised, which means that a substance can not get any energy value (translational or rotational or vibrational or electronic) it likes. It has only certain specified energy values allowed for it. When the molecule is excited or (demoted from excited state) it will absorb (or emit) only that energy which is the difference between energy values of the present state and the state where it is going by excitation (or demotion). Excitation of a molecule (or demotion) takes place by absorption (or emisson) of light quanta of electromagnetic radiation, and Planck's quantum theory suggests that energy of a photon (light quanta) is $h\nu$ where h = Planck's constant and ν = frequency of radiation. So if ΔE is the energy difference between two states then $\Delta E = h\nu$. As 'ΔE' values are fixed, so 'ν' --values are also fixed. It means during absorption (or emission) of electromagnetic radiation, the substance will absorb only certain selected frequencies. The complete frequency chart of absorption (or emission) is called spectra of the substance. when absorption is studied it is called absorption spectra and when emission is studied it is emission spectra. The two are virtually same.

19.9 DIFFERENT TYPES OF SPECTRA ATOMIC; MOLECULAR

We have just now discussed that there are some selected energy levels for a substance. In case of translational energy gaps between successive energy levels are so small that the change appears to be continuous and that is why spectral studies can not be made. For rotational energy gaps are somewhat more and the energy content of microwave - light (far I.R) can cause rotational transition. So rotational spectra are called far I.R. spectra. For vibrational transition gaps between successive energy levels are larger and more energy is needed for such transition. I.R. light can provide this — thus vibrational spectra are I.R. spectra and for electronic transition gaps are so large that we need visible (or to some extent UV) light. The following table shows this.

Table 19.2

	MICRO-WAVE	INFRA-RED	VISIBLE/U. V.
Frequency (Hz)	$3\times10^{12} \leftrightarrow 3\times10^{14}$	$3\times10^{14} \leftrightarrow 3\times10^{16}$	$3\times10^{16} \leftrightarrow 3\times10^{18}$
Wave length (A^0)	$10^8 \leftrightarrow 10^6$	$10^6 \leftrightarrow 10^4$	$10^4 \leftrightarrow 10^2$
Energy Joules/mole	$10 \leftrightarrow 10^3$	$10^3 \leftrightarrow 10^5$	$10^5 \leftrightarrow 10^7$

Now in case of an atom there are no rotational or vibrational motion so only electronic transition can occur and we get some selected lines in the spectral chart showing the frequencies of absorption. But a polyatomic molecule has all these (rotational vibrational and electronic) energies. Thus in the spectral chart of molecule we get a thick line, which is actually a cluster of so many fine lines, indicating rotational and vibrational transition. Thus molecular spectra is band spectra.

19.10 ROTATIONAL SPECTRA

Let us consider a diatomic molecule AB consisting of dissimilar atoms (heteronuclear). It will have some dipole moment.

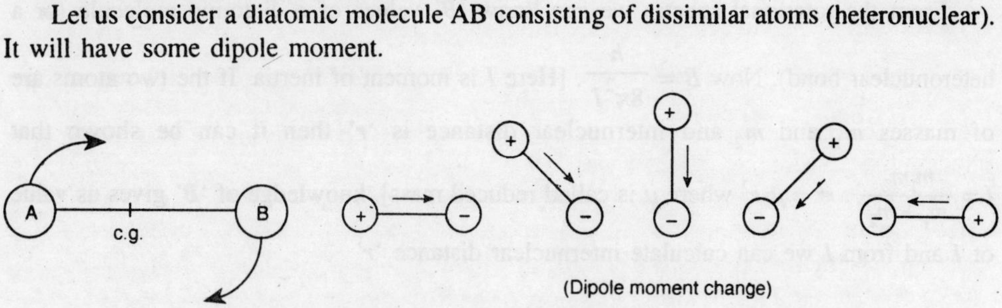

Fig. 19.6 Change of dipole moment during rotation.

The two nuclei A and B will rotate around their centre of gravity (c.g.). During rotation direction of dipole moment will continuously change with time i.e. the molecule will have some oscillating electrical moment inside it. It can then interact with another oscillating electrical field e.g. light. Such an interaction may lead to rotational energy increase when they are in phase i.e. when the light is absorbed. Thus a permanent dipole can show rotational spectra. From quantum mechanics it can be shown that allowed rotational energy levels are —

$$E = \frac{h^2}{8\pi^2 I} J(J+1) = BhJ(J+1)$$

where B is a constant and is $\frac{h}{8\pi^2 I}$ [I = Moment of inertia of the molecule].

J is an integer having values 0, 1, 2, 3, etc. J is called rotational quantum number. Quantum mechanical calculations allow only those rotational transitions for which $\Delta J = \pm 1$. So a molecule at J^{th} rotational level can go to $(J+1)^{th}$ rotational level. In J^{th} level its energy is

$$E_J = BhJ(J+1)$$

and in $(J+1)^{th}$ level $E_{J+1} = Bh(J+1)(J+2)$.

The energy difference is $\Delta E = 2Bh(J+1)$. If this energy is supplied by microwave radiation of frequency ν then —

$$h\nu = 2Bh(J+1)$$

or $$\nu = 2B(J+1). \qquad .....(19.5)$$

This is the frequency of absorption. So —

for transition from $J = 0$ to $J = 1$ frequency absorbed $\nu_0 = 2B$,

" " " $J = 1$ to $J = 2$ " " $\nu_1 = 4B$,

" " " $J = 2$ to $J = 3$ " " $\nu_2 = 6B$.

So the molecule will absorb $2B$, $4B$, $6B$, $8B$ etc. frequency. If a graph in drawn showing absorbance (see chapter - 20) vs frequency, then the graph will show peak points at $2B$, $4B$, $6B$ etc. frequencies. This graph will be called rotational spectra of the molecule. We say that rotational lines are equi-spaced, spacing between successive lines is $2B$.

19.11 APPLICATIONS OF ROTATIONAL SPECTRA

From the rotational spectra we can know 'B' values of a diatomic molecule (or a heteronuclear bond). Now $B = \frac{h}{8\pi^2 I}$. [Here I is moment of inertia. If the two atoms are of masses m_1 and m_2 and internuclear distance is 'r' then it can be shown that $I = \frac{m_1 m_2}{m_1 + m_2} \cdot r^2 = \mu\, r^2$ where μ is called reduced mass]. knowledge of 'B' gives us value of I and from I we can calculate internuclear distance 'r'

19.12 VIBRATIONAL SPECTRA

Let us consider a heteronuclear diatomic molecule AB. The bond between A and B can be regarded as a spring and the two atoms as two rigid spheres attached to the two ends of the spring. If now the two spheres are pressed and the pressure subsequently released they will oscillate and show a to and fro motion. In the absence of any resistance, this vibration will go on continuing. If the molecule is heteronuclear then it has some dipole moment. But as bond length is continuously changing due to eternal vibration so dipole moment is also oscillating. This oscillating electrical moment can interact with such oscillating electrical moment arising due to propagation of light. Thus the molecule can absorb light of selected frequencies. This absorption chart is vibrational spectra. Since here frequencies absorbed fall in the infra-red range, the spectra is also called I. R. spectra. Some vibrations are shown :

Table 19.3

Molecule	Vibration	Comment
AB		Dipole moment changes during vibration; I. R active.
AB_2	Symmetrical - Stretching Vibration	Dipole moment does not change. Vector sum is always zero; I. R inactive.
AB_2	Anti-symmetrical - Stretching Vibration	Dipole moment changes; I. R. active.
AB_2	Bending vibration	Dipole moment changes; I. R. active.

19.13 POTENTIAL ENERGY DIAGRAM

We have just learnt about vibrations of nuclei in a molecule. Now this vibration is to some extent comparable with oscillation of a pendulum. But we know that during oscillation potential energy (P.E.) of the pendulum changes with amplitude. It is highest at the two extreme positions and lowest at the mean position. Similar will be the case of P.E. of a diatomic molecule; during vibration internuclear distance continuously changes with time and so P.E. will also change. It will be highest when the atoms are either at shortest or largest distance. But it will be minimum at some distance, which is generally regarded as bond length of the molecule. Variation of P.E. with internuclear distance is shown by the following diagram known as P.E. diagram.

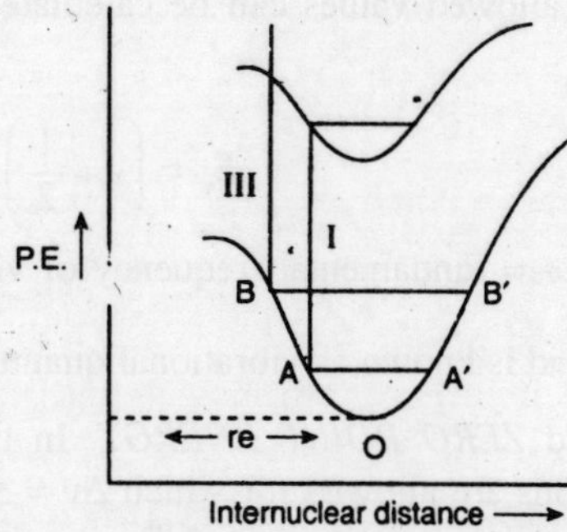

Fig. 19.7 P. E. diagram.

In the above figure let AA′ is the vibrational state of a diatomic molecule in the ground electronic state. It means during vibration the minimum and maximum internuclear distances correspond to the points A and A′. So this distance continuously changes from A to A′ in a simple harmonic motion. With this variation of internuclear distance P. E. of the molecule changes along the line AOA′. At O (internuclear distance v_e) P. E. is minimum. This is called bond length. Similar will be the case when the molecule is promoted to a higher vibrational level BB′. Here the internuclear distance changes from B to B′ while P. E. changes along the path BOB′. In the given figure (19.7) the lower curve shows vibrational states when the molecule is in the ground electronic state and the upper curve shows vibrational state in the excited electronic state. Now when a molecule absorbs visible or UV light its electronic state is promoted. Subsequently vibrational state will be promoted. Since this promotion takes some time (may be very small) and since atoms are continuously oscillating, so a question, which may come in our mind is that whether internuclear distance will remain the same just after transition i.e whether transition will be shown by a vertical line or inclined line ? According to Franck and Condon, since nuclei are much heavier than electron cloud, their movement will be sluggish and during the span of rearranging electron cloud i.e. electronic transition, internuclear distance will be almost same. So transition will be shown by vertical line. This is Franck-Condon principle, In the chapter of photochemistry we shall discuss in detail the results of such promotion.

19.14 VIBRATIONAL SPECTRA

When a substance executes a simple harmonic periodic motion, then according to Hooke's law the restoring force (F) which tries to bring it back, is proportional to its displacement (x). So $F = kx$, 'k' is a constant known as force constant. For a diatomic molecule where atoms execute such sort of oscillation, restoring force constant is a measurement of bond strength. Also for an oscillator the fundamental frequency of vibration (ω) is

$$\omega = \frac{1}{2\pi}\sqrt{\frac{k}{m}}.$$

where 'm' is mass of the oscillator. For a diatomic molecule behaving as a simple harmonic oscillator 'm' is reduced mass (μ)

$$\text{or } m = \mu = \frac{m_1 m_2}{m_1 + m_2}.$$

where m_1 and m_2 are massess of individual atoms.

Now like other forms of energies, vibrational energy of a molecule is also quantised and its allowed values can be calculated using quantum-mechanics. The allowed values are —

$$E_v = \left(v + \frac{1}{2}\right) h\omega, \quad .....(19.6)$$

where ω = fundamental frequency of vibration and v is an integer having values 0, 1, 2, 3, etc and is known as vibrational quantum number. When $v = 0$; $E_v = \frac{1}{2} h\omega$. This energy is called *ZERO POINT ENERGY*. In this case also there is selection rule. Only those transitions are allowed for which $\Delta v = \pm 1$. So if a molecule is in the vth vibrational state it can be promoted to $(v + 1)^{th}$ vibrational state. Corresponding energy changes can be calculated.

$$E_v = \left(v + \frac{1}{2}\right) h\omega,$$

$$E_{v+1} = \left(v + \frac{3}{2}\right) h\omega,$$

$$\Delta E = (E_{v+1} - E_v) = h\omega.$$

If this energy is supplied by light of frequency ν , then $h\nu = h\omega$ or $\nu = \omega$. Hence frequency absorbed will be equal to fundamental frequency of vibration of the bond. Now each and every bond has its own characteristic fundamental frequency of vibration. So from these frequencies nature of bond present can be known. The following table gives fundamental frequencies of vibrations of some common bonds.

Table 19.4

Bond	Mode	Wave No. (cm^{-1})
C-H	Stretch	2700—3300
C-H	Bend in phase	1300—1500
C-H	Bend out phase	800—830
O-H	Stretch	3000—3700
O-H	Bending	1200—1500
N-H	Stretch	3000—3700

19.15 VIBRATION-ROTATION SPECTRA

In an assembly of molecules different molecules are present in different energy state. Let we find some molecules in the ground vibrational ($v = 0$) and ground rotational ($J = 0$) state. If it absorbs ω frequency its vibrational state will be promoted and we get pure vibrational spectra. But it absorbs $(\omega + 2B)$ frequency its both vibrational and rotational states will be promoted and we shall get vibration-rotation spectra. Some cases are shown in the next page :

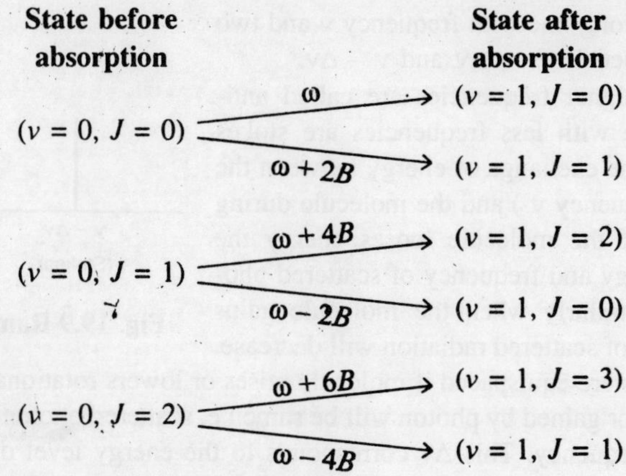

So the molecule will absorb ω ; ω + 2*B*; ω + 4*B*; ω + 6*B*; etc and ω – 2*B*; ω – 4*B*; ω – 6*B*........etc. frequencies. Now 'ω' is in the I. R. range; but '*B*' is in the far I.R range so ω, ω + 2*B*; ω + 4*B*;.........; ω – 2*B*; ω – 4*B*.......etc. are very close. They will appear as a cluster of lines and we shall call it band - spectra. The expected curves will be as follows —

The missing line at the middle is for pure vibrational transition corresponding to absorbance of ω frequency (Q - line). On its right are ω + 2*B*; ω + 4*B*; ω + 6*B* etc. (R-lines) and on the left are ω – 2*B*; ω – 4*B*; ω – 6*B* etc. (P – lines). This is vibration-rotation spectra.

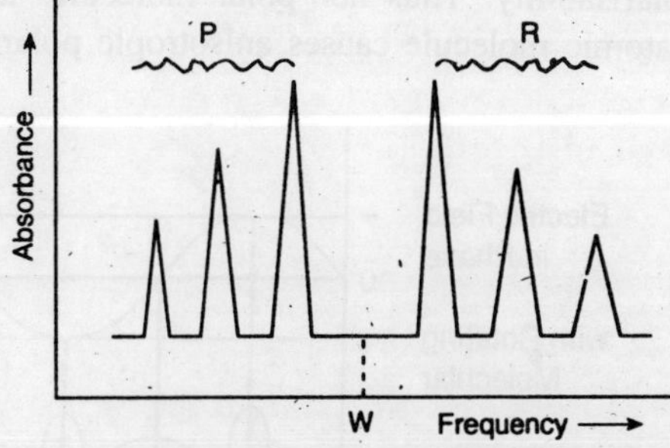

Fig. 19.8 Vibration-rotation spectra for HCl.

19.16 APPLICATIONS

From the study of vibrational spectra we can know fundamental frequency of vibration (ω). This helps us to identify the nature of bond present. Also $\omega = \frac{1}{2\pi}\sqrt{\frac{k}{\mu}}$. Thus force constant '*k*' can be known and *k* is a measurement of bond strength.

19.17 RAMAN SPECTRA

When light falls on a substance then it causes electronic polarisation within the molecule. The polarised electron cloud sends back a portion of the light from which it was coming. The phenomena is known as scattering of light. Rayleigh observed that frequency of scattered radiation is the same as that incident radiation. Such elastic scattering is called Rayleigh scattering. But Raman found that during scattering while most of the incident photons are scattered with the incident frequency (Rayligh scattering) a few may be scattered with different frequencies. Such inelastic scattering of electromagnetic radiation, out side the x-ray region is called Raman scattering; Raman spectra or simply Raman effect.

19.18 STOKES AND ANTISTOKES LINES

Let a monochromatic radiation of frequency ν is incident on a substance capable of Raman scattering. If the scattered radiation is passed through a grating (prism like) we shall get three lines; one strong line with frequency ν and two weak-lines with frequencies ν + Δν and ν − Δν.

The lines with higher frequencies are called anti-stokes line and those with less frequencies are stokes lines. This is due to the exchange of energy between the incident photon (frequency ν) and the molecule during in-elastic collision. If the molecule looses energy the photon will gain energy and frequency of scattered photon will increase. Similarly when the molecule gains energy the frequency of scattered radiation will decrease.

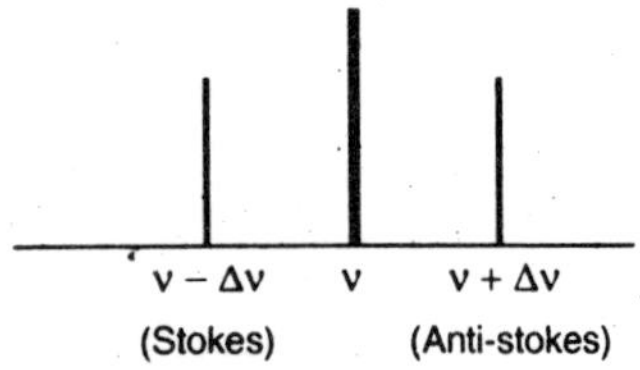

Fig. 19.9 Raman spectra.

Since rotational levels are equispaced if molecule raises or lowers rotational energy during collision, energy lost or gained by photon will be same i.e. scattered photon will have either ν − Δν or ν + Δν frequency. This Δν corresponds to the energy level difference of the molecule and thus it is the characteristic of the molecule.

19.19 APPLICATIONS

The condition that a molecule will be Raman-active is that it must have anisotropic polarisability. Thus non-polar molecules are Raman-active. Rotation of a homonuclear diatomic molecule causes anisotropic polarisability.

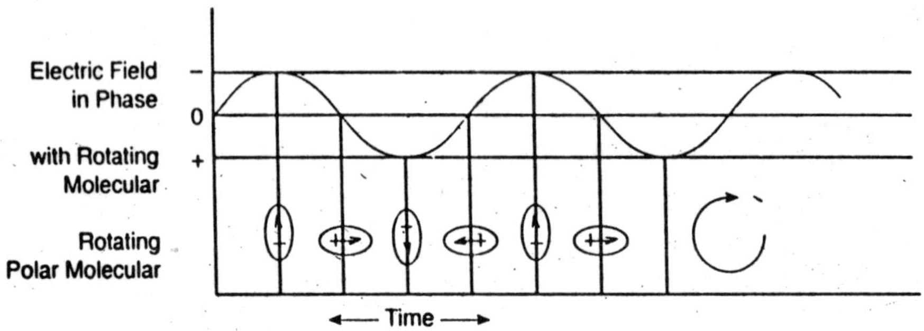

Fig. 19.10

Thus molecules, which are I.R. inactive due to absence of polar nature are Raman-active. That is why Raman spectra and I. R. spectra are said to be complementary to each other.

POINTS TO REMEMBER

Structure of molecules can be known from study of both physical and chemical properties. Study of physical properties involve molar polarisation and dipole moment measurement and molecular spectra analysis. Any heteronuclear link will act as a tiny bar magnet due to non-coincidence of c.g.'s of negative charge cloud and positive nuclei. Moment of this bar magnet is link moment and vector summation of all link moments in a molecule is dipole moment. Its measurement helps us to know percentage of ionic character; bond angle and correct structure. While dipole moment selects the correct structure among a number of proposed structures, molecular spectra gives us the actual structure. Rotational spectra tells about moment of inertia, bond length and presence of isotope. Vibrational spectra (I. R. spectra) identifies the nature of bonds and calculates bond strength. They are found in case of molecules having permanent dipole moments. For non-polar molecules we use Raman spectra.

QUESTIONS

SHORT TYPE

1. What is polarisation ? What is the unit of molar polarisation ?
2. Define link moment and group moment.
3. Why dipole moment of CO_2 is zero ?
4. What structural information can be obtained from O-H link moment and dipole moment of water.
5. 'Distortion polarisation does not depend on temperature but orientation polarisation depends' – explain.
6. Whose polarisation will change with temperature and why — H_2 ; Cl_2 ; HCl ?
7. Why atomic spectra are line spectra ?
8. How many types of energy a molecule can possess ?
9. Which of the following molecules is I.R. active ? H_2 ; HCl ; Cl_2.
10. Will H_2 show Raman effect and I.R. spectra ?

ESSAY TYPE

1. State and explain molar polarisation. Explain why this value for some molecules is independent of temperature but for some other is dependent on temperature.
2. Explain bond moment and dipole moment; show with some examples, how dipole moment values help to elucidate structure of molecules.
3. What are additive and constitutive properties? Name the most important constitutive property and show how it helps us in getting structural information.
4. Assuming that rotational energy of a diatomic molecule $E_J = BhJ\,(J+1)$ in the Jth level, show that rotational lines are equi-spaced. Find out the spacing and explain its a· plication.
5. What are r ıre vibrational and vibrational-rotational spectra? Explain their applications.
6. Distinguish between Rayleigh scattering and Raman scattering. What are the characteristics of Raman spectra?

PROBLEMS

1. The dielectric constant of CH_4 (g) at 0°C and 1atm. pressure is 1.00094. Assuming that methane behaves as an ideal gas, calculate (a) the induced polarisation and (b) the polarisability of the molecule.

   [*Hints* : $P_i = \frac{4}{3}\pi N\alpha = \frac{D-1}{D+2}\cdot\frac{M}{\rho}$ α = Polarisability].

2. A compound AB having partly ionic partly co-valent character has the dipole moment 1.03 debye. Show how the per cent ionic character may be calculated on the basis of charge (4.8×10^{-10} e.s.u.) separated by equilibrium bond length 1.3 Å.

   [*Hints* : % of ionic character = $\frac{1.03 \times 10^{-18}}{4.8 \times 10^{-10} \times 1.3 \times 10^{-8}} \times 100$].

3. The dipole moment of HCl is 1.03 debye and its bond length is 1.27 Å. Calculate (i) the charge on the constituent atoms and (ii) the percentage of ionic character.

[*Hints* : (i) If charge = δe then $\delta e \times 1.27 \times 10^{-8} = 1.03 \times 10^{-18}$
(ii) Same as before].

4. The moment of inertia of NH radical is 1.68×10^{-40} gm cm^2. At what frequency in the microwave region you expect to detect transition from J = 2 to J = 3.

[*Hints* : $J = 2$ to $J = 3$; $\Delta E = 6Bh$; $\upsilon = 6B = \frac{6h}{8\pi^2 I}$].

5. In the near I.R. spectrum of CO there is an intense bond at 2144 cm^{-1}. Calculate (a) the fundamental vibration frequency of CO (b) the period of vibration (c) the force constant (d) the zero point energy in cal/mole.

[*Hints* : (a) $\upsilon = c\bar{\upsilon} = 3 \times 10^{10} \times 2144$ sec^{-1}, (b) $t = \frac{1}{\upsilon}$,

(c) $\upsilon = \frac{1}{2\pi}\sqrt{\frac{k}{\mu}}$, $\mu = \frac{12 \times 16}{12 + 16} \times 1.66 \times 10^{-24}$, (d) Zero point energy $= \frac{1}{2} h\upsilon \times N$].

OBJECTIVE TYPE

1. Write √ mark over the correct word.

(a) Volume is (additive/constitutive) property.
(b) For non-polar molecules molar polarisation is (dependent/independent) of temperature.
(c) A molecule AB_2 having zero dipole moment is (linear/ angular).
(d) Molecular spectra is shown by (polar/non-polar) molecules.
(e) Raman spectra requires (isotropic/anisotropic) polarisability.

2. Write T against true and F against false statements.

(a) Link moments are vector quantities.
(b) H_2O is polar so it will be linear.
(c) p-dichlorobenzene has zero dipole moment.
(d) o-dichlorobenzene has no dipole moment.
(e) Rotational spectra is shown by I. R. light.
(f) Atomic spectra is line spectra.
(g) Raman spectra and I. R. spectra are complimentary.

3. Match terms of column A with those of column B.

A	B
(a) Clausius-Massotti.	(a) Anisotropic polarisability.
(b) Franck-Condon.	(b) Unit for dipole moment.
(c) Debye.	(c) Elastic scattering.
(d) Rayleigh.	(d) Molar polarisation.
(e) Raman spectra.	(e) P. E. diagram.

Answers :

1. (a) additive (b) independent (c) linear (d) polar (e) anisotropic.

2. (a) T (b) F (c) T (d) F (e) F (f) T (g) T.

3. a – d; b – e; c – b; d – c; e – a.

Chapter 20

PHOTOCHEMISTRY

20.1 INTRODUCTION — PHOTOCHEMICAL REACTION

During study of chemical kinetics, we have learnt that occurrence of a chemical reaction requires activation energy. Sometimes light energy can supply the necessary activation energy and can initiate a reaction. Such reactions will be termed as photochemical reaction. A very common example is the combination between H_2 and Cl_2. They do not combine in the dark, but when exposed to light they combine with explosion. The chapter of photochemistry discusses different aspects of such reactions i.e. reactions initiated by photons (light particles).

20.2 THERMAL AND PHOTOCHEMICAL REACTIONS — A COMPARISON

Reactions where activation energy comes from molecular collisions (thermal in nature) are called dark reaction or thermal reaction. But reactions where light supplies activation energy are photochemical. Generally visible and U. V. radiation carry necessary energy to break a chemical bond and thus these lights can cause photochemical reaction.

Rate of thermal reaction increases according to Arrhenius equation (vide chapter of chem. kinetics). However, rate of photochemical reactions do not depend on nature - but depends on light intensity.

The course of a thermal reaction is guided by thermodynamics and thus all thermal reactions are associated with a decrease in free energy. But photochemical equations follow quantum theory and such reactions may be associated with an increase in free energy — example is photosynthesis.

20.3 SOME DEFINITIONS

Let us acquiant ourselves with some terms often used in photochemistry.

The term 'intensity' of light at some place means number of photons striking unit area at that place in unit time. No./area/sec is the unit. If frequency of photon is known then energy received by the unit area per second can be calculated. And so sometimes intensity is expressed as ergs/area/sec.

If frequency of a photon is υ , then its energy $E = h\upsilon$. Putting $\frac{c}{\lambda}$; c is velocity of light; λ is wave length we get

$$E = \frac{hc}{\lambda}.$$

The total energy of N (N = Avogadro number) photons is called one einstein. Thus energy of one einstein is

$$E = \frac{Nhc}{\lambda}.$$

In c. g. s. unit $$E = \frac{6.023\times10^{23}\times6.624\times10^{-27}\times3\times10^{10}}{\lambda} \text{ ergs/mole}$$

F-20

$$\text{or} \qquad E = \frac{1.196 \times 10^8}{\lambda} \text{ ergs/mole.}$$

Here wave length of light λ is in c. m. unit. If it is expressed as A° unit then as 1 A° ≡ 10^{-8} cm.

$$E = \frac{1.1696 \times 10^{16}}{\lambda} \text{ ergs/mole.}$$

The smaller the wave length the greater is the energy of einstein. Thus one blue einstein carries more energy than a red einstein.

20.4 LAWS OF ABSORPTION-LAMBERT'S LAW; BEER'S LAW

Since photochemical reactions are carried out by absorption of light, it is first necessary to know about the laws of absorption. There are two such laws, which are (i) Lambert's law and (ii) Beer's law.

When light falls on a substance then a part of it may be absorbed and a part is transmitted. Naturally intensity of the transmitted light is less than that of incident light. Lambert's law speaks about this decrease in intensity. **The law states that when light is transmitted through a substance then decrease in intensity is directly proportional to the thickness of the medium and also the intensity of incident radiation.** Thus if I is the intensity of radiation incident on a substance of thickness dl, then decrease in intensity $(-dI)$ of the transmitted light will be —

$$-dI \propto I \cdot dl$$

$$\text{or} \qquad -dI = KI \cdot dl.$$

Here K is a constant.

$$\text{or} \qquad -\frac{dI}{I} = K\, dl.$$

If the substance is of thickness 'l' and I_0 and I are intensities of incident and transmitted radiation, then on integration —

$$\int_{I_0}^{I} -\frac{dI}{I} = K \int_0^l dl$$

$$\text{or} \qquad \ln \frac{I_0}{I} = K \cdot l$$

$$\text{or} \qquad \log \frac{I_0}{I} = \frac{K}{2.303} \cdot l = \varepsilon l. \qquad \text{.....(20.1)}$$

Here $\varepsilon = \frac{K}{2.303}$ is a new constant, known as extinction co-efficient. The term $\log \frac{I_0}{I}$ is called absorbance or optical density (abbreviated as *O. D.*) of the medium. 'ε' is a measurement of absorbing capacity i.e. extinguishing capacity and hence the name extinction co-efficient has been chosen.

If the absorbing substance is a solution (or gas) then extent of absorption will depend on concentration of solute (or pressure of gas as $P = CRT$) also. For such cases **Beer extended Lambert's law as Decrease in intensity of light during transmisson through a medium is directly proportional to thickness of the medium, intensity of incident light and molar concentration of solution.** Thus here —

$$-dI \propto I.\ dl.\ C$$

or $$-dI = K.\ I.\ dl.\ C.$$

Proceeding as before we shall get —

$$\text{O. D.} = \log \frac{I_0}{I} = \varepsilon\ C.\ l. \qquad .....(20.2)$$

20.5 MORE ABOUT EXTINCTION CO-EFFICIENT

We have already stated that extinction co-efficient measures absorbing capacity (or extinguishing capacity) of the substance. Since any substance can not absorb any frequency of light equally, so extinction co-efficient depends not only on nature of substance but also on frequency of light employed. Extinction co-efficients of a substance at various frequencies can be measured by an instrument, known as spectrophotometer and if we plot ε-values against frequency then the curve that we obtain is the 'spectra' of the substance. 'ε'-values have several utilities. Since these are highly sensitive towards impurity, therefore purity of a sample can be tested by measuring its ε-value at some selected frequency. Since O.D. values are additive in nature, we can employ ε-values for accurate quantitative estimation. To illustrate let we have a solution containing two substances A and B at concentration C_A and C_B. We are to evaluate C_A and C_B. Now O. D. of solution = O. D. due to A + O. D. due to B. Let d_1 and d_2 are experimentally measured O.D. values of the substance at two different frequencies ν_2 and ν_2. Then if 'l' is the length of the cell in which solution was taken during O. D. measurement and if $(\varepsilon_{A_1}\ ;\ \varepsilon_{A_2})$ and $(\varepsilon_{B_1}\ ;\ \varepsilon_{B_2})$ are extinction co-efficients of A and B at frequencies ν_1 and ν_2 respectively then

at ν_1 $$d_1 = \varepsilon_{A_1}.C_A.l + \varepsilon_{B_1}.C_B.l$$

and at ν_2 $$d_2 = \varepsilon_{A_2}.C_A.l + \varepsilon_{B_2}.C_B.l$$

All terms of above equations excepting C_A and C_B are known. Thus C_A and C_B can be calculated. The result is highly accurate.

20.6 LAWS OF PHOTOCHEMISTRY

There are two laws of photochemistry. These are (a) Grothus -Draper law and (b) Stark -Einstein's law for photochemical equivalence.

We have seen in section 20.4 that when light falls on a substance a part is absorbed and a part is transmitted. Now if some photochemical reaction occurs, which part of the light is responsible for this ? Grothus and Draper observed that **'it is the absorbed light which can cause chemical reaction, the transmitted light has no role to play.'** This observation is called Grothus-Draper law. The law though simple and self-evident is purely qualitative in nature. It does not give any idea about extent of chemical reaction that may occur. For this purpose we need another law and that law was given by Stark and Einstein.

According to Stark and Einstein - **'In the primary step of a photochemical reaction, one photon can activate only one molecule.'** In photochemical reaction, the primary step is the step where light quanta i.e. photon is absorbed. This step is followed by secondary steps, which are generally thermal in nature. Stark-Einstein's law is regarding this first step.

20.7 QUANTUM YIELD - MISUNDERSTANDING OF EINSTEIN'S LAW

Einstein's law predicts that one photon activates only one molecule. This prediction is misunderstood sometimes as — 'if 'n' photons are absorbed then 'n' molecules will react.'

Thus the ratio —

$$\frac{\text{No. of molecules reacted}}{\text{No. of photons absorbed}} = \text{let } \phi,$$

was expected to be unity always. 'ϕ' is called **quantum efficiency** or **quantum yield.** In actual practice we find reaction with $\phi = 1$ [e. g. $SO_2 + Cl_2 \xrightarrow{h\upsilon} SO_2Cl_2$] ; with $\phi > 1$ [e. g. $H_2 + Cl_2 \xrightarrow{h\upsilon} 2HCl$] and with $\phi < 1$ [e. g. $H_2 + Br_2 \xrightarrow{h\upsilon} 2HBr$]. Reactions with high quantum yield ($\phi > 1$) or low quantum yield ($\phi < 1$) are not violations of Einstein's law. Einstein's law is for the primary process i.e. absorption of light process and there it is very much correct. But because of secondary processes total number of molecules reacted may not always be equal to total number of products. Following examples will clarify.

20.8 EXAMPLES OF HIGH AND LAW QUANTUM YIELD

Photochemical reaction between H_2 and Cl_2 is believed to follow following mechanism —

1. $Cl_2 \xrightarrow{h\upsilon} Cl + Cl$
2. $Cl + H_2 \longrightarrow HCl + H$ (exothermic)
3. $H + Cl_2 \longrightarrow HCl + Cl$
4. $Cl + Cl + X \longrightarrow Cl_2 + X$.

In the primary step one photon decomposes only one chlorine molecule into two chlorine atoms. These atoms may start a chain reaction via path '2' or terminate the chain by recombination (path -4). Since path '2' is exothermic its probability of occurrence at ordinary temperature is more. Path '4' is a three body collision process; its probability is less. Thus number of Cl_2 molecules reacted becomes much greater than number of photons absorbed and ϕ becomes > 1

On the contrary the reaction between H_2 and Br_2 proceeds as —

1. $Br_2 \xrightarrow{h\upsilon} Br + Br$
2. $Br + H_2 \longrightarrow HBr + H$ (endothermic)
3. $H + Br_2 \longrightarrow HBr + Br$
4. $H + HBr \longrightarrow H_2 + Br$
5. $Br + Br \longrightarrow Br_2$.

Here step '2' is endothermic, so its probability of occurrence at room temperature is less. Step '5' is a two - body collision, so its probability is more. So number of Br_2 molecules reacted with H_2 is less than number of Br_2 molecule decomposed by light. Thus here quantum yield is less than one.

20.9 ONE EXAMPLE

The decomposition of HI takes place by the absorption of light of wave lengths 2070 A° ; 2530 A°; 2820 A° etc. The proposed mechanism is —

$$HI \xrightarrow{h\upsilon} H + I \quad \text{(primary step : } I_0 \text{ is intensity)}$$

$$H + HI \longrightarrow H_2 + I \quad \text{Thermal } (k_1)$$

$$I + I \longrightarrow I_2 \quad \text{Thermal } (k_2).$$

Rate of decomposition of HI is

$$-\frac{d}{dt}[HI] = I_0 + k_1 [H] [HI]. \qquad .....(1)$$

According to steady-state theory concentration of intermediates i.e. [H] or [I] do not change with time. So

$$\frac{d}{dt}[H] = 0$$

But $\frac{d}{dt}[H] = I_0 - k_1[H][HI]$.

So $I_0 = k_1[H][HI]$.

Hence equation (1) becomes $-\frac{d}{dt}[HI] = 2I_0$.

$$\text{So quantum yield} = \frac{\text{No. of } HI \text{ molecules decomposed in a given time}}{\text{No. of photons absorbed in that time}}$$

$$= \frac{\text{Rate of decomposition of HI}}{\text{Intensity}}$$

$$= \frac{2I_0}{I_0} = 2.$$

So quantum yield = 2.

As time passes another reaction $H + I_2 \rightarrow HI + I$ begins to be prominent. So quantum yield decreases.

20.10 PHOTOSTATIONARY STATE OR PHOTOCHEMICAL EQUILIBRIUM

Let us consider a reversible reaction, whose forward reaction is both thermal and photochemical but backward reaction is only thermal. If such a reaction is carried out both in absence and presence of light then equilibrium yield will be different in two cases. In presence of light rate of forward reaction will increase due to its occurrence in two paths (thermal and photochemical) and rate of backward reaction will also increase due to accumulation of greater amount of product at any moment. Ultimately a new equilibrium will be established. This is known as photochemical equilibrium. Many authors do not prefer to call it an equilibrium, because it is a temporary state created in presence of light. It will go back to original equilibrium when brought in to dark. They therefore call it a photostationary state. Dimerisation of anthracene (A) is an example of such a process. The mechanism is —

$$A \xrightarrow{h\upsilon} A^* \quad \text{(activation)}$$

$$A^* \xrightarrow{-h\upsilon'} A \quad \text{(fluoroscence)}$$

$$A^* + A \longrightarrow A_2 \quad \text{(thermal)}$$

$$A_2 \longrightarrow A + A \quad \text{(thermal)}$$

The interesting thing to note is that while in case of thermal reaction, equilibrium yield depends on concentration of reactant, for photochemical equilibrium equilibrium yield is independent of concentration of reactant, but depends only on intensity of radiation.

20.11 EFFECT OF ABSORPTION OF LIGHT DIFFERENT PHENOMENA

When a substance absorbs visible or U.V. light its electronic state will be promoted. This promotion is also associated with change in vibrational state (Vide : P.E. diagram, chapter-19) Depending on the nature of vibrational state in the upper electronic state following changes may occur.

If the vibrational state in the upper electronic state is unstable, the molecule will break down in to nascent atoms or free radicals which will then carry out the reaction. Thus a photochemical reaction will start.

If however the vibrational state in the upper electronic state is stable, then the molecule will spend sometime here and then will come back to the original state by emitting radiation. The important thing is the life period of the excited state. If this state is a singlet then its life period is very small (10^{-8} sec) - thus the molecule returns to the ground state almost immediately after excitation. This is a case of **fluoroscence**. If frequencies of absorbed and emitted radiations are same it is **résonance fluoroscence.** Thus Hg absorbs light of wave length 2537A° and emits the same. But sometimes the molecule returns to the ground state in steps emitting various frequencies. This is **ordinary fluoroscence** ZnS; mineral-oils show this.

If the excited vibrational state is a triplet then its stability and hence life period is more. In such cases molecule stays in that state for quite a long time and emits radiation even after the source of incident radiation is removed. This phenomena is called **phosphorescence** sulphates of Ca, Ba and Sr exhibit this.

Light of wavelength 2537 A° carries necessary energy for dissociation of H_2 - molecule. But H_2 - does not absorb this and so dissociation will not occur. But Hg-vapour absorbs this light and fluoroscese. If Hg-vapour is present with H_2 gas and this light is made incident Hg-vapour will absorb it and pass necessary energy to H_2 by thermal collision to cause its dissociation. Such a process is called **photosensitization.** Hg-vapour is a **photo-sensitizer.** There are many such examples. The best example is photosynthesis in plants. Chlorophyl acts as a photosensitizer for the reaction between CO_2 and H_2O to give HCHO, which ultimately forms glucose. UO_2^{2+} acts as a photosensitizer for the decomposition of oxalic acid by light.

20.12 CHEMILUMINESCENCE

Certain chemical reaction are associated with emission of visible light. Such phenomena are called **chemi-luminescence**. Light emitted is called cold-light because it does not create any heating effect. A popular example is glow of firefly. This is due to oxidation of luciferon.

POINTS TO REMEMBER

Photochemical reaction are initiated by light. Laws regarding absorption are given by Lambert and Beer. According to Lambert-Beer **log** $\frac{I_0}{I} = \varepsilon\ C.\ l.$ 'ε' is extinction co-efficient. Its value depends on wavelength of light and nature of substance. It is used for checking purity of compound and also for quantitative estimation. Laws of photochemistry are (a) Grothus-Draper law and (b) Stark-Einstein's law.

$$\textbf{Quantum yield} = \frac{\textbf{No. of molecules reacted}}{\textbf{No. of photons absorbed}}.$$ It may be equal to, greater than or less than unity. Absorption of light by a substance may cause (i) fluoroscence (ii) phosphorescence (iii) photosensitization and (iv) photochemical reaction. What will happen that depends on vibrational state of the molecule achieved after absorption.

QUESTIONS

SHORT TYPE

1. What is an einstein ? Prove that blue einstein carries more energy than red einstein.
2. What are O. D. and extinction co-efficient ? On which factors does the latter depend ?
3. What is 'quantum yield' ?
4. Give examples of low and high quantum yield.
5. Distinguish between phosphoresce and fluoroscence.
6. What is photosensitization ? Give example.
7. What is important difference between a chemical equilibrium and photochemical equilibrium?
8. What is chemi-luminescence?

ESSAY TYPE

1. Compare a photochemical reaction with a thermal reaction.
2. State Lambert-Beer's law and discuss some of the utilities of measuring extinction co-efficient.
3. Write down the laws of photochemistry. What are apparent deviations from Einstein's law-how can you explain them?
4. Write down the mechanism for photo-dissociation of HI and calculate the quantum yield.
5. Briefly explain photostationary state citing an example.
6. Predicts the different things that may happen when a molecule absorbs visible light.

PROBLEMS

1. A 0.003 M solution of $Co(NH_3)_6^{3+}$ transmits 75% of incident light of 500 mμ if the path length is 1cm. Calculate the extinction co-efficient and the percent absorption for a 0.01.M. solution.

[*Hints* : O. D. = $\log\frac{100}{75} = \varepsilon \times 0.003 \times 1$; ε is same for 0.01 M sol^n.

Here O. D = $\log\frac{100}{1} = \varepsilon \times 0.01 \times 1$, find I which is transmittance.

(100 – I) is percentage of absorbance].

2. A 0.03M. solution of $Co(C_2O_4)_3^{3-}$ has on O.D.2.0 at 660 mμ using a 1cm cell. Calculate (a) the value of extinction co-efficient (b) the % absorption of a 0.051 M solution in the same cell.

[*Hints* : same as '1'.]

3. A 2cm. cell filled with certain gas at STP absorps 60% of incident light of wave length 580 mμ,. Give the % transmission for a 2 cm. cell with the gas at 2 atm using the same wavelength at some temperature.

[*Hints* : For gas $C = \frac{P}{RT}$; so O. D. = $d = \frac{\varepsilon P}{RT} \cdot l$.

$$\frac{d_1}{d_2} = \frac{P_1}{P_2}$$ when others are constant].

4. A spectrophotometer cell when filled with liquid. A transmits 50% and when filled with another liquid B transmits only 25% of the incident light of a certain wave length. What would be the optical dencity at this wavelength when the same cell is filled with a mixture of equal volumes of two liquids.

[*Hints :* In a mixture of equal volumes concentrations of each is halved; so O. D. values will also be halved. For liquid A it is $\frac{1}{2}\log\frac{100}{50}$ for B it is $\frac{1}{2}\log\frac{100}{25}$. As O. D. is additive property. So O. D. of mixture $= \frac{1}{2}\left(\log\frac{100}{50} + \log\frac{100}{25}\right)$].

OBJECTIVE TYPE

1. Fill in the gaps with appropriate word.

(a) In photochemical reaction free energy ——— increase. (must not / may)
(b) Laws of absorption is given by ——— (Einstein / Beer)
(c) Extinction co-efficient ——— on concentration. (depends / does not depend)
(d) Absorbance ——— on cocentration. (depends / does not depend)
(e) $H_2 + Cl_2$ reaction is an example of ——— quantum yield. (low / high)
(f) Photosynthesis is an example of ——— .
(phosphorescence / photo-sensitization)

2. Write T against true and F against false.

(a) Blue einstein has less energy than red einstein.
(b) Einstein's law predicts that one photon will make only one molecule react.
(c) Quantum yield for photo dissociation of *HI* is generally 2.
(d) For fluoroscence excited state is triplet.
(e) During photosynthesis Hg acts as a photo-sensitizer.

3. Match terms of column A with those of column B.

A	B
(a) Fluoroscence.	(a) Fire-fly.
(b) Phosphorescence.	(b) Singlet.
(c) Chemiluminescence.	(c) Chlorophyl.
(d) Absorbance.	(d) Triplet.
(e) Photosensitizer.	(e) Optical-density.

Answers :

1. (a) may (b) Beer (c) does not depend (d) depends (e) high (f) photosensitization.

2. (a) F (b) F (c) T (d) F (e) F.

3. a – b; b – d; c – a; d – e; e – c.

APPENDIX

TABLE 1 : Traditional & S.I. Units of some common parameters

Parameter	Traditional unit	S.I. unit	Conversion factor
Pressure	(i) m.m. (ii) c.m. (iii) torr (iv) bar 1 atm ≡ 76 c.m. 1 atm ≡ 1.013 bar	Pascal (Kg. m^{-1} s^{-2}) (or N m^{-2})	1 atm ≡ 1.013 × 10^5 pascal
Volume	c.c. or litre	$(metre)^3$	1 $metre^3$ ≡ 10^6 c.c. ≡ 10^3 liter
Force	Dyne	Newton	1 newton ≡ 10^5 dyne
Energy	Calorie/Ergs	Joule	1 cal. ≡ 4.2 J
Molar mass	gm/mole	Kg/mole	1 gm/mole ≡ 10^{-3} Kg/mole
Density	gm/c.c.	Kg/$metre^3$	Kgm/$metre^3$ ≡ (gm/c.c.) × 10^{-3}
Surface Tension	dyne cm^{-1}	Joule $metre^{-2}$	1 dyne cm^{-1} ≡ 10^{-3} Joule $metre^{-2}$
Viscosity	Poise i.e. dyne-sec/cm^2	Newton-sec/$metre^2$	1 newton-sec/$metre^2$ ≡ 10 poise

TABLE 2 : Values of some universal constants

Constant	Traditional unit	S.I. unit
Velocity of light	3 × 10^{10} cm. sec^{-1}	3 × 10^8 metre sec^{-1}
Electronic charge	4.8 × 10^{-10} e.s.u.	1.60 × 10^{-19} coulomb.
Avogadro Number	6.023 × 10^{23} $mole^{-1}$	6.023 × 10^{23} $mole^{-1}$
Rest mass of electron	9.11 × 10^{-28} gm.	9.11 × 10^{-31} Kg.
Rest mass of proton	1.67 × 10^{-24} gm.	1.67 × 10^{-27} Kg.
Faraday	96,486 coulomb.	96,486 coulomb.
Planck's constant	6.626 × 10^{-27} erg-sec.	6.626 × 10^{-34} Joule sec.
Boltzmann Constant	1.38 × 10^{16} erg-deg^{-1}	1.38 × 10^{-23} Joule deg^{-1}
Gas Constant →	8.31 × 10^7 erg. deg^{-1} $mole^{-1}$ 1.987 cal deg^{-1} $mole^{-1}$ 0.0821 litre-atm. deg^{-1} $mole^{-1}$	8.31 Joule deg^{-1} $mole^{-1}$

Subject Index

RECENT UNIVERSITIES QUESTIONS

PART I

CHAPTER 1

Atomic Structure Classical and Wave Mechanical Approach

1. Write notes on : Pauli's exclusion principle. **[C.U. '92, '94, '96 '98]**

2. What are the quantum numbers required to describe the properties of a rotating electron? State and explain their significance. **[C.U. '91, '94, '99]**

3. Sketch the three *p*-orbitals. **[C.U. '94, '99]**

4. Write down the electronic configuration of element having at. no. 15. How many electrons are unpaired? Write down all the quantum numbers of these unpaired electrons. **[C.U. '98]**

5. Show the probability distribution curves of *s*- and *p*- orbitals. **[C.U.H. '98]**

CHAPTER 3

Radioactivity and Nuclear Chemistry

1. What do you mean by half-life period and average life period of radio element? Show their inter-relationship. **[C.U. '90, '92, '98]**

2. Write notes on : Radioactive equilibrium. **[C.U. '92, '94, '96, '98]**

3. Distinguish between fission and fusion. **[C.U. '93, '97, '99]**

4. State and explain the theory of radioactive disintegration. Half-life period of a radio element is 231 min. How long it would take for $\frac{9}{10}$th fraction of the element to decay?

(Ans. t = 767.5 min) [C.U. '98]

5. Write notes on artificial radioactivity. **[C.U. '99]**

PART II

CHAPTER 1

Ideal gas—Kinetic Theory

1. State the postulates of kinetic theory of gases. Discuss the cause of deviation of real gases from ideal behaviour. **[C.U. '97]**

2. State the postulates of kinetic theory of gases and deduce $PV = \frac{1}{3}mnc^2$. Derive an expression of total K. E. per mole of an ideal gas. **[C.U. '96]**

3. Show that the time of diffusion of O_2 is four times that of H_2 at identical temperature and premine.. **[C.U. '97]**

CHAPTER 2

Real Gas

1. Deduce $\frac{RT_c}{P_cV_c}$. [C.U. '91, '93, '95, '98]

2. Calculate the pressure of the 3 moles of a Van der Waals gas in a container of 20 litre at 27°C. [$a = 6.5$ atm.lit^2.mole^{-2}, $b = 0.056$ lit.mole^{-1}] **(Ans. 3.58 atm.) [C.U. '98]**

3. Write down Van der Waal's equation for 'n' moles and state the units of 'a' and 'b' **[C.U. '99]**

4. Express the constant 'a', 'b' in terms of critical constants. **[C.U.H. '99]**

CHAPTER 4

The Liquid State

1. Explain : CH_4 is non-polar but $CHCl_3$ is polar. **[C.U.H. '98]**

2. Explain : Arrange in order of increasing d.m. H_2O, o-dichlorobenzene, Br_2, CCl_4, nitrobenzene. **[C.U.H. '97]**

CHAPTER 5

The Solid State

1. What is a 'Unit cell' of a cyrstal? State the number of atoms in a unit cell of a face-centred and a body-centred cube. What are Miller indices of a crystal? **[C.U.H. '95]**

CHAPTER 6

Thermodynamics—First law

1. Joule-Thomson effect. **[C.U. '93, '95, '97, '99]**
2. State the First law of Thermodynamics. **[C.U. '91, '98]**
3. Deduce the relation $C_p - C_v = R$. **[C.U. '92, '96, '99]**
4. Distinguish between isothermal and adiabatic process. **[C.U. '91, '98]**
5. Deduce the temperature-volume relation of an ideal gas in an adibatic process. **[C.U. '94, '98]**

CHAPTER 7

Thermochemistry

1. Write notes on : Hesses law. **[C.U. '96, '98]**

2. Define heat of reaction at constant pressure (ΔH) and at constant volume (ΔE) and deduce a relation between them. **[C.U. '94, '97]**

CHAPTER 8

Thermodynamcs—Second Law

1. What is a cyclic process? Deduce the expression of efficiency of a reversible Carnot cycle. Hence show that efficiency of a reversible isothermal cycle is zero. **[C.U. '90, '92, '95, '97, '99]**

2. Write notes on : Entropy. **[C.U. '90, '99]**

3. State the 2nd Law of thermodynamics. **[C.U. '91, '98]**

4. 2 moles of an ideal gas expands from 1 litre to 10 litre at 100°C. Calculate the reversible work done in ergs. **(Ans. 1.43×10^{11} ergs.) [C.U. '98]**

CHAPTER 9

Chemical Equilibrium

1. Write notes on : Le Chatelier's principle and its application. **[C.U. '94, '97]**

2. State law of mass action. What is active mass? Relate K_p with α and P for $PCl_5 (g) \rightleftharpoons PCl_3 (g) + Cl_2 (g)$.

CHAPTER 11

Dilute Solution—Colligative property

1. Human blood is isotonic with 0.1 (N) NaCl solution. The Van't Hoff factor of NaCl is 1.82. Calculate the osmotic pressure of human blood at 37°C. **(Ans. 4.62 atm.)** **[C.U. '93, '97]**

2. State and formulate Raoult's law of relative lowering of vapour pressure. State its limitations. How the law can be used for determination of molecular weight. **[C.U. '94, '98]**

3. State the principle and draw a lebelled diagram for the measurement of osmotic pressure by Berkeley-Heartly's method. **[C.U. '97]**

4. What is plasmolysis? **[C.U. '97]**

5. What is the boiling point of solution containing 5 gm. urea (M = 60) in 100 gm water? ($K_b = 0.513$) **(Ans. 100.4275°C)** **[C.U. '98]**

6. What are colligative properties? Why are they so called? Define Van't Hoff factor. Find out the Osmotic pressure of a 0.001(M) K_2SO_4 solution at 27°C. **[C.U.H. '98]**

7. Calculate the osmotic pressure at 25°C of a solution containing 1 gm. glucose ($C_6H_{12}O_6$) and 1 gm. sucrose ($C_{12}H_{22}O_{11}$) per litre. **[C.U. '97]**

CHAPTER 12

Heterogeneous Equilibrium—Phase Rule

1. Write notes on : Phase diagram of water. **[C.U. '97]**

2. Write notes on : Steam distillation. **[C.U. '97]**

3. Explain : Degree of freedom of one component system at triple point is zero. **[C.U. '99]**

4. Draw the phase diagram of CO_2 system. What is dry ice? **[C.U.H. '99]**

5. What is consolute temperature? Give examples of upper and lower consulate temperature. **[C.U.H. '93, '95, '97]**

6. Discuss the phase diagram of nicotine-water system. **[C.U.H. '97]**

CHAPTER 13

Chemical Kinetics

1. Write notes : Half-life period of a 1st order reaction. **[C.U. '98]**

2. Assuming the initial concentrations of both the reactants to be equal to '*a*', derive the expression for the specific rate constant of the reaction. Show that half-life period of this reaction is inversely proportional to '*a*'. **[C.U. '99]**

CHAPTER 14

Catalysis—Adsorption

1. Write notes : Auto catalysis, catalyt proproter. **[C.U. '94, '96, '99]**

2. Write notes : Catalysis.

CHAPTER 15

Colloids

1. Write note : Isoelectric point. **[C.U. '90, '99]**
2. Write note : Dialysis. **[C.U. '93, '95, '99]**
3. Electrophoresis. **[C.U. '94, '98]**
4. Write note : Gold number. **[C.U. '97, '99]**
5. Write note : Zeta-potential. **[C.U.H. '98]**
6. What are electro-kinetic phenomenon? Discuss these qualitatively. **[C.U.H. '95, '99]**
7. Explain why $Al_2(SO_4)_3$ is more active than Na_2SO_4 in coagulating a sol. **[C.U.H. '99]**
8. Why lyophilic colloids are more stable than lyophobic Colloids? **[C.U.H. '99]**

CHAPTER 16

Electrolytic Conduction

1. Write note : Conductance ratio. **[C.U. '97]**
2. Equivalent conductance at infinite dilution. **[C.U. '93, '95, '97, '99]**

CHAPTER 18

Electromotive Force

1. Standard electrode potential, E° of Zn^{2+}/Zn and Cu^{2+}/Cu are –0.763 and 0.337 volts. Design an electrochemical cell and write the cell reaction. Calculate the E° of the cell. **[C.U. '98]**

2. Discuss the principle of determination of pH of a solution by a glass electrode. **[C.U.H. '98]**

CHAPTER 17

Ionic Equilibrium

1. Write note : Ostwald's dilute on law. **[C.U. '90, '92, '98]**
2. Lewis definition of acid and bases. **[C.U. '93, '98]**
3. Deduce an expression for the degree of hydrolysis of a salt of weak acid and strong base. **[C.U. '92, '96, '97]**
4. Describe the e.m.f. method for determination of pH of solution. **[C.U. '89, '92, '95, '97]**
5. Calculate the pH of a mixture containing 0.01 (M) acetic acid and 0.03 (M) sodium acetate. ($pk_a = 4.8$) **(Ans. 5.28)** **[C.U. '94, '97]**

6. Deduce the expression of hydrolysis constant of a salt of strong acid and weak base. Will the solution be acidic or alkaline? **[C.U. '95, '99]**

7. What is meant by buffer solution. **[C.U. '97]**
8. Which of the following mixture will acts as buffer-why?

   CH_3COOH + NaOH (2 : 1 mole ratio)

   NH_4Cl + HCl.

   NH_4OH + HCl (2 : 1 mole ratio) **[C.U. '98]**

9. An aqueous solution of 0.1 (M) KCN is 2% hydrolysed. Find the value of dissociation constant of HCN. **[C.U. '99]**

10. Discuss the principle of determination of pH of a solution using a glass electrode. [C.U.H. '98]

11. The solubility of AgCl in water is 0.001435 gm/lit. Find its solubility in 0.1(M) NaCl solution. [Ag = 108, Cl = 35.5] [C.U. '99]

12. Write notes on common ion effect and its application in analytical chemistry. [C.U. '99]

CHAPTER 20

Photochemistry

1. Write notes on : Fluorescence and phosphorescence. [C.U.H. '96, '98]

2. State Einstein's Law of photochemical equivalence. What is quantum yield? Can the latter be both less or greater than unity? [C.U.H. '96, '99]

3. How can it be experimentally determined? [C.U.H. '96, '99]

4. What is an actinometer? [C.U.H. '98]

5. What is 'One Einstein'? [C.U.H. '98]

APPENDIX-2

[UNIVERSITY QUESTIONS]
(with Hints & Answers)

GAS - IDEAL - REAL

1. What is Boyle Temp^r ? **[K.U.'12; B.U.'11; N.B.U'2006]**

Ans. For Ideal gas PV = constant; when temperature is kept constant. This is Boyle's law. For real gas PV is not constant under isotheramal condition. However real gases obey Boyle's law only at a particular temperature. This temperature is called Boyle temperature. At this temperature $\left[\frac{d(PV)}{dp}\right]_T = 0$. Its value is $\frac{a}{Rb}$ ['a' & 'b' are van-der waals constant].

2. What is compressibility factor? **[K.U.'12]**

Ans. Real gases do not follow ideal gas equation PV = nRT. Here we write PV = ZnRT, where $Z = \frac{V_{real}}{V_{ideal}}$. This factor 'Z' is called compressibility factor. For ideal gas Z=1. For real gas Z may <1 [more compressible] or >1 [less compressible].

3. What is the value of 'R' in S.I. unit. **[K.U.'12]**

Ans. Value of 'R' in S.I. unit is joules deg^{-1} mole^{-1}.

4. Why real gases show deviations from ideal behaviour. **[K.U.'12]**

Ans. Real gases do not obey postulates of kinetic theory properly. This is the reason for showing deviation from ideal gas equation.

Thus real gases posses intermolecular attraction. Due to this their pressure becomes less that ideal gas pressure. Van-der-wall showed $P_{ideal} = P_{real} + \frac{a}{V^2}$.

Molecules of real gases are not point mass. They occupy some volume. Thus entire volume (V) of the vessel is not available for movement. Available volume is (V–b); 'b' is co-volume. Thus while for ideal gases PV=RT for one mole; for real gases it is $\left(P + \frac{a}{V^2}\right)\left(V - b\right) = RT$.

5. From the equation deduced from kinetic Theory arrive at Boyle's law and Avogadro's law. **[K.U. 2012: N.B.U 2000; V.U. 2004]**

Ans. From the postulates of kinetic theory expression for pressure is $PV = \frac{1}{3}mnc^2$, where symbols have their usual meaning. We can arrive at Boyle's law form this.

We write $PV = \frac{1}{3} mnc^2$

or, $PV = \frac{2}{3} \times n \times \frac{1}{2} mc^2$

$\frac{1}{2} mc^2$ is K.E. of one molecule. So $n \times \frac{1}{2} mc^2$ = Total K.E. Hence

$$PV = \frac{2}{3} \times \text{Total K.E.}$$

From the postulates K.E. & T. So K.E. = constant ´ T. Hence

PV = Constant × T

It 'T' becomes constant, then 'PV' is cosntant. This is Boyle's law.

To prove Avogadro's law we consider one mole of two gases at same P.V. and T. For 1st gas $PV = \frac{1}{3} m_1.n_1.c_1^2$ and for second $PV = \frac{1}{3} m_2.n_2.c_2^2$; where '$n_1$' and '$n_2$' are number of molecules of gases '1' and '2' in this case. So $\frac{1}{3} m_1.n_1.c_1^2 = \frac{1}{3} m_2.n_2.c_2^2$ ....(1).

Since both gases have same temperature, so K.E./molecule is also same. So $\frac{1}{2} m_1.c_1^2 = \frac{1}{2} m_2.c_2^2$ ....(2) compairing '1' & '2' we get $n_1 = n_2$. So equal volumes of two gases at same 'P' & 'T' will have same number of molecules. This is Avogadro's law.

6. **Mention two postulates of Kinetic theory.** **[K.U. 2011]**

Ans. See section 1.6

7. **What do you understand by average velocity and root mean square velocity of gas molecules?** **[K.U. 2011]**

Ans. If C_1, C_2, C_3 .... C_n etc. be velocities of different molecules present in an assembly of 'n' molecules, then average velocity is —

$$C_{av} = \frac{C_1 + C_2 + C_3 + \ldots\ldots\ldots\ldots + C_n}{n}$$

and r.m.s. velocity is —

$$C_{r.m.s} = \left[\frac{C_1^2 + C_2^2 + C_3^2 \ldots\ldots\ldots\ldots + C_n^2}{n}\right]^{\frac{1}{2}}$$

considering vector nature and isotropic behaviour Çav = 0; but $C_{r.m.s.}$ is not zero.

8. Mention the SI units of 'a' and 'b' for van-der-waals gas. **[B.U. 2011; N.B.U. '96; V.U. 2005]**

Ans. S.I. unit of van-der-waals constants are —

$(\text{metre})^6$ pascal mole^{-2} for 'a'

and metre^3 mole^{-1} for 'b'.

9. At room temperature 'N' number of molecules of an ideal gas, each of mass 'm' are kept in a cubical container of side length L. Derive necessary expression for the pressure of the gas.
Using the expression $PV = \frac{1}{3} m\ Nc^2$, verify Boyle's law and Avogradro's law.

[B.U. 2011]

Ans. See section 1.8 for the first part and section 1.12 for the second part.

10. Instead of the simple average speed why root mean square is used in kinetic theory? [B.U. 2011]

Ans. See section 1.10.

11. Draw the representative Andrew's curve for CO_2 system, and give proper explanation.. **[B.U. 2011]**

Ans. The adjacent figure shows representative Andrew curves for CO_2. Curve III is below critical temp^r with three seperate parts, AB (representing gas); BC (gas-liquid transition) and CD (liquid). Cure II is at critical temp^r. It has a hump at which gas is liquified and curve I is above critical temp^r where liquid state is absent.

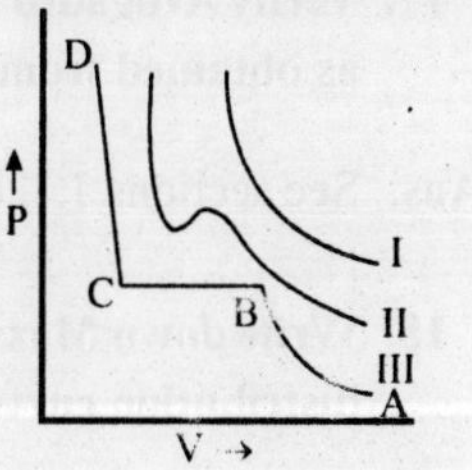

12. Write the van-der-waals equation for 'n'-mole real gas. **[B.U. 2012]**

Ans. For 'n' mole real gas van-der-waals equation is —

$$\left[P + \frac{n^2a}{V^2}\right]\left[V - nb\right] = nRT$$

13. Under what condition CO_2 gas can be liquified? **[B.U. 2012]**

Ans. CO_2 gas can be liquified by isothermal compression, if its temperature is below its ertical temperature, which is 31.1°C.

14. Using the appropriate reasons obtain the equation of state for one-mole of a van-der-waals gas from the equation of state (PV=RT) for one mole of an ideal gas. [B.U. 2012]

Ans. See section 2.3.

15. What are meant by critical temperature critical pressure and critical volume of a real gas? Write the relationship among them for a van-der-waals gas.

[B.U.2012]

Ans. Critical temperature (T_C) is the temperature at which or below which a gas can be liquified by isothermal compression. Above T_C application of pressure can not liquify it. A gas below T_C is also called vapour.

Critical pressure (Pc) is the minimum pressure required to liquify the gas at critical state.

Critical volume (V_c) is volume occupied by one mole gas at T_c temperature and P_c pressure. The relation among them for a van-der-waals gas is —

$$\frac{RT_c}{P_cV_c} = 2.66$$

16. Under what condition the average kinetic energy of the gaseous molecules of ammonia and oxygen becomes equal? [B.U. 2012]

Ans. Temp is a measurement of average K.E. If T is same, average kinetic energy of molecules of ammonia and oxygen will be same.

17. Verify Avogadro's law and Graham's law of diffusion from the equation of pressure as obtained from kinetic theory of gas. [B.U. 2012]

Ans. See sections 1.12(b) and 1.12(c)

18. Write down Maxwell's velocity distribution equation and explain the terms. Draw the distribution curves at two temperatures T_1 and T_2 (T_2>T_1).

[C.U. 2006; N.B.U. 2006]

Ans. See section 1.17

19. What will be the expression for r.m.s. speed in terms of pressure and density?

[N.B.U. 2006]

Ans. Root means square velocity $C_{r.m.s.}$ is

$$C_{r.m.s.} = \left[\frac{C_1^2 + C_2^2 + \dots\dots\dots\dots + C_n^2}{n}\right]^{\frac{1}{2}}$$

Now, $PV = \frac{1}{3} m.n.c^2_{r.m.s.}$ or, $C^2_{r.m.s.} = 3P.\frac{V}{mn}$

But $\frac{mn}{V} = \frac{\text{Total mass}}{\text{Total volume}} = \text{density 'd'}$

So, $C^2_{r.m.s.} = \frac{3P}{d}$

Hence, $C_{r.m.s.} = \sqrt{\frac{3P}{d}}$

20. $C_P - C_V$ value is same for all gases (ideal) but $\frac{C_P}{C_V}$ values are different — explain.

[N.B.U 2006]

Ans. It can be shown that for one mole ideal gas $C_p = \frac{5}{2}R + a$ and $C_V = \frac{3}{2}R + a$ where 'a' is increase in intramolecular (rotational and vibrational) energy of the gas molecule for 1°C rise in temperature. So $C_p - C_V = \left(\frac{5}{2}R + a\right) - \left(\frac{3}{2}R + a\right) = R$; which is same for all gases. But $\frac{C_P}{C_V} = \frac{\left(\frac{5}{2}R + a\right)}{\left(\frac{3}{2}R + a\right)}$ As 'a' value depends on atomicity; $\frac{C_P}{C_V}$ value will depend on nature of gas e.g. for monoatomic gases a = 0; $\frac{C_P}{C_V} = \frac{5}{3} = 1.66$; for polyatomic gases, 'a' has some values.

21. Find out van-der-waals constant 'b' for a gas having molecular diameter 3A°. [N.B.U. 2006]

Ans. Van-der-waals constant $b = 4N\frac{4}{3}\pi r^3$.

Here r = radius of the molecule.

In this case $r = \frac{3}{2}$ A° = 1.5×10^{-8} c.m.

So, $b = 4 \times 6.023 \times 10^{23} \times \frac{4}{3} \times 3.14\,(1.5 \times 10^{-8})^3$

or, $b = 34.042$ c.c. mole$^{-1} \equiv 0.034$ litre mole^{-1}

22. Using van-der-waals equation calculate pressure exerted by 2 mole nitrogen taken in a 2 litre vessel at 21°C. [Given a = 1.39 atm. litre^2 mole^{-2} and b = 0.039 litre mole^{-1}]

[N.B.U. 2006]

Ans. For 'n' mole van-der-waals gas

$$\left(P + \frac{n^2a}{v^2}\right)\left(v - nb\right) = nRT$$

$$\text{or, } P = \frac{nRT}{v - nb} - \frac{n^2a}{v^2}$$

Here, n = 2; R = 0.082 litre atm; $mole^{-1}$; deg^{-1}; T = 300K

v = 2 litre; b = 0.039 litre $mole^{-1}$ and a = 1.39 atm $litre^2$ $mole^{-2}$

$$\text{So, } P = \frac{2 \times 0.082 \times 300}{2 - 2 \times 0.039} - \frac{4 \times 1.39}{4}$$

$$\text{Or, } P = 25.598 - 1.39 = 24.208 \text{ atm.}$$

LIQUID

1. Define surface tension of a liquid. How does it change with temperature? **[K.U. 2012]**

Ans. Surface of a liquid is in a state of strain due to unbalanced backward force exerted by liquid molecules, just below its surface. This force tends to decrease the surface area. Amount of force acting perpendicular to the surface per unit length is called surface tension (γ). Its unit is dyne cm^{-1} (or newton $metre^{-1}$). According to Eotovos susface tension is related with temperature as —

$$\gamma\left(\frac{M}{\rho}\right)^{\frac{2}{3}} = \kappa\,(T_C - T)$$

γ = Surface tension ; M = Molar mass ; ρ = Density; T_c = Critical temp.

and T = $Temp^r$; K = Constant.

A more accurate relation is

$$\gamma\left(\frac{M}{\rho}\right)^{\frac{2}{3}} = \kappa\,(T_C - T - 6)$$

2. Why viscosity of water is greater than that of methyl alcohol? **[K.U. 2012]**

Ans. Viscosity of a liquid is generally increased when degree of association increases. Due to hydrogen bonding water is more associated than methyl alcohol. That is why its viscosity is more.

3. Give two examples of surface tension in everyday life. **[K.U. 2011]**

Ans. A needle floats on water; if water is thrown in the air, the water drops that fall are spherical. These are examples of surface temsion in every-day life.

4. Define viscosity co-efficient. What is its S.I. unit. **[K.U. 2011]**

Ans. Viscosity co-efficient is the frictional force between two liquid layers moving side by side, where area of contact and velocity gradient are unity. Its S.I. unit is newton sec $metre^{-1}$.

5. What is the effect of temperature on the viscosity of a liquid and gas? **[K.U. 2011]**

Ans. Viscosity of a liquid decreases with increase in temperature; but viscosity of a gas increases with increase in temperature.

6. Water rises in a capillarly but mercury falls — why ? **[N.B.U. 2011]**

Ans. If meniscus of a liquid is curved then atmospheric pressure is more on the concave side. Excess pressure is inversely proportional to radius of the tube. For capillary radius is very small, so excess pressure is large. For water in a capillary upper meniscus is concave (contact angle < 90°). So due to pressure difference water level rises. For mercury upper meniscus is convex (contact angle > 90°), so here pressure difference acts downwards and the level falls.

THERMODYNAMICS — THERMO CHEMISTRY

1. What is a reversible process? **[K.U. 2012; N.B.U. '98; C.U./V.U./K.U.'2000]**

Ans. If during a process difference between driving and opposing force is infinitesimally small, then the proess may occur in both forward and backward direction. System is always in equilibrium. Such a process is reversible process. Expansion of a gas against a continuously decreasing pressure in an example.

2. Under what condition efficiency of a Carrot's engine will be unity? **[K.U. 2012]**

Ans. Efficiency of carnot's engine (n_t) is $\eta = \frac{T_1 - T_2}{T_1}$; T_1 and T_2 are temperatures of source and sink. If T_2 = O.K. i.e. temperature of sink is absolute zero, then efficiency will be one (1).

3. State and explain Hess's law. **[C.U. 2011; K.U. 2012; B.U. 2012]**

Ans. Hess's law states, 'if a reaction takes place in a single step or in a number of steps then total heat of reaction in each path is same. To explain let us consider oxidation of C to CO_2 via following paths :—

Path I : $C(S) + O_2\,(g) = CO_2(g);\ \Delta H$

Path II : $C(S) + \frac{1}{2}O_2\,(g) = CO(g) + \Delta H_1$

$CO(g) + \frac{1}{2}O_2(g) = CO_2(g)\ ;\ \Delta H_2$

Here $\Delta H = \Delta H_1 + \Delta H_2$

4. Write down Clausius – Clapeyron equation for Solid $\rightleftharpoons$ Liquid transition; explain each term **[K.U. 2012; N.B.U '06]**

Ans. For solid $\rightleftharpoons$ liquid transition Clausius — Clapeyron equation is —

$$\frac{dP}{dT} = \frac{L_f}{T(V_l - V_s)}$$

Here dT is change in melting point for dP unit change of pressure. L_f = latent heat of fusion per mole at melting point T; V_l and V_S are molar volumes of liquid and solid phases.

5. One mole of an ideal gas expands isothermally reversibly from 1 litre to 10 litre at 27°C calculate ΔU, q, w and ΔH. **[K.U. 2012]**

Ans. We known for ideal gases $\Delta U = nC_V dT$ and $\Delta H = nC_P dT$, [symbols have usual meaning]. In isothermal process dT=0; so ΔH = ΔU = 0.

$$w = nRT\ln\frac{V_2}{V_1}$$

Here n = 1 T = (273 + 27) = 300K ; V_2 = 10 litre

V_1 = 1 litre R = 8.314 joule |K|mole

$$\text{So } W = 8.314 \times 300 \times 2.303 \log\frac{10}{1}$$

$$= 5744.14 \text{ joule}$$

Now q = ΔU + w

as ΔU = 0 so q = w = 5744.14 joule.

6. How will you explain that entropy is a measurement of chaos? **[K.U. 2011; V.U. 2001; C.U. 2005]**

Ans. Calculation shows that entropy of the universe (system + surrounding) increases in spontaneous and irreversible processes. Now all natural processes are spontaneous and irreversible. So entropy of the nature or universe is always increasing. But if we look carefully we shall see that chaos and randomness are continuously increasing in nature. So entropy is a measurement of chaos.

7. Give an example of a spontaneous process. How is ΔG related with such process at constant pressure and temperature? **[K.U. 2011]**

Ans. Flow of gas from high to low pressure is an example of spontaneous process; free mixing of two different gases at same P and T is also an example. Here $\Delta G_{P.T.}$ = –ve

8. Explain why U, H, S and G all are called state functions in thermodynamics. **[B.U. 2011]**

Ans. Values of U, H, S and G of a system depend only on the state of the system; they are independent of path or previous history. In cyclic process they do not change. That is why they are called state functions.

9. For one mole of a gas prove that $C_P - C_V = \left[P + \left(\frac{\delta U}{\delta V}\right)_T\right]\left(\frac{dV}{dT}\right)_P$ **– symbols have usual meaning.** **[B.U. 2011]**

Ans. See section 6.12.

10. One mole of an ideal gas system undergoes reversible expassion to some volume by isothermal and by adiabatic pathway. Draw the necessary P–V diagram and explain why adiabatic curve is steeper. **[B.U. 2011]**

Ans. See section 6.17.

11. The heat energy can be completely converted in to mechanical work in isothermal reversible expansion — comment. **[B.U. 2011]**

Ans. Mathematical form of first law of thermodynamics is $dq_{rev} = dU + PdV$; symbols have their usual meaning. Now $dU = nC_V dT$. In isothermal process temperature does not change, so $dT = 0$; hence $dU = 0$. So $dq_{rev} = PdV$. Hence in isothermal revessible process heat absorbed is completely converted to mechanical work.

12. Calculate the work done in calorie when 4 gm. He gas (ideal gas) is expanded isotherimally reversibly exponentially 'e' times of its initial volume at 127°C. **[B.U. 2011]**

Ans. Expression for work done during isothermal reversible expansion for 'n' mole ideal gas

is $W = nRT\ln\frac{V_f}{V_i}$

For He (mol.wt = 4) 4 gm ≡ 1 mole

$R = 1.987$ Cal deg^{-1} mole^{-1} $T = 273 + 127 = 400K$

and $\frac{V_f}{V_i} = e$ (given)

So, $W = 1 \times 1.987 \times 400 \ln e$

$= 794.8$ Cal.

13. Write notes on, efficiency of Carnot's engine. [B.U. 2011]

Ans. Efficiency of Carnot's engine means fraction of the heat (absorbed by the engine) converted to work. Thus if a Carnot's engine absorbs Q_1 heat from a source at T_1 and perform W work per cycle. Then efficiency is $\frac{W}{Q_1}$. From an imaginary cycle consideration it was shown that

$$\frac{W}{Q_1} = \frac{T_1 - T_2}{T_1}$$; T_2 = Temp^r of the sink.

Theoretically efficiency will be 100% i.e. $W = Q_1$ only when T_2= O K.

14. Explain why the final temperature of a gas – system falls during adiabatic expansion. [B.U. 2012]

Ans. During adiabatic expansion the system does not receive any heat from surrounding (dq = 0); but it performs work. So its internal energy (dU) decreases. Hence temperature falls. dq = dU + W. When dq = 0; w = –dU. During expansion w is positive, so dU is negative. But $dU = C_V (T_f - T_i)$. So, $T_f < T_i$

15. Establish the relation between the heat of reaction at constant pressure and the heat of reaction at constant volume. [B.U. 2012]

Ans. See Sections 7.3 and 7.4.

16. Calculate the difference in energy (in cals) between the heat of reaction at constant pressure and constant volume for the following reaction at 27°C. $2H_2S(g) + SO_2(g) = 3S(g) + 2H_2O(l)$. [B.U. 2012]

Ans. Difference between two types of heats of reaction is ΔnRT. [ΔH = ΔU + ΔnRT]. Here Δn = Difference in number of moles of GASEOUS products and reactants. In this case Δn = 3 – (2+1) = 0. So the two heat of reactions have same value; no difference.

17. Derive the mathematical expression for the efficiency of a carnot-engine. [B.U. 2012; C.U. 2010]

Ans. See section 8.4.

18. Under what condition a process will take place spontaneously? [B.U. 2012]

Ans. Conditions for spontaneity of a precess are $dG_{P,T}<0$ and $dS_{T,V}>0$. [G = Gibbs potential; S = Entropy]

19. Write notes on Joule-Thomson effect. [B.U.2012]

Ans. See section 6.13.

20. One mole of an ideal gas undergoes a reversible isothermal expansion at 27°C, then the final volume becomes ten times of its initial volume. Calculate the work done and heat absorbed in e.g.s. unit. [B.U. 2012]

Ans. Expression for isothermal reversible work is

$$W = nRT \ln \frac{V_f}{V_i}$$

Here n =1 (no. of mole); R = 1.987 cal. deg^{-1} $mole^{-1}$

$$T = 273 + 27 = 300 \text{ and } \frac{V_f}{V_i} = 10$$

So, W = 1.987 × 300 ln 10 = 1372.82 Cal

In isothermal expansion heat absorbed = work done.

So, here heat absorbed = 1372.82 Cal

21. What are meant by isothermal change and adiabatic change? Deduce the relation PV^r = constant for adiabatic reversible change of an ideal gas. [C.U. 2010]

Ans. If during some change temperature of the system remains constant, then it is an isothermal change.

If during some change the system does not exchange energy (heat) with the surrounding, then it is an adiabatic change.

(For derivation of the relation PV^r = Const. See section 6.16)

22. 2 mole of ideal gas is expanded reversibly and isothermally from a volume of 1 litre to 10 litres at a temperature of 27°C calculate ΔS_{sys}. [C.U. 2010]

Ans. For isothermal change expression for entropy change of an ideal gas is —

$$\Delta S_{sys} = nR \ln \frac{V_2}{V_1}$$

In this case n = 2 moles; R = 8.314 $JK^{-1}mole^{-1}$

V_2 = 10 litre ; V_1 = 1 litre ;

$$\text{So } \Delta S_{sys} = 2 \times 8.314 \times \ln \frac{10}{1} = 38.29 \text{ J.K}^{-1}$$

23. Why is the Joule-Thomson Co-effecient of an ideal gas is zero? [C.U. 2010]

Ans. Expression for Joule-Thomson co-efficient is $\mu_{J-T} = \frac{1}{C_P}\left[\frac{2a}{RT} - b\right]$; 'a' and 'b' are van-der-waals constant. As ideal gas does not have attraction or repulsion, a = 0 and b = 0. So $\mu_{J-T} = 0$.

24. Derive Kirchoff's equation showing the variation of heat of reaction with temperature. **[C.U. 2010]**

Ans. See section 7.12.

25. The heat of neutralization of a strong acid by a strong alkali is always the same. Explain **[C.U. 2010]**

Ans. See section 7.9.

26. Deduce $\Delta G = \Delta H + T\left[\frac{\delta(\Delta G)}{\delta T}\right]_P$; where the terms have usual meaning.

[C.U. 2010]

Ans. See section 8.13.

27. Draw a Carnot cycle on a P-V diagram. **[C.U. 2010, N.B.U'06]**

Ans. See section 8.4

28. What is inversion temperature? **[C.U. 2010]**

Ans. Inversion temperature is that temperature, at which Joule-Thomson co-efficient is zero. Below this temperature throttling produces cooling, but above this it produces heating.

29. What are meant by closed system and isolated system? Derive expression for maximum work done in isothermal reversible expansion of 'n'-moles of an ideal gas. [C.U.2011]

Ans. Closed system is a system whose mass can not be altered but energy can be changed. Gas enclosed in a cylinder is an example.

Isolated system is a system whose neither mass nor energy can be altered. Gas enclosed in a cylinder covered by an adiabatic wall is an example.

[For derivation of expression for isothermal reversible work see section 6.5]

30. At constant pressure of 2 atmosphere a gas absorbs 300 cal of heat for expansion from a volume of 10 litre to 20 litre. Calculate the change in internal energy.[1 litre atm ≡ 24.2 cal] **[C.U. 2011]**

Ans. From first law dq = dU + dw

Here dq = heat absorbed = 300 cal.

dw = work done = $P(V_2 - V_1) = 2\,(20 - 10)$

$= 20$ litre. atm ≡ 484 cal

So, 300 = dU + 484

or, dU = – 184 cal

31. Write down the significance of Joule-Thomson co-efficient. **[C.U. 2011, N.B.U 2006]**

Ans. See section 6.14

32. Give an example of spontaneous process. State how Gibb's free energy at constant temperature and pressure is related to it. **[C.U. 2011]**

Ans. Flow of heat from high to low temperature is a spontaneous process.

Condition of spontaneity is $dG_{P,T}$ = negative.

[G = Gibbs free energy]

33. What is meant by stantement – "The efficiency of a Carnot's engine is 1". Is it really possible? **[N.B.U '06 ; C.U. 2011]**

Ans. Efficiency of Carnot's engine (η) is $\eta = \frac{T_1 - T_2}{T_1}$. T_1 and T_2 are temperatures of source and sink. So if T_2 = OK then efficiency = 1. It is not really possible as we can never reach OK.

34. What is the physical meaning of entropy? **[C.U. 2011]**

Ans. See section 8.10.

35. Define thermodynamically heat capacity at constant volume (C_V) and heat capacity at constant pressure C_p. Deduce theirmodynamically $C_P - C_V = R$ for 1 mole idal gas. **[C.U. 2011]**

Ans. Molar heat capacity at constant volume (C_V) can be definied thermodynamically as the increase in internal energy (dU) of one mole of a system per degree rise in temperature when heated at constant volume. So $C_V = \left(\frac{dU}{dT}\right)_V$ Similarly molar heat capacity at constant pressure (C_p) can be defined as increase in enthalpy (dH) of one mole a system per degree rise in temperature, when heated at constant pressure. So $C_P = \left(\frac{dH}{dT}\right)_P$.

[For derivation $C_P - C_V = R$ for 1 mole see section 6.12]

36. Show that heat absorbed by a system at constant volume is equal to its internal energy increase. **[N.B.U.'95]**

Ans. From first law of thermodynamics we know dq = dU + W; where dq = heat absorbed; dU = increase in internal energy and W = work done. But W = PΔV. At constant volume ΔV = 0; so W = 0. Hence dq = dU. So heat absorbed at constant volume is equal to internal energy increase.

37. A Cannot's engine works between O°C and 127°C. How much work will be done if it absorbs 400 K.cal heat? [V.U. 2003]

Ans. From Carnot's cycle $\frac{W}{Q} = \frac{T_1 - T_2}{T_1}$

W = Workdone; Q = heat absorbed = 400 K. cal

T_1 = Source temp^2 = (273 + 127) = 400K; T_2 = Sink temp^2 = 273K

SU $\frac{W}{400} = \frac{400-273}{400}$; $W = \frac{400 \times 127}{400}$

or W = 127 K. Cal

CHEMICAL EAUILIBRIUM

1. What do you mean by K_C and K_P for a hyprothetical reaction $aA + bB \rightleftharpoons mM + nN$? K_C for the reaction $SO_2 + \frac{1}{2}O_2 \rightleftharpoons SO_3$ at 600°C is 61.7. Calculate K_P.

[B.U. 2011]

Ans. For a hypothetical reaction —

$aA + bB \rightleftharpoons mM + nN$

$K_C = \frac{(C_M)^m . (C_N)^n}{(C_A)^a . (C_B)^b}$; C = molar concentration

and $K_P = \frac{(P_M)^m . (P_N)^n}{(P_A)^a . (P_B)^b}$; P = Partial pressure.

For the reaction $SO_2 + \frac{1}{2}O_2 \rightleftharpoons SO_3$

$\Delta n = 1 - \left(1 + \frac{1}{2}\right) = -\frac{1}{2}$

Now $K_P = K_C . (RT)^{\Delta n}$

Here $K_C = 61.7$ R = 0.082 litre atm. K^{-1} mole^{-1} and $\Delta n = -\frac{1}{2}$

T = 273 + 600 = 873

So, $K_P = 61.7(0.082 \times 873)^{-\frac{1}{2}} = 7.29\ (atm)^{-\frac{1}{2}}$

2. Discuss the effect of T and P on the following reaction on the basis of Le-chatclier's principle. $N_2(g) + 3H_2(g) = 2NH_3(g)$; $\Delta H = -92$ K.J. mole^{-1}

[B.U. 2012]

Ans. For the given reaction $N_2(g) + 3H_2(g) = 2NH_3(g)$ 'ΔH' is negative. So it is an exothermic reaction. According to Le-chatclier's principle exothermic reactions are favoured by lowering the temperature. So here yield will be more at low T. If T is raised percentage of NH_3 will decrease.

In this reaction four moles of reactant give two moles of product ($\Delta n = -2$). So forward reaction is associated with decrease in volume. Hence if 'P' is raised yield will increase.

3. Under what condition K_P and K_C of a gaseous reaction become equal.

[B.U. 2012, C.U. 2011]

Ans. $K_P = K_C (RT)^{\Delta n}$; Δn = Difference in number of moles of gaseous products and reactants. If $\Delta n = 0$ then $K_P = K_C$. Example is $H_2(g) + I_2(g) = 2HI(g)$ reaction.

4. Is state of chemical equilibrium dynamic? **[C.U. 2010]**

Ans. Equilibrium is always dynamic in name. At equilibrium forward and backward reaction occur with exactly same speed. Here $\Delta G_{P,T} = 0$. This can be experimentally verified by radio–isotope.

5. Derive an expression for K_P in terms of degree of dissociation (α) and total pressure (P) for the following reaction.

Ans. For the reaction $COCl_2(g) = CO(g) + Cl_2(g)$; if we start with 1 mole $COCl_2$ and if α be degree of dissociation, then at equilibrium number of moles of reactant and products are — $n_{COCl_2} = 1-\alpha$; $n_{CO} = \alpha$ and $n_{Cl_2} = \alpha$

So total moles $\Sigma n = 1 - \alpha + \alpha + \alpha = 1 + \alpha$.

As partial presures = mole fraction × total pressure

So partial pressure are —

$$P_{COCl_2} = \frac{1-\alpha}{1+\alpha}.P\ ;\ P_{CO} = P_{Cl_2} = \frac{\alpha}{1+\alpha}.P$$

$$\text{or,}\quad K_P = \frac{P_{CO}\cdot P_{Cl_2}}{P_{COCl_2}} = \frac{\frac{\alpha}{1+\alpha}.P \times \frac{\alpha}{1+\alpha}.P}{\frac{1+\alpha}{1+\alpha}.P}$$

$$\text{or,}\quad K_P = \frac{\alpha^2.P}{1-\alpha^2}$$

6. Will equilibrium constants for the equations $\frac{1}{2}N_2 + \frac{3}{2}H_2 = NH_3$ and $N_2 + 3H_2 + 2NH_3$ be same? **[B.U. 2000; K.U. 2003]**

Ans. For the equation $\frac{1}{2}N_2 + \frac{3}{2}H_2 = NH_3$; equilibrium constant $K_1 = \frac{a_{NH_3}}{a_{N_2}^{1/2} . a_{N_2}^{3/2}}$; but for the equation $N_2 + 3H_2 + 2NH_3$; it is $K_2 = \frac{a_{NH_3}^{2}}{a_{N_2} \cdot a_{H_2}^{3}}$ (a = active mass). It is clear $K_2 = K_1^2$. So the numerical values are different.

PHASE RULE

1. State Henry's law regarding the effect of pressure on the solubility of gas in a liquid at a fixed temperature. **[B.U. 2011]**

Ans. See section 10.5.

2. State Nernst's distribution law corresponding to the solubility of a solute in the composite system containing two different immiscible solvents.
[B.U. 2012; C.U. 2010]

Ans. See section 10.6

3. Draw the lebelled phase diagram of sulphur system and discuss its salient features. [C.U. 2010]

Ans. See section 12.7 (c)

4. Write notes on :
(a) Gibbs phase rule (b) Eutectic mixture (c) steam distillation. **[C.U. 2010]**

Ans. (a) See section 12.5.

(b) Eutectic (means easy-melting) mixture is a mixture (of definite composition) of two solids, whose melting point is less than the melting point of either of the two solids, or their mixture with any other composition. It was initially misunderstood as a compound. But it is actually a mixture as its composition becomes different if 'P' is changed. For eutectic point degree of freedom is zero.

(c) See section 10.12.

5. Draw the labelled phase diagram of water system and explain it briefly.
[C.U. 2011]

Ans. See section 12.7(a).

6. 'Azeotrope is not a compound' — explain. **[C.U. 2011]**

Ans. Azcotrope is a liquid-liquid mixture of a particular composition, which boils with unchanged composition. For this peculiarity it was assumed that it may be a compound. But it is not a compound as its composition is pressure dependent. Had it been a compound composition should remain same.

COLLIGATIVE PROPERTIES

1. Write down the relation between osmotic pressure and concentration of a dilute solution and show that a dilute solution behaves as an ideal gas. **[K.U. 2012]**

Ans. If 'π' is the osmotic pressure of a dilute solution (concentration = C M/L) of non-volatile non-electrolyte solid solute at a temperature T, then the relation among them is

$\pi = CRT$

Now if 'n' mole solute is present in 'V' litre solution then $C = \frac{n}{V}$. So $\pi = \frac{n}{V}$, RT or

$\pi V = nRT$. This is similar to ideal gas equation PV=nRT, so a dilute solution behaves as an ideal gas.

2. 20 gm solute is dissolved in 1 litre solution. If osmotic pressure is 3 atm at 127°C, find out molar mass of solute, given R=0.082 litre atm, mole^{-1} K^{-1}. **[K.U. 2012]**

Ans. If M is molar mass then $20g \equiv \frac{20}{M}$ mole. It is present in 1 litre, so molar concentration

$C = \frac{20}{M}$ Now

$\pi = CRT$

Here, π = 3atm ; R = 0.082 litre atm mole^{-1} K^{-1} T = 273 + 127 = 400K

So, $3 = \frac{20}{M} \times 0.082 \times 400$

or, $M = \frac{20 \times 0.082 \times 400}{3} = 218.67$

3. Between 0.1 M urea solution and 0.1 (M) NaCl solution whose osmotic pressure is more and why? **[K.U. 2012]**

Ans. Urea is a non-electrolyte, so one molecule gives one particle. NaCl is a strong electrolyte, so one molecule gives two particles (Na^+ and Cl^- ions). Since osmotic pressure is a colligative property i.e. particle dependent, so osmotic pressure is more for 0.1(M) NaCl.

4. What is vant Hoff factor? **[K.U. 2012]**

Ans. For solultes, which undergo dissociation or association in solution, experimental values of colligative propertis are greater or less than the theoretically calculated values. Ratio of experimental value and theoretical value is called vant Hoff factor (i). So,

$$i = \frac{\text{Exptl value of colligative property}}{\text{Theo. value of that colligative property}}$$

For disociation cases i>1 and for association cases i<1.

5. Write down mathematical form of Raoult's law for lowering of vapour pressure. [K.U. 2012]

Ans. For dilute solution of non-volatile, non-electrolyte solid solutes, equation for Raoult's law for lowering of vapour pressure is

$$\frac{P^o - P}{P^o} = x_2$$

where P^o and P are vapour pressure of pure solvent and solution respectively and 'x_2' is mole fraction of solute.

6. Define ebullioscopic constant. What is its unit?

Ans. For dilute solution of non-volatile, non-electrolyte solid solute, Raoult's law for elevation of boiling point is $\Delta T_b = \kappa_b . C_m$. Here ΔT_b is elevation of boiling point; C_m = mola concentration. The constant κ_b is ebullioscopic constant. It is defined as elevation of boiling point for one molal (C_m=1) solution. Its unit is deg. kg. $mole^{-1}$.

7. Why boiling point of a solution is more than the solvent. **[K.U. 2011]**

Ans. Vapour pressure of a solution is less than that of solvent. A liquid boils at a temperature where its vapour pressure becomes equal to atmospheric pressure. So boiling point of solvent is less (of solution is more).

8. A dilute solution is prepared by dissolving w_2 gm of a non-volatile, non-eletrolyte solid solute (molecular weight M_2) in w_1 gm solvent (Mol. wt M_1). Express the concentration in molality (m_2) and mole-fraction (x_2) unit. **[B.U. 2011]**

Ans. Weight of solute = W_2 gm ;

Molecular weight of solute = M_2.

So mole of solute = $W_2/M_2 = n_2$

This is dissolved in W_1 gm solvent.

So, number of moles present in 1000 gm is $\frac{w_2 \times 1000}{w_1 \times M_2}$. This is molal concentration (m).

So $m = \frac{w_2 \times 1000}{w_1 \times M_2}$.

Again mole of solvent $= \frac{w_1}{M_1} = n_1$.

So, mole - fraction of solute $x_2 = \frac{n_2}{n_1 + n_2}$.

Or, $x_2 = \frac{w_1/M_1}{w_1/M_1 + w_2/M_2}$.

9. Draw the necessary vapour pressure (P) versus temperature (T) diagram for pure solvent and dilute solution. Hence show graphically how the boiling point of the solution will change. [B.U. 2011]

Ans. See section 11.3(a)

10. A solution of 1.73 gm 'A' in 100 C.C. water is found to be isotonic with 3.42% solution of sucrose. Calculate mol. wt. of A. [B.U. 2011]

Ans. We know expression for osmotic pressure (π) is

$$\pi V = \frac{w}{M} . RT \quad \text{or,} \quad \pi = \frac{w}{MV} . RT$$

(π)= osmotic pressure of V litre solution with 'w' gm solute (mol. wt = M) at temp. T]

In the given problem for A $\pi = \frac{1.73}{M \times 0.1} . RT$

and for sucrose $\pi = \frac{3.42}{342 \times 0.1} \times RT$

since they are isotonic, π is same and 'M' f or sucrose ($C_{12}H_{22}O_{11}$) is 342.

So, $\frac{1.73}{M \times 0.1} = \frac{3.42}{342 \times 0.1}$ or, $M = 173$ gm mole^{-1}

11. Mention two important limitation's of Raoult's law. [B.U. 2011]

Ans. Raoult's laws for colligative properties viz. lowering of vapour pressure, elevation of boiling point and depression of freezing point are valid for *dilute* solutions of *non-volatile* and *non-eletrolyte* solid solutes. So these are the limitations.

12. What is meant by osmosis? Define osmotic pressure of a dilute solution with a diagram. [B.U. 2012]

Ans. See section 11.5 (b)

13. State the vant Hoff's law relating to osmotic pressure and write the mathematical equation of the osmotic pressure for a dilute solution. [B.U. 2012]

Ans. See section 11.5(e) and 11.5(f).

14. Calculate the osmotic pressure of 0.1(M) KCl, solution at 27°C in S.I. Unit. [B.U. 2012]

Ans. Expression for osmotic pressure of an electrolytic solution is $\pi = iCRT$

π = osmotic pressure; i = vant Hoff factor;

C = Molar concentration; T = absolute temp.

KCl is a strong (1–1) electrolyte; it dissociates completely. So i = 2

Hence $\pi = 2 \times CRT$

In S.I. unit 'C' should be in mole/(metre)^3.

Here C = 0.1 mole/litre i.e. mole.(deci-metre)^3

So, C = 0.1×10^3 mole/(metre)^3 in S.I. unit.

$$\therefore \pi = 2 \times 0.1 \times 10^3 \times 8.314 \times 300 \text{ pascal}$$
$$= 4.988 \times 10^5 \text{ pascal}$$

CHEMICAL KINETICS – CATALYSIS – ADSORPTION

1. What is the unit of specific reaction rate for a second order reaction? [K.U. 2012]

Ans. Rate equation for a second order reaction is $-\frac{dc}{dt} = kc^2$; 'k' is specific reaction rate so unit of 'R' is (conc.)$^{-1}$ time^{-1}. If 'concentration is in moles/litre and 'time' in second unit, then unit of 'k' is litre mole^{-1} sec^{-1}.

2. Derive an expression for rate constant for a first order reaction. Show that in this case half-life does not depend on initial concentration. [K.U. 2012]

Ans. See section 13.9 and 13.10

3. Rate constants of a reaction are 2.45×10^{-5} sec^{-1} and 16.2×10^{-4} sec^{-1} at 273K and 303K respectively. Calculate activation energy of the reaction. [K.U. 2012]

Ans. From Arrhenius equation

$$ln = \frac{k_2}{k_1} = \frac{E}{R} = \frac{T_2 - T_1}{T_1 T_2}$$

Here, $T_1 = 273K$ $k_1 = 2.45 \times 10^{-5} sec^{-1}$ $R = 1.987$ Cal K^{-1} $mole^{-1}$

$T_2 = 303$ K $k_2 = 16.2 \times 10^{-5}$ sec^{-1}

$$\text{So, } ln\frac{16.2 \times 10^{-4}}{2.45 \times 10^{-5}} = \frac{E}{1.987} \times \frac{303 - 273}{303 \times 273}$$

Or, E = 22964.2 Cal = 22.9642 K Cal

4. Describe a process for determining order of a reaction. [K.U. 2012]

Ans. See section13.21

5. Derive an expression for rate constant of a second order reaction with two reactants with different initial concentration [K.U. 2011]

Ans. See section13.14

6. How can you know whether a reaction is of first order or second order? [K.U. 2011]

Ans. For first order reaction half life is constant and independent of initial concentration For second order reaction half-life is inversely proportional to initial concentration.

7. Write down the integrated form of rate equations for first and second order chemical reactions considering same initial concentration of the reactants. Using these forms obtain the units of the corresponding rate constants. [B.U. 2011]

Ans. If 'a' be initial concentration(s) of the reactant(s) and 'x' amount decomposes in time 't', then expression for rate constants for first order (k_1) and second order (k_2) are as follows —

$$k_1 = \frac{2.303}{t} \log \frac{a}{a-x}$$

and $k_2 = \frac{1}{t} \cdot \frac{a}{a(a-x)}$

unit of k_1 is $time^{-1}$ e.g. sec^{-1}

unit of k_2 is $(conc)^{-1}$ $time^{-1}$ e.g. litre $mole^{-1}$ sec^{-1}

8. If temperature of the system undergoing first order reaction is increased how the value of half-life period change. [B.U. 2011]

Ans. For 1st order reaction half-life ($t_{1/2}$) is $t_{1/2} = \frac{0.693}{k_1}$; k_1 = rate constant. Now from Arrhenius eq^n $k_1 = A.e^{-E/RT}$. [A = frequency factor; E = activation energy]. So if T is increased k_1 will increase, hence $t_{1/2}$ will decrease.

9. Calculate the half-life of the decomposition reaction at 300K of a substance with initial concentration 0.05 moles/litre. Given rate constant = 6.8×10^{-4} litre mole^{-1} sec^{-1} **[B.U. 2011]**

Ans. Given rate constant (k) = 6.8×10^{-4} litre mole^{-1} sec^{-1}. Unit of 'k' indicates that it is a second-order reaction. So its half-life $t_{1/2} = \frac{1}{k.a}$ [a = initial concentration]

$$t_{1/2} = \frac{1}{6.8 \times 10^{-4} \times 0.05} = 2.94 \times 10^{4} \text{ secs.}$$

10. Explain the significance of the terms 'k', 'A' and 'E_a' in the following Arrhenius equation $k=A.e^{-Ea/RT}$. **[B.U. 2011; B.U.2012; C.U. 2011]**

Ans. In the Arrhenius equation $k = A.e^{-E/RT}$, 'k' is rate constant or specific reaction rate. It is actually rate of the reaction, when concentration of reactants are specified as unity.

'A' is frequency factor. It is a masurement of number of collisions among reactant molecules occuring /c.c./sec.

'Ea' is activation energy i.e. minimum energy required by a reactant molecule to undergo chemical reaction.

11. Prove that $t_{75\%} = 2\times t_{50\%}$ for a first order reaction. **[C.U. 2012]**

Ans. For first order reaction expression for rate constant is $k = \frac{2.303}{t} \log \frac{a}{a-x}$

So, $t = \frac{2.303}{k} \log \frac{a}{a-x}$

When $t = t_{50\%}$ $x = \frac{a}{2}$

So, $t_{50\%} = \frac{2.303}{k} \log \frac{a}{a/2} = \frac{2.303}{k} \log 2$....................(1)

When $t = t_{75\%}$ $x = \frac{3}{4}a$

$t_{75\%} = \frac{2.303}{k} \log \frac{a}{a - \frac{3}{4}a} = \frac{2.303}{k} \log 4$....................(2)

$\therefore \frac{t_{75\%}}{t_{50\%}} = \frac{\log 4}{\log 2} = \frac{2\log 2}{\log 2} = 2$

$\therefore t_{75\%} = 2 \times t_{50\%}$

12. A given reaction is either first order or second order. How the order of the said reaction can be identified by a simple method. [B.U. 2012]

Ans. We start the reaction with initial concentration C. If time taken for conversion of conc. from C to $\frac{C}{2}$ is same as the time taken for conversion form $\frac{C}{2}$ to $\frac{C}{4}$ then it is first order.

But for second order time taken in the second case $\left(\frac{C}{2} \rightarrow \frac{C}{4}\right)$ will be twice the time taken in the first case $\left(C \rightarrow \frac{C}{2}\right)$.

13. Write down the rate equation of a zero-order reaction and derive mathematical expression for rate constant. Show that half-life period of zero-order reaction is directly proportional to initial concentration of the reaction. [C.U. 2010]

Ans. Sec section 13.17

14. Show that for a first order reaction, the time required for completion of 99.9% is about 10 times the time required for 50% completion of the reaction. [C.U. 2010]

Ans. Expression for rate constant of a first order reaction is —

$$k = \frac{2.303}{t} \log \frac{a}{a-x} \text{ [a = Initial conc}^n\text{; x = amount decomposed in time 't']}$$

So, $t = \frac{2.303}{k} \log \frac{a}{a-x}$

If a = 100 then time for 50% decompostion is $t_{50} = \frac{2.303}{k} \log \frac{100}{100-50} = \frac{2.303}{k} \log 2$

$= \frac{2.303}{k}.0.3$ and time for 99.9% decompositon is, $t_{99.9} = \frac{2.303}{k} \log \frac{100}{100-99.9}$

$= \frac{2.303}{k} \log 10^3 = \frac{2.303}{k}.3.$

$\therefore \frac{t_{99.9}}{t_{50}} = \frac{3}{0.3} = 10$

15. Write notes on energy of activation [C.U. 2010]

Ans. See section 13.23.

16. Write notes on enzyme catalysis. [C.U. 2010]

Ans. See section 14.8

17. What do you mean by molecularity and order of a reaction? [C.U. 2010]

Ans. Sec. section 13.8.

18. Define autocatalysis, citing an example. **[C.U. 2010]**

Ans. If one of the products of a reaction begin to catalyse the reaction then it is called autocatalysis. Thus during oxidation of $C_2O_4^{2-}$ ion by MnO_4^-, Mn^{2+} produced, catalyses the reaction. That is why rate is slow at the start (no Mn^{2+} ion) but increases later due to Mn^{2+} ion produced.

19. What do you mean by half life period of a reaction? Show that for a first order reaction half-life period ($t_{1/2}$) does not depend on initial concentration of the reactant. **[C.U. 2011]**

Ans. See section 13.1.

20. A first order reaction completes 50% in 1 hour. Calculate the time for 90% completion of the same reaction. **[C.U. 2011]**

Ans. For first order reaction, rate constant (k) is $k = \frac{2.303}{t} \log \frac{a}{a-x}$; where 'a' is initial concentration and 'x' amount reacts in time 't'. Now if a = 100 and x = 50 then t = 1 hr.

$$\text{So, } k = \frac{2.303}{1} \log \frac{100}{100-50} = 2.303 \log 2 = 0.693$$

So for 90% reaction —

$$t = \frac{2.303}{k} \log \frac{a}{a-x} = \frac{2.303}{0.693} \log \frac{100}{100-90}$$

$$t = \frac{2.303}{0.693} = 3.32 \text{ hour}$$

21. Write notes on catalyst promotor. **[C.U. 2011]**

Ans. See section 14.6

22. What is adsorption? On what factors does it depend? **[N.B.U 1999]**

Ans. Sticking of foreign materials on the surface of solid or liquid (surface has excess free energy due to unbalanced backward attraction) is called adsorption.

It depends on (i) nature of adsorbate (ii) nature of adsorbent (iii) nature of surface (iv) temperature and (v) pressure or cencentration.

23. Distinguish between adsorption and absorption. **[N.B.U 1999]**

Ans. Sticking on the surface is adsorption. It is a surface phenomena. Thus charcoal adsorbs ammonia; it (ch coal) also adsorbs acetics acid from its solution.

Penetration inside the bulk is absorption. Thus blotting paper absorbs ink; sponze absorbs water.

24. Compare between physical and chemical adsorption. **[N.B.U 1999]**

Ans. See section 14.12

25. Derive Langmuir's adsorption isotherm and arrive at Freundlich's isotherm. **[N.B.U. 2000; 2001; 2006]**

Ans. See section 14.16

COLLOID

1. To transform a fixed amount of muddy water in to clear water, why potash-alum is more effective than lime. **[B.U. 2011]**

Ans. Muddy water is a sol (colloid) of clay in water. To transform it to clear water sol-particles are to be co-agulated. Co-agulation is done by adding electrolyte. According to Schulze-Hardy rule ions with higher valency are more effective. In potash alum we have Al^{3+} besides K^{+} and SO_4^{2-}; but in lime we get Ca^{2+} and OH^{-} ions. So potash alum is more effective.

2. Define with example : lyophillic colloids and association colloids. **[B.U. 2011]**

Ans. Lyophillic colloids are those where sol-particles (dispersed phase) have a strong affinity for the solvent (dispersion medium). Sols of organic substances e.g. starch; blood etc. are examples. They are highly stable.

In a true solution, solute particles are of 1–10A° dimension. If some how, the particles can be associated and dimension is raised to 10 to 1000A°, then a colloid will be produced. This is association colloid. One example of such process is solvent change. Thus if to an alcoholic solution of phenolphthalien water is added we get a sol.

3. 'The gold number of haemoglobin is 0.03' – explains. **[B.U.2012]**

Ans. Gold number is a measurement of protective capacity of lyophillic colloids. When we say gold number of haemoglobin is 0.03, we mean, a minimum 0.03 mg. of haemoglobin should be added to 10 c.c. gold sol (~ 0.0053 %) to check its co-agulation during addition of 1 c.c. 10% NaCl solution.

4. Write notes on —
(i) Brownian motion (ii) Iso-electric point (iii) Electrodialysis (iv) Tyndall effect. **[B.U. 2012]**

Ans. See sections 15.8 (d) (e) ; 15.6 and 15.8 (c)

5. Write down the difference between the process co-agulation and peptization. **[B.U. 2012]**

Ans. Co-agulation is the process of destroying a colloid by addition of electrolyte. Thus clay particles present in muddy water can be co-agulated by adding alum (electrolyte).

Peptization is the process of forming a colloid by striking a fresh ppt. by solvent. Thus if a freshly precipitated $Fe(OH)_3$ is struck by water in the form of a jet from a wash battle, it is peptized to form a colloid.

6. Write note on dialysis. **[C.U. 2010]**

Ans. Sec section '15.6

7. Write note on Schulze-Hardy rule. **[C.U. 2010]**

Ans. See section 15.12

8. What is understood by electrophoresis of colloid? What is the origin of such property? **[C.U. 2010]**

Ans. Sec sections 15.8(e) and 15.9

9. Define gold number. **[C.U. 2010; C.U. 2011]**

Ans. See section 15.13 written in bold letters.

10. $Al_2(SO_4)_3$ is more effective than K_2SO_4 in co-agulating a sol— why? **[C.U. 2011]**

Ans. See answer to question no '1'.

11. Write notes on Brownian motion. **[C.U. 2011]**

Ans. See section 15.8(d).

ELECTROLYTIC CONDUCTANCE

1. What is conductometric titration? Draw and explain the conductometric titration curve for the titration of a strong acid by strong base. **[K.U. 2012 ; C.U. 2010]**

Ans. Sec section 16.20(c)

2. Define equivalent conductivity of a solution. How dos it change with dilution.
[K.U. 2012]

Ans. See section 16.12 (for definition) and 16.16

3. Define transport number of an ion. **[K.U. 2012]**

Ans. Transport number of an ion is fraction of the total charge carried by that ion. Thus if q_+ and q_- are charges carried by cation and anion, then respective transport numbers are $t_+ = \frac{q_+}{q_+ + q_-}$ and $t_- = \frac{q_-}{q_+ + q_-}$

4. Equiv. conductance at infinite dilution for Na-butyrate; NaCl and HCl are 83, 127 and 426 $ohm^{-1}\ cm^2\ equiv^{-1}$ at 23°C. Find out equiv conductance at infinite dilution for butyric acid. **[K.U. 2012]**

Ans. From kohtrausch's law —

$$\wedge^o_{HA} = \wedge^o_{HCl} + \wedge^o_{NaA} - \wedge^o_{NaCl}$$

So $\wedge^o_{\text{butyric acid}} = 426 + 83 - 127 = 382\ ohm^{-1}\ cm^2\ equiv^{-1}$

5. In KCl solution transport numbr of K^+ is 0.49; What is the transport number of Cl^-? **[K.U. 2011]**

Ans. If 't' represents transport number, then in KCl solution $t_{K+} + t_{Cl-} = 1$ [Here $t_{k+} = 0.49$]
So, $t_{Cl} = 1 - 0.49 = 0.51$

6. Establish the reaction between specific and equivalent conductance.
[K.U. 2011, B.U.2011]

Ans. We know conductance (C) of a solution ; is $C = \kappa \times \frac{A}{l}$ ....(1) where κ = specific conductanc :A = cross-sectional area of each electrode and l = distance between two electrodes. Now euqivalent conductance (∧) of a solution is the conductance of a solution containing one equivalent of electrolyte placed between two electrodes, which are one centimetre apart. If C(N) is concentration, then C equiv. of electrolyte is present in 1000 c.c. sol^n. Hence 1 equiv is present in $\frac{1000}{C}$ c.c. sol^n. If we are to keep it between two electrodes one centimetre apart (l = 1cm). then area of each electrode (A) should be $\frac{1000}{C}$ Cm^2. So eq^n (1) becomes

$$\wedge = \kappa \times \frac{10^3}{C}$$

This is the relation.

7. What are the units of specific and equiv. conductances. **[K.U. 2011]**

Ans. Unit of specific conductance is ohm^{-1} cm^{-1} (in S.I. seimens $metre^{-1}$)

Unit of equivalent conductance is ohm^{-1} cm^2 eq^{-1} (in S.I. seimens $metre^2$ $equivalent^{-1}$)

8. State and explain Kohltausch's law **[K.U. 2011; C.U. 2010]**

Ans. See chapter 16.18

9. At 25°C equivalent conductance at infinite dilution for NaOH, NH_4Cl and NaCl are 248, 149.5 and 126.45 mho cm^2 $equiv^{-1}$. Calculate equivalent conducatance at infinite dilution of NH_4OH. **[K.U. 2011]**

Ans. From Kohtraush's law –

$$\wedge^o_{NH_4OH} = l^o_{NH_{4+}} + l^o_{OH\_} \quad [\wedge^o = \text{Eq. conductance at infinite dil}^n.$$

$$l^o = \text{Eq. ionic conductance}]$$

$$\text{or} \quad \wedge^o_{NH_4OH} = (l^o_{NH4+} + l^o_{Cl^-}) + (l^o_{Na+} + l^o_{OH^-}) - (l^o_{Na+} + l^o_{Cl^-})$$

$$= \wedge^o_{NH_4Cl} + \wedge^o_{NaOH} - \wedge^o_{NaCl}$$

$$= 149.5 + 248 - 126.45 = 271.05 \text{ ohm}^{-1} \text{ cm}^2 \text{ eq}^{-1}.$$

10. Why direct current (D.C.) is never used for accurate measurement of electrical conductivity of an electrolyte solution? **[B.U. 2011]**

Ans. See section 16.18; last part marked 'Q'.

11. How could you differentiate a strong electrolyte from a weak electrolyte by the measurement of equivalent conductance at different concentration? **[B.U. 2011]**

Ans. Equivalent conductance of both strong and weak electrolyte increase with increase in dilution. But for strong electrolytes successive dilution very soon gives a constant value. which does not change further. But for weak electrolyte equivalent conductance does not become constant; it increases at a very very slow rate with dilution.

12. Explain why the observed conductance of 1(M) aqueous solution of HCl is much higher than corresponding NaCl solution. **[B.U. 2011]**

Ans. Conductance of HCl is due to conductance of H^+ and Cl^- ions; whereas conductance of NaCl is due to conductance of Na^+ and of Cl^- ions. Now due to Grothus type of conductance, conductance value of H^+ (~350 in appropriate unit) is higher than that of Na^+ (~50 at same unit at 25°C) . So conductance of HCl is much higher.

13. The specific conductance of 0.1 (M) NaOH is 0.22 ohm^{-1} cm^{-1}. When equal volume of 0.1(M) HCl is added, the specific conductance falls to 0.0056 ohm^{-1} cm^{-1}. Find the equivalent conductance of NaCl solution. [B.U. 2011]

Ans. When equal volume 0.1(M) HCl is added to 0.1(M) NaOH, the resultant mixture is a solution of NaCl (due to complete neutralisation) with conc. $\frac{0.1}{2} = 0.05$ (M) (volume is doubled, so concentration is halved. Its opecific conducatance (κ) is 0.0056; So equiv. conductance '^' is $\wedge = \kappa \times \frac{1000}{C} = 0.0056 \times \frac{1000}{0.05} = 112$ ohm^{-1} cm^{-1} $equiv^{-1}$.

14. Explain why we do not find the equivalent conductance of a salt as exactly equal to the sum of ionic conductance of the components at moderate concentration. [B.U. 2011]

Ans. In a salt solution of moderate concentration, large number of ions are present in small space. So there will be strong inter-ionic attraction between ions of opposite charge. An ion-atmosphere of opposite ion is built surrounding an ion. This atmosphere reduces ionic velocity. That is why equivalent conductance becomes less than theoretically calculated value from ion conductance.

15. Write notes on abnormal transport numbers in concentrated CdI_2 solution. [B.U. 2011]

Ans. See section 16.10.

16. "The specific conductance of a solution containing a definite weak electrolyte increases with dilution" — Justify the statement. [B.U. 2011]

Ans. We know specific conductance of a solution is its conductance enclosed in a unit cube. For strong electrolytes (which are fully dissociated) dilution decreases number of ions present in a unit cube. So specific conductance decreases with dilution.
for weak electrolytes dilution increases degree of dissociation. So number of ions per unit cube may be more in case of dilute solution due to greater extent of dissociation. Hence specific conductance increases with dilution in such cases.

17. Calculate the equivalent conductance of a 0.01 (N) KCl solution at 18°C, whose specific conductance is 0.001225 mho cm^{-1} [B.U. 2012]

Ans. Equivalent conductance (^) and specific conductance (κ) are related as

$$\wedge = \kappa \times \frac{10^3}{C} \quad [C = Conc^n \text{ in normality unit}]$$

So here $\wedge = 0.001225 \times \frac{10^3}{0.01} = 122.5$ ohm^{-1} cm^2 $equiv^{-1}$

IONIC EQUILIBRIUM

1. In every acid-base titration pH at end point is not necessarily seven — why? **[K.U. 2012]**

Ans. When an acid reacts with a base it produces a salt. Salts of strong acid-weak base, weak acid-strong base and weak acid—weak base undergo hydrolysis. During hydrolysis either excess H^+ or OH^- is produced. So the solution becomes acidic or alkaline. Hence pH is not 7 (at 25°C)

2. What will be the pH of the mixture when 10 ml. 0.2(N) KOH is mixed with 30 ml. 0.1 (N) CH_3COOH. [$K_a = 2 \times 10^{-5}$] **[K.U. 2012]**

Ans. We know 1000 C.C. 1(N) sol^n contain 1 equiv. solute

So, 10 ml. 0.2 N KOH contains $\frac{10 \times .2}{1000} = \frac{2}{1000}$ eq. KOH

Similarly 30 ml 0.1 N CH_3COOH contains $\frac{30 \times 0.1}{1000} = \frac{3}{1000}$ eq. CH_3COOH

When they are mixed the mixture will be a buffer with $\frac{3-2}{1000} = \frac{1}{1000}$ equiv. weak acid

and $\frac{2}{1000}$ equiv. salt.

So, $\frac{C_{salt}}{C_{acid}} = \frac{2}{1}$, we know pH of a buffer is $pH = P_{ka} + \log \frac{C_{salt}}{C_{acid}}$

$= - 10g\ (2 \times 10^{-5}) + \log 2 = 5$

3. What is the pH of a 0.05 (M) H_2SO_4? **[K.U. 2012]**

Ans. H_2SO_4 is a strong acid; it dissociates completely. So a 0.05 (M) H_2SO_4 will produce $2 \times 0.05 = 0.10$(M) H^+ is sol^n.

Now $pH = -\log C_{H^+} = -\log(0.1) = 1$

So its pH will be 1.

4. What is Buffer solution? **[K.U. 2012; B.U. 2012, C.U. 2010]**

Ans. Buffer solutions are those solutions, whose pH remain more or less same during addition of little amount of acid or alkali. These are made by mixing (i) weak acid and its salt (ii) weak base and its salt (iii) two acid salts of same acid.

5. What will happen if HCl(g) is passed through saturated NaCl soluion? [K.U. 2012]

Ans. NaCl is a highly soluble salt. So its solubility product value [$K_s = C_{Na^+}, C_{Cl^-}$ in saturated solution] is very high. If we send HCl(g) in a saturated solution of NaCl,

C_{Na+} does not change, (solution is not diluted as gas is passed) but C_{Cl-} increases. So the ionic product (C_{Na^+}, C_{Cl^-} exceeds solubility product, Hence pure NaCl is precipitated.

6. Cations of group-II are precipitated as sulphides in acid medium but cations of group-III B are precipitated in alkaline medium — why? [K.U. 2012]

Ans. Sulphides of group II have very very small solubility product value. So very small amount of S^{2-} ion is required for their precipitation. If we take them in acid medium and pass H_2S (weak dibasic acid), then due to common ion effect we get very small S^{2-} which is sufficient for precipitation of group II sulphides. But group IIIB metal sulphides have relatively high solubility product value. So they require large amount of S^{2-} ion. This is done by passing H_2S in NH_4OH medium where the salt $(NH_4)_2S$ is produced, which dissociates to give large amount of S^{2-} ion.

7. What are pH of 10^{-3} (N) HCl and 10^{-3} (N) NaOH? [K.U. 2012; B.U. 2011]

Ans. In 10^{-3} NHCl $\quad C_{H+} = 10^{-3}$

So, pH = $-\log(10^{-3}) = 3$

In 10^{-3} N NaOH ; $C_{OH-} = 10^{-3}$ so $C_{H+} = \frac{K_w}{C_{OH^-}}$

[As ionic product of water $K_w = C_{H^+} \cdot C_{OH^-}$]

But $K_w = 10^{-14}$ so $C_{H+} = \frac{10^{-14}}{10^{-3}} = 10^{-11}$

Hence its $p_H = -\log(10^{-11}) = 11$

8. Dissociation constant and ionic product of water are not same — explain. [K.U.'12; B.U.'11]

Ans. Dissociation constant (K) of water is $K = \frac{C_{H^+} \cdot C_{OH^-}}{C_{H_2O}}$;

ionic product $K_w = C_{H^+} + C_{OH^-}$

So $K \times C_{H_2O} = K_w$

$C_{H_2O} = 55.5$ M/L

So $K \times 55.5 = K_w$

9. Why a solution of sodium acetate is alkaline? [K.U. 2012]

Ans. Sodium acetate is a salt of strong base (NaOH) and weak acid (CH_3COOH). In solution anion of the salt (CH_3COO^-) will hydrolyse as —

$CH_3COO^- + H_2O = CH_3COOH + OH^-$

Due to formation of OH^- ; The solution will be alkaline.

10. At 25°C dissociation constant of acetic acid is 1.778×10^{-5}; Calculate pH of its ..... solution. (neglect degree of dissociation in comparison to concentration).
[K.U. 2012; B.U. 2012]

Ans. Expression for pH of a solution of weak acid is $pH = \frac{1}{2} pk_a - \frac{1}{2} \log C$ During derivation of this equation it was assumed that degree of dissociation is negligible.

In this case $K_a = 1.778 \times 10^{-5}$

So $p_{ka} = -\log K_a = 4.75$

$C = \frac{1}{10}(N) \quad \log C = -1$

$\therefore \quad pH = \frac{1}{2} pk_a - \frac{1}{2} \log C$

$= \frac{1}{2} \times 4.75 - \frac{1}{2}(-1) = 2.38 + 0.5 = 2.88$

11. State with reason whether a $FeCl_3$ solution will be acidic or alkaline.
[K.U. 2011; B.U. 2011]

Ans. $FeCl_3$ is a salt of weak base $Fe(OH)_3$ and strong acid HCl. So in solution its nature will be acidic due to cationic hydrolysis —

$Fe^{3+} + 2H_2O \rightleftharpoons Fe(OH)_3 + 3H^+$

12. If solubility product of Ag_2CO_3 is 8×10^{-12}, what will be its solubility?
[K.U. 2011]

Ans. In a saturated sol^n of Ag_2CO_3, following equilibrium $Ag_2CO_3(S) = 2Ag^+ + CO_3^{2-}$ exists. If its solubility is 'S' moles/litres, then $C_{Ag+} = 2S$ and $C_{CO_3^{2-}} = S$. So solubility product $K_s = (C_{Ag+})^2 . CO_3^{2-} = (2S)^2 \times S = 4S^3$

So, $4S^3 = 8 \times 10^{-12}$

$S^3 = 2 \times 10^{-12}$

$\therefore \quad S = 1.26 \times 10^{-4}$ M/L

13. Why CH_3COONH_4 solution behaves as a buffer? [K.U. 2011]

Ans. CH_3COONH_4 is a salt. So in solution it dissociates completely producing large amount of CH_3COO^- & NH_4^+ ions. To this solution it we add few H^+ ions it will be captured by CH_3COO^- to produce undissociated CH_3COOH. On the contrary if we add few OH^- ions; NH_4^+ will capture these to give undissociated NH_4OH. In both cases pH remain same. So it will behave as a buffer solution.

14. Derive Ostwald's dilution law. Express it in terms of equivalent conductance of a weak electrolyte. **[K.U. 2011]**

Ans. See section 17.1 and last few lines of section 17.3.

15. Dissociation constant of aqueous solution of ammonia is 2.5×10^{-5}; calculate its degree of dissociation in a 0.01 M solution. **[K.U. 2011]**

Ans. We know for a C(M) solution of NH_3, dissociation constant $K_b = \frac{\alpha^2 C}{1-\alpha}$; α = degree of dissociation. Usually $\alpha << 1$. So $K_b \approx \alpha^2 C$ or $\alpha = \sqrt{\frac{K_b}{C}}$

Putting the given values $\alpha = \sqrt{\frac{2.5 \times 10^{-5}}{0.01}} = 5 \times 10^{-5} = 0.05$.

16. What is the characteristic of an indicator? **[K.U. 2011]**

Ans. Indicators are those substances which indicate end-point of titration processes by means of colour change. Acid-base indicators have two tautomeric structures with different colours; which colour will be visible, that depends on p_H of the medium.

17. What is pH of blood? **[K.U. 2011]**

Ans. Blood is alkaline in nature having pH value 7.4.

18. Calculate degree of hydrolysis of a 0.05 (N) aqueous sodium acetate solution. Dissociation constant of acetic acid is 1.74×10^{-5}. **[K.U. 2011]**

Ans. Sodium acetate is a salt of weak acid and strong base. So its hydrolysis constant (K_h) is

$$K_h = \frac{Kw}{Ka}$$

Kw = Ionic product of water = 10^{-14} (at 25°C)

Ka = 1.74×10^{-5}

$$\therefore\ K_h = \frac{10^{-14}}{1.74 \times 10^{-5}} = 5.7 \times 10^{-10}$$

It 'h' is degree of hydrolysis, then

$$K_h = \frac{h^2 C}{1-h} \approx h^2 C \quad [\text{as } h<<1]$$

$$\therefore\ h = \sqrt{\frac{K_h}{C}} = \sqrt{\frac{5.7 \times 10^{-10}}{0.05}} = 1.07 \times 10^{-4}$$

19. Derive Henderson equation. State with reason which indicator is to be used during titration of CH_3COOH with NaOH. **[B.U. 2011; 2012]**

Ans. [For derivation of Henderson equation – see 'pH of buffer in section 17.9]

CH_3COOH is a weak acid, whereas NaOH is a strong base. Hence during titration pH at end point will be greater than 7 due to hydrolysis. Here pH near end point shows a sharp change from ~7 to ~11. So indicators, whose P_{KIn} value fall in this range are to be selected. Phenolphthalien is such an indicator.

20. What is the pH of pure water at normal tem^{r} (25°C)? **[B.U. 2011]**

Ans. pH of water (pure) at 25°C is 7.

21. Establish the relation $K_h = \frac{Kw}{Ka}$ for hydrolysis of a salt sol^n of weak acid and strong base. **[B.U. 2011]**

Ans. See section 17.10(b).

22. What is common ion effect? Give an example. **[B.U. 2011]**

Ans. See section 17.4.

23. Write notes on Ostwald's dilution law. **[B.U. 2011]**

Ans. Weak electrolytes are partially dissociated. There is always an equilibrium between undissociated and dissociated part. The equilibrium constant for this dissociation equilibrium is called dissociation constant. If concentration of the weak electrolyte is C(M) and 'α' be its degree of dissociation, then dissociation constant (K) is $K = \frac{\alpha^2 C}{1-\alpha}$ Dilution will decrease 'C', so to keep 'K' constant, 'α' must increase. Hence there will be increase in dissociation with increase in dilution". This is Ostwald's dilution law.

24. 10^{-8} HCl does not show pH =8. Why? **[B.U. 2012]**

Ans. In any HCl solution.

$$C_{H+} = C_{H+} \text{ (from HCl)} + C_{H+} \text{ (from } H_2O)$$

In H_2O at 25°C. $C_{H+} = 10^{-7}$, that is why its pH is 7. When concentraiton of HCl is appreciable, we neglect C_{H+} from water during pH calculation. Thus for 0.1(N) HCl we take pH = –log 10^{-1} = 1. But in this case (10^{-8} N HCl) C_{H+} coming from HCl is less than C_{H+} coming from water. So we are to take it in to account. Hence pH will be little less than 7.

25. Aqueous solution of NH_4Cl is acidic. Explain. **[B.U. 2012]**

Ans. NH_4Cl is a salt of strong acid HCl and weak base NH_4OH. So it will undergo cationic hydrolysis as $NH_4^+ + H_2O = NH_4OH + H^+$. Due to production of H^+, it will be acidic.

26. Why is Ostwald dilution law not applicable for strong electrolyte? [C.U. 2010]

Ans. Ostwald's diluiton law states that extent of dissociation will increase will dilution. So it is valid for those electrolytes, which are partially dissociated; strong electrolytes are completely dissociated. So here the law is not applicable.

ELECTROMOTIVE FORCE

1. What do you understand by the statement. "Redox potential of Fe^{3+}/Fe^{2+} couple is +0.77 volt at 25°C. [K.U. 2012; C.U. 2010]

Ans. The statement — "redox potential of Fe^{3+}/Fe^{2+} couple is +0.77 volt" means the e.m.f. of a cell whose left hand electrode is standard hydrogen electrode and right hand electrode is Pt/Fe^{3+}, Fe^{2+} electrode (with unit activities) is +0.77V. i.e. for the cell—

(Pt) H_2(1 atm) | H^+ (a=1)|| Fe^{3+} (a=1); Fe^{2+}(a=1) | P; E_{cell} = 0.77 volt at 25°C.

2. Derive Nernst's equation for e.m.f. of a cell. [K.U. 2012]

Ans. Inside an electrochemical cell the reaction that occurs is nothing but a redox reaction. Let the reaction is —

rR = pP

Then from vant Htoff's isotherom free energy change ΔG is —

$$\Delta G = \Delta G^\circ + RT \ln \frac{a_P{}^p}{a_R{}^r}$$

Where ΔG° is standard free energy change. But if during reaction number of electrons exchanged = n and 'E' is the e.m.f. of the cell inside which the reaction occurs, then ΔG = –nFE and ΔG° = – nFE°.

$$\text{So } -nFE_{Cell} = -nFE^\circ_{Cell} + RT \ln \frac{a_P{}^p}{a_R{}^r}$$

$$\text{or , } E_{Cell} = E^\circ_{Cell} + \frac{RT}{nF} \ln \frac{a_R{}^r}{a_P{}^p}$$

This is Neinst's equation for e.m.f.

3. How can you measure pH of a solution by e.m.f. method? [K.U. 2012]

Ans. See chapter 18.19(a).

4. Standard reduction potentials for Zn^{2+}/Zn and Ag^{+}/Ag (E°) are –0.77 V and +0.799V. Construct a cell with these electrodes and calculate its standard e.m.f.

[K.U. 2012]

Ans. If we construct a cell with $Zn|Zn^{2+}$ & $Ag|Ag^{+}$ electrodes it will be $Zn|Zn^{2+}(a_0)||Ag^{+}(a)|Ag$. Its $E^{\circ}_{cell} = E^{\circ}_{Ag^{+}/Ag^{-}} - E^{\circ}_{Zn^{2+}|Zn} = 0.799 + 0.77 = 1.569$ V.

5. Write down electrode reaction of a calomel electrode and derive an expression for its potential. **[K.U. 2011]**

Ans. See section 18.11.

6. What is calomel electrode? How will you determine the potential of calomel electrode? **[C.U. 2010]**

Ans. (For first part see section 18.11)

Pontential of a calomel electrode can be known by constructing the following cell and measuring its e.m.f.

(Pt) H_2(1atm) | H^{+} (a=1)|| satd. calomel.

For this cell left hand electrode is a standard hydrogen electrode, whose potential is zero. So $E_{Cell} = E_{Calomel}$. Measuring E_{Cell} by a potentiometer $E_{Calomel}$ can be known.

PHOTOCHEMISTRY

1. What do you mean by 'Quantum yield' of a photochemical reaction? Give an example of low and high quantum yield. **[C.U. 2010]**

Ans. See sections 20.7 and 20.8.

2. Write notes on 'fluoroscence'. **[C.U. 2010]**

Ans. See section 20.11.

3. State Stark-Einstein law of photochemistry. **[C.U. 2010]**

Ans. See section 20.6.

4. What is molar extinction co-efficient? **[C.U. 2010]**

Ans. From Lambert Beer's law $\log\frac{I_0}{I} = \varepsilon.cl$. Here I_0 and I are intensities of incident and transmitted light. C is molar concentration of a solution taken in a photochemical cell of length l and ε is molar extinction co-efficient. Now if $\frac{I_0}{I} = 10$ and C = 1(M) ; then

$\varepsilon = \frac{1}{l}$ So molar extinction co-efficient is the reciprocal of cell length (filled with a 1(M) solution). required to reduce the intensity of transmitted light by ten times.

5. State Lambert - Beer's law and establish the integrated form of the mathematical expression of the law. **[C.U. 2011]**

Ans. See section 20.4

6. Calculate the value of one einstein corresponding to a radiation of wavelength 4000A°. **[C.U. 2012]**

Ans. Energy content of one einstein is $E = Nh\frac{C}{\lambda}$

N = Avogadro number = 6.023×10^{23}

h = Planck's constant = 6.627×10^{-27} erg. sec.

C = velocity of light = 3×10^{10} c.m. sec^{-1}

λ = Wavelength of radiation = 4000×10^{-8} c.m.

$$E = \frac{6.023 \times 10^{23} \times 6.627 \times 10^{-27} \times 3 \times 10^{10}}{4000 \times 10^{8}} \text{ ergs}$$

$$= 2.99358 \times 10^{11} \text{ ergs} \equiv 2.99358 \times 10^{4} \text{ Joule.}$$

7. Write notes on phosphorescence. **[C.U. 2011]**

Ans. See section 20.11.